Utilization of Waste Biomass in Energy, Environment and Catalysis

Novel Biotechnological Applications for Waste to Value Conversion

Series Editor:

Neha Srivastava and Manish Srivastava

Utilization of Waste Biomass in Energy, Environment and Catalysis
Dan Bahadur Pal and Pardeep Singh

For more information about this series, please visit: www.routledge.com/our-products/book-series

Utilization of Waste Biomass in Energy, Environment and Catalysis

Edited by
Dan Bahadur Pal and Pardeep Singh

CRC Press
Taylor & Francis Group
Boca Raton London New York

CRC Press is an imprint of the
Taylor & Francis Group, an **informa** business

First edition published 2022
by CRC Press
6000 Broken Sound Parkway NW, Suite 300, Boca Raton, FL 33487-2742

and by CRC Press
2 Park Square, Milton Park, Abingdon, Oxon, OX14 4RN

© 2022 selection and editorial matter, Dan Bahadur Pal and Pardeep Singh; individual chapters, the contributors

CRC Press is an imprint of Taylor & Francis Group, LLC

ISBN: 978-1-03-205162-8 (hbk)
ISBN: 978-1-03-205163-5 (pbk)
ISBN: 978-1-00-319635-8 (ebk)

DOI: 10.1201/9781003196358

Typeset in Times
by Newgen Publishing UK

Contents

Preface

In the past few decades, researchers around the world have confirmed the potential of different biomass as feed-stocks for the production of not only bio-fuels but also other value-added products such as energy, environmental pollution and bio-catalysts. Since the 1970s, the focus was mainly on bioenergy production from different biomass and much less attention was given to high-volume, medium-value industrial products. However, the situation has been drastically changed in the last few years after realizing that biomass industries can survive only if they co-produce other value-added products along with energy production. The concept of integrated bio-refinery was developed and adapted by many industries especially in the developed countries. Now, most of the developing countries are also centering on the biomass-based industries with targeting multiple products along with bioenergy. Biomass synthesis and its applications act as feed-stocks for the production of not only biofuels but also other value-added products such as energy, environmental pollution, catalysts and others. These are the focus mainly on bioenergy production from different biomass and much less attention was given to high-volume, medium-value industrial products. This book investigates problems of controlled synthesis of these materials and the effect of their morphological, physical and chemical characteristics on their adsorption or desorption capacity and recent progress on green catalysts derived from biomass for various catalytic applications. Socioeconomic impacts in environment/climates of waste biomass. This book's 15 chapters deal with the following topics: Chapter 1 (agricultural waste biomass utilization as a bio-adsorbent: activation carbon for dye removal), Chapter 2 (agricultural waste biomass utilization in waste water treatment), Chapter 3 (phytochemical extraction from waste biomass), Chapter 4 (biomass [agricultural waste] as sustainable reinforce in polymer composite), Chapter 5 (biomass accretion and control strategies in gas biofiltration), Chapter 6 (enzymatic biodiesel production from biomass), Chapter 7 (catalytic cracking of jatropha curcas non-edible oil to hydrocarbons of gasoline fraction: optimization studies through the Box-Behenken method), Chapter 8 (production of hydrogen from waste biomass), Chapter 9 (microbial mediated waste management and bioenergy production), Chapter 10 (use of waste biomass as remediator for environmental pollution), Chapter 11 (recent trends in biomass conservation and management), Chapter 12 (revalorization of waste biomass for preparing biodegradable composite materials), Chapter 13 (biomass of microalgae as potential biodiesel source for future energy needs), Chapter 14 (waste biomass pretreatment using novel materials) and Chapter 15 (corporate social accountability in waste production and management).

Editors

Dan Bahadur Pal Assistant Professor, Department of Chemical Engineering, Birla Institute of Technology, Mesra, Ranchi-835215, Jharkhand, India. His research areas are nanotechnology, catalysis, energy, environment, biomass application and wastewater treatment. He has published around 25 research papers in journals, 15 book chapters and 18 conferences proceedings.

Pardeep Singh Assistant Professor, Department of Environmental Science, PGDAV College, University of Delhi, New Delhi, India. His research areas are nanotechnology, energy, environment, biomass application and wastewater treatment. He has published around 60 research papers in journals, edited 25 books and written 39 book chapters.

Contributors

Chapters	Authors	Corresponding authors
1	**Agricultural Waste Biomass Utilization as a Bio-Adsorbent: Activated Carbon for Dye Removal** Avdesh Singh Pundir[1*], Kailash Singh[2] and Sunil Rajoriya[1*] [1]Department of Chemical Engineering, Meerut Institute of Engineering and Technology, Meerut, Uttar Pradesh, India [2]Department of Chemical Engineering, Malaviya National Institute of Technology, Jaipur, Rajasthan, India *Corresponding author Email ID: 2013rch9513@mnit.ac.in	**Avdesh Singh Pundir** 2013rch9513@mnit.ac.in Sunil Rajoriya 2013rch9519@mnit.ac.in Kailash Singh Ksingh.mnit@gmail.com
2	**Agricultural Waste Biomass Utilization in Waste Water Treatment** Amit Kumar Tiwari*, Dan Bahadur Pal and Nirupama Prasad Department of Chemical Engineering, Birla Institute of Technology, Mesra, Ranchi – 835215, Jharkhand, India	**Amit Kumar Tiwari** amit_tiwari20@yahoo.com Nirupama Prasad nirupama.2k3@gmail.com Dan Bahadur Pal danbahadur.chem@gmail.com
3	**Phytochemical Extraction from Waste Biomass** Bidyut Mazumdar*, Amit Keshav, Alok Sharma Department of Chemical Engineering, National Institute of Technology Raipur Chhattisgarh, India	**Bidyut Mazumdar** bmazumdar.che@nitrr.ac.in
4	**Biomass (Agricultural Waste) as Sustainable Reinforce in Polymer Composite** Nirupama, Dan Bahadur Pal and Amit Kumar Tiwari Department of Chemical Engineering, Birla Institute of Technology Mesra, Ranchi (JH)-India	**Nirupama** nirupama.2k3@gmail.com Dan Bahadur Pal danbahadur.chem@gmail.com Amit Kumar Tiwari atiwari@bitmesra.ac.in

Chapters	Authors	Corresponding authors
5	**Biomass Accretion and Control Strategies in Gas Biofiltration** Rahul[1*], Vivek Kumar[2], Amarendra Kumar Dash[3] and Jay Mant Jha[4] [1]Chemical Engineering Department, Jaipur National University, Jaipur, Rajasthan – 302017, India [2]Chemical Engineering Department, Rajiv Gandhi University of Knowledge Technologies, R.K. Valley, Andhra Pradesh – 516330, India [3]English Department, Rajiv Gandhi University of Knowledge Technologies, Nuzvid, Andhra Pradesh – 521202, India [4]Chemical Engineering Department, Maulana Azad National Institute of Technology, Bhopal, Madhya Pradesh – 462003, India	**Rahul** rahuliitr2004@gmail.com **Vivek Kumar** **vivekkumar@rguktrkv.ac.in** **Amarendra Kumar Dash** **dash_amarendra@rguktn.** **ac.in** **Jay Mant Jha** **04ch20@gmail.com**
6	**Enzymatic Biodiesel Production from Biomass** Nisha Gupta[1], Dilip Kumar[2] and Bhawna Verma[3] [1,3] Department of Chemical Engineering, Indian Institute of Technology, Banaras Hindu University, Varanasi, Uttar Pradesh, India [2]Department of Biochemical Engineering, Harcourt Butler Technical University, Kanpur Uttar Pradesh, India	**Bhawana Verma** bverma.che@itbhu.ac.in Nisha Gupta - nishagupta.che19@itbhu.ac.in - Dilip Kumar dilipbe@hbtu.ac.in
7	**Catalytic Cracking of Jatropha Curcasa Non-Edible Oil to Hydrocarbons of Gasoline Fraction: Optimization Studies through the Box-Behenken Method** Saurabh Yadav[1], Neeru Anand[2], Dinesh Kumar[2] and Suantak Kamsonlian[1*] [1]Department of Chemical Engineering, Motilal Nehru National Institute of Technology Allahabad, Prayagraj – 211004, India [2]University School of Chemical Technology, GGS Indraprastha University Delhi, Delhi – 110078	**Suantak Kamsonlian** suantakk@mnnit.ac.in Saurabh Yadav indiansy@gmail.com Neeru Anand neeruanand@ipu.ac.in Dinesh Kumar dinesh.usct@gmail.com

Chapters	Authors	Corresponding authors
8	**Production of Hydrogen from Waste Biomass** Saptarshi Das[1], Satarupa pattanayak[2] and Sumit Kumar Jana[1]* [1]Department of Chemical Engineering, Birla Institute of Technology, Mesra, Ranchi, India [2]Department of Chemistry, Birla Institute of Technology, Mesra, Ranchi, India	**Sumit Kumar Jana** sumitkrjana@bitmesra.ac.in Saptarshi Das saptarshidas822@gmail.com Satarupa Pattanayak Satarupapattanayak1987@gmail.com
9	Microbial mediated waste management and bioenergy production Janeeshma E.[1], Gurudatta Singh[2] and Jos T. Puthur[1]* [1,*] Plant Physiology and Biochemistry Division, Department of Botany, University of Calicut, C.U. Campus P.O., Kerala – 673635, India, Tel.: +91-9447507845, Fax +91-494-2400269 [2] Institute of Environment and Sustainable Development, Banaras Hindu University, Varanasi, India – 221005.	**Jos T. Puthur** jtputhur@yahoo.com Gurudatta Singh gurudatta.singh2@bhu.ac.in Jos T. Puthur jtputhur@yahoo.com
10	**Use of Waste Biomass as Remediator for Environmental Pollution** Amit Kumar Tiwari*, Nirupama Prasad and Dan Bahadur Pal Department of Chemical Engineering, Birla Institute of Technology, Mesra, Ranchi – 835215, Jharkhand, India	**Amit Kumar Tiwari** amit_tiwari20@yahoo.com Nirupama Prasad nirupama.2k3@gmail.com Dan Bahadur Pal danbahadur.chem@gmail.com
11	**Recent Trends in Biomass Conservation and Management** Amarendra Kumar Dash[1], Vivek Kumar[2*], Dipti Sahoo[3], Jay Mant Jha[4], Balgovind Tiwari[5] and Rahul[6] [1]Department of English, Rajiv Gandhi University of Knowledge Technologies, Nuzvid, AP – 521202, India [2]Department of Chemical Engineering, Rajiv Gandhi University of Knowledge Technologies, RK Valley, AP – 516330, India [3]Department of Management, Rajiv Gandhi University of Knowledge Technologies, Nuzvid, AP – 521202, India [4]Department of Chemical Engineering, Maulana Azad National Institute of Technology, Bhopal, MP – 462003, India	**Vivek Kumar** **vivekkumar@rguktrkv.ac.in** Amarendra Kumar Dash dash_amarendra@rguktn.ac.in Jay Mant Jha 04ch20@gmail.com Balgovind Tiwari balgovindtiwari@rguktrkv.ac.in Dipti Sahoo sahood9@rguktn.ac.in Rahul rahuliitr2004@gmail.com

Chapters	Authors	Corresponding authors
	[5]Department of Physics, Rajiv Gandhi University of Knowledge Technologies, RK Valley, AP – 516330, India [6]Department of Chemical Engineering, Jaipur National University, Jaipur, RJ – 302017, India	
12	**Revalorization of Waste Biomass for Preparing Biodegradable Composite Materials** K. R. Srivastava[1*], D. B. Pal[3], P. K. Mishra[1] and P. K. Srivastava[2] [1]Department of Chemical Engineering and technology, Indian Institute of Technology, (BHU), Varanasi – 221005, Uttar Pradesh, India [2]School of Biochemical Engineering, Indian Institute of Technology, (BHU), Varanasi – 221005, Uttar Pradesh, India [3]Department of Chemical Engineering and Technology, Birla Institute of Technology, Mesra, Ranchi – 835215, Jharkhand, India	**Kumar Rohit Srivastav** krsrivastava6@gmail.com
13	**Biomass of Microalgae as Potential Biodiesel Source for Future Energy Needs** Deen Dayal Giri[1], Juhi Khan[3], Ajay Giri[2], Dan Bahadur Pal[4] and Amit Kumar Tiwari[4] [1]Department of Botany, Maharaj Singh College, Saharanpur (UP), India 247001 [2]Department of Basic Education, Uttar Pradesh, India [3]Department of Botany, IFTM University, Moradabad (UP), India [4]Department of Chemical Engineering, BIT Mesra, Ranchi (Jharkhand), India	**Deen Dayal Giri** ddgiri1@gmail.com
14	**Waste Biomass Pretreatment Using Novel Materials** [1*]Vivek Kumar, [2]Jay Mant Jha, [3]Amarendra Kumar Dash, [4]Balgovind Tiwari and [5]Rahul, [1]Department of Chemical Engineering, Rajiv Gandhi University of Knowledge Technologies, RK Valley, AP – 516330, India	**Vivek Kumar** vivekkumar@rguktrkv.ac.in Amarendra Kumar Dash dash_amarendra@rguktn.ac.in Jay Mant Jha 04ch20@gmail.com Balgovind Tiwari balgovindtiwari@rguktrkv.ac.in Rahul rahuliitr2004@gmail.com

Chapters	Authors	Corresponding authors
	[2]Department of Chemical Engineering, Maulana Azad National Institute of Technology, Bhopal, MP – 462003, India [3]Department of English, Rajiv Gandhi University of Knowledge Technologies Nuzvid, AP – 521202, India [4]Department of Physics, Rajiv Gandhi University of Knowledge Technologies RK Valley, AP – 516330, India [5]Department of Chemical Engineering, Jaipur National University, Jaipur, RJ – 302017, India	
15	**Corporate Social Accountability in Waste Production and Management** K. N. Ajoykumar,[1] Gurudatta Singh[2] and A. M. Shackira AM[3*] [1]Department of Botany, Nirmalagiri College, Kannur, Kerala, India – 670701. [2]Institute of Environment and Sustainable Development, Banaras Hindu University, Varanasi, India – 221005 [3]Department of Botany, Sir Syed College, Kannur, Kerala, India – 670142	**Shackira AM** shackimajeed@gmail.com KN Ajoykumar knajoykumar@gmail.com Gurudatta Singh gurubhu89@gmail.com

1 Agricultural Waste Biomass Utilization as a Bio-Adsorbent

Activated Carbon for Dye Removal

Avdesh Singh Pundir[1], Kailash Singh[2] and Sunil Rajoriya[1]
[1]Department of Chemical Engineering, Meerut Institute of Engineering and Technology, Meerut, Uttar Pradesh, India
[2]Department of Chemical Engineering, Malaviya National Institute of Technology, Jaipur, Rajasthan, India

CONTENTS

DOI: 10.1201/9781003196358-1

1

1.1 INTRODUCTION

Industrialization and urbanization are essential for enhancing human living standards. This is one side of the coin. On the other side, these have resulted in polluted wastewater in the environment. It is well known that effluent dyes have been identified as one of the polluted water resources (Somasekhara Reddy et al., 2017). Wastewater effluent from pulp and paper, tannery, textile, food, pharmaceutical and electroplating industries have high levels of synthetic dye pollutants. This high level of dye pollutants must be reduced within the range of regulatory limits. Literature shows that there are more than a thousand synthetic dyes have been used in industries (Ukanwa et al., 2019).

Despite the existence of a large number of dye wastewater treatment scheme, challenges in treating wastewater still exist in respect to their economic aspects. Major difficulties out of them include improper dye removal and COD reduction (Şayan, 2006). Industrial applied technologies consist of various biological and chemical treatment, adsorption, and membrane-based filtration techniques. One way to enforce the wastewater problem is to introduce a number of policies at the state and central level. The challenges to monitoring its implementation are mainly due to associated treatment costs (Boumchita et al., 2017).

The word "adsorption" was coined in 1881 by German physicist Heinrich Kayser. He described this word to make the difference between the surface phenomena vs intermolecular penetration. Bio-adsorption is an important process in which the waste water is treated on the surface of biological material, which may contain live or dead micro-organism, agricultural waste, industrial waste, etc. However, chemical adsorption is strictly reserved for chemically attachment of adsorbate, needing to be removed to solid a waste surface, called adsorbent. In broad classification, the adsorption process can be divided into physical and chemical processes. The physical adsorption process, as its name suggests, is controlled by the physical forces available at the surface, in general. The major physical forces available are dipole-dipole interaction, van der Waals forces, hydrogen bond, polarity, static interaction hydrophobicity, etc. When some foreign species come in contact with the surface, adsorption takes place and the contacted species attaches to the surface by means of strong chemical interactions called chemisorption. The extent to which the adsorption process takes place strongly depends on the physical and chemical properties of the adsorbent such as molecular structure, molecular size, polarity, etc.

The primary artificial dye, Mauveine, was searched with the aid of Perkin in 1856. As a result, the dyestuffs industry can rightly be defined as mature. Figure 1.1 shows the chemical structure of Mauveine. It remains a colorful, challenging industry requiring a continuous stream of new products due to the fast converting global wherein we live. Environmental regulations are getting stronger day by day in the developing countries such as India. Effluent from the dye industries is restricted to treating within allowable limits, especially to color removal and chemical oxygen demand (COD). These are changeling issue to handle dye industry wastewater. However, many dyes have microbial resistance and natural light

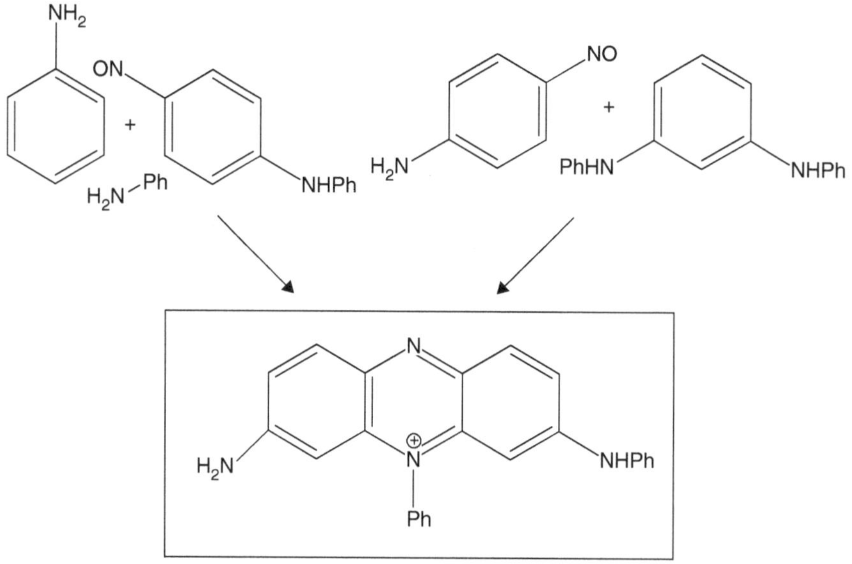

FIGURE 1.1 Synthetic chemical structure of Mauveine.

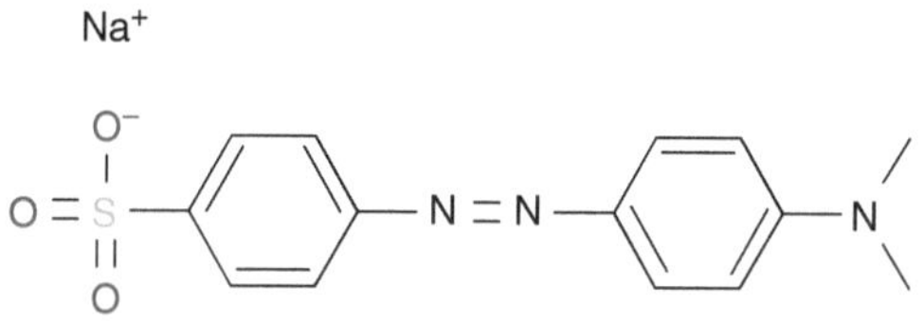

FIGURE 1.2 Synthetic chemical structure of methyl oranges.

resisting characteristic which make them difficult to treat, in general. Therefore, the treatment of dye wastewater is necessary to comply the national environmental regulations (Wu et al., 2019).

Subbaiah and Kim (2016) have investigated the adsorption of methyl orange from aqueous solution by aminated pumpkin seed powder. Figure 1.2 shows the chemical structure of methyl orange. They have reported that aminated pumpkin seed powder can efficiently remove the dye (maximum adsorption capacity of 143.7 mg/g) at an optimum temperature of 298 K. This study has shown the high efficiency of natural bio-adsorbent based on aminated pumpkin seed powder for the treatment of dye wastewater.

Yu et al. (2018) have investigated the utilization of biochar adsorbent prepared from chicken manure for the adsorption of methyl orange dye. They have reported that almost 100% dye removal was achieved in 30 min when the biochar adsorbent dosage of 1.0 g/L was employed.

Overall, it can be concluded that natural bio-adsorbents as well as activated carbon prepared from bio-adsorbent may be a better option for the treatment of dye wastewater.

It can be seen from the Table 1.1 that many researchers have reported the utilization of natural bio-adsorbent as well as AC prepared from the biomass towards the treatment of wastewater. Some of the authors have applied the natural bio-adsorbent for the removal of different type of dyes from the aqueous solution, whereas a few researchers have employed the AC produced from biomass. Table 1.2 depicts the advantages and disadvantages of the few major conventional techniques of wastewater.

1.2 DYE

Many kinds of impurities usually found in wastewater may be grouped as suspended solids, dissolved mineral salts, organic and coloring matters, dissolved gases, bacteria, and different microorganisms. A few parameters which might be responsible for exchange in inherent characteristics of wastewater because of pollution are color, turbidity, dissolved solids, pH, dissolved oxygen (DO), absorbable organic halogens (AOX), biodegradability of wastes in terms of biochemical oxygen call for (BOD) and chemical oxygen call for (COD), and oils and grease. Dyes manufacture unit observed of the structural aspect liable for color ingredient in dye called chromogens, except one or two cases, nowadays, all the sorts of dye have been discovered in the 1800s. The great alternate in this field was discovered in the period 1930–1950 by the production of synthetic fiber nylon, polyester, and polyacrylonitrile. Dyeing of the cotton leapt forward with the development of reactive dye in 1954 and 1956. In that era, a variety of in-depth studies regarding reactive dyes turned into endured over the following a long time and, this segment continues to be persevering with today. The most important problem was encountered with the oil crisis in the early 1970s, which reasons a steep gradient hike in the prices of basic uncooked substances for dyes, therefor emphasis is to find a process for alternative cost-effective approaches to produce dyes. The performance of the dye manufacturing plant, by means of converting vulnerable structure chromogens, consisting of anthraquinone, with stronger structure chromogens, together with (heterocyclic) azo and benzodifuranone, has grown plenty but alternatively push the restriction to address the extra strong dye wastewater to deal with. The decolorization trouble is nonetheless crucial and ongoing, as are the modern-day subject matters of social protection and safety of our environment. There are numerous important dyes used for superior technology applications, particularly in the subject of electronics and non-impact printing industries. Dye classification is based on their chemical structure, utilization, and application technique. The structure-based classification method is generally followed by means of dye chemists, who use terms which includes azo dyes, anthraquinone dyes, and phthalocyanine dye (Hunger, 2002). The utilization or application-based approach is predominantly followed by using the dye technologist, and that they talk of reactive dyes for cotton and disperse dyes for polyester.

TABLE 1.1

Overview of Work Done on Dye Removal from Wastewater Using Different Types of Adsorbents

S. No.	Name of pollutant	Type of adsorbent	Treated Volume (mL)	Ranges of Process parameters studied	Optimum parameters	Removal efficiency (%)/ adsorption capacity (mg/g)	References
1.	Acid green 25 (AG25)	Calcined oyster shells	100 mL	pH = 2.0, 4.0, 6.0, 8.0, 10.0, and 11.0; Adsorbent dosage = 1.0, 2.0, 3.0, 5.0, and 7.0 g/L; Temperature (K) = 288.15, 293.15, 303.15, and 315.15 K; Initial dye concentration = 10, 20, 30, 50, and 70 mg/L	Contact time of 60 min; AG25 concentration of 50 mg/L; Adsorption temperature of 303.15°C; adsorbent dosage of 2 g/L; pH of 11	96.5%	(Inthapanya et al., 2019)
2.	Methylene blue	Activated carbon prepared from coconut husk	100 mL	Temperature (K) = 303 to 323K; Initial dye concentration = 50 to 500 mg/L	Contact time of 120 min; initial dye concentration of 500 mg/L; Adsorption temperature of 303 K; adsorbent dosage of 0.10 g/100 mL;	405 mg/g	(Tan et al., 2008)

(continued)

TABLE 1.1 (Continued)
Overview of Work Done on Dye Removal from Wastewater Using Different Types of Adsorbents

S. No.	Name of pollutant	Type of adsorbent	Treated Volume (mL)	Ranges of Process parameters studied	Optimum parameters	Removal efficiency (%)/ adsorption capacity (mg/g)	References
3.	Acid Orange 7 (AO7)	Zeolite-activated carbon macro composite		pH = 2.0, 7.0, and 12.0; For each experiment, the dye solution was agitated with MC at a ratio of 2:1; Initial dye concentration = 100 to 400 mg/L	Contact time of 360 min; Initial dye concentration of 100 mg/L; Adsorption temperature of 25°C; pH of 2	99%	(Lim et al., 2013)
4.	Direct Red 23 (DR23) and Direct Blue (DB78)	Poly (vinyl alcohol) (PVA) nanofiber	250 mL	pH = 2.1, 3, 4, 5, 6, 7 and 8; Temperature (K) = 298, 308, 318 and 325 K;	Contact time of 60 min; initial dye concentration of 30 mg/L; Adsorption temperature of 298 K; adsorbent dosage of 0.02 g/250 mL; pH of 2	370 and 400 mg/g for DR23 and DB78, respectively.	(Mahmoodi et al., 2017)
5.	Methyl Orange	Biochar adsorbent prepared from chicken manure	200 mL	pH = 4.5–10.5; Adsorbent dosage = 0.2 to 1.4 g/L; Temperature (K) = 298 K to 318 K; Initial dye concentration = 25 to 75 mg/L	Contact time of 30 min; initial dye concentration of 25 mg/L; Adsorption temperature of 298 K; adsorbent dosage of 1 g/L; pH of 4.5	89.19% and 22.30 mg/g	(Yu et al., 2018)

No.	Dye	Adsorbent	Volume	Range	Optimized conditions	Efficiency	Reference
6.	Methyl Orange	Aminated pumpkin seed powder	30 mL	pH = 3–11; Temperature (K) = 298 K to 318 K; Initial dye concentration = 100 to 1000 mg/L	Contact time of 110 min; initial dye concentration of 300 mg/L; Adsorption temperature of 298 K; adsorbent dosage of 0.05 g/30 mL; pH of 4.5	143.7 mg/g	(Subbaiah and Kim, 2016)
7.	Acid Orange 7 (AO7)	Activated carbon	200 mL	Adsorbent dosage = 3 to 6 g/L; Initial dye concentration = 50 to 200 mg/L	Contact time of 120 min; initial dye concentration of 100 mg/L; adsorbent dosage of 5 g/L;	98%	(Li et al., 2017)
8.	Methylene blue	Garlic straw (GS)	100 mL	pH = 2.0–10.0; Adsorbent dosage = 0.005 to 0.20 g/10 mL; Temperature (K) = 303.16 K to 323.16 K; Initial dye concentration = 20 to 200 mg/L	Contact time of 200 min; initial dye concentration of 100 mg/L; Adsorption temperature of 303.16 K; adsorbent dosage of 0.04 g/10 mL; pH of 7	85%	(Kallel et al., 2016)
9.	Methyl red	Sugarcane bagasse	100 mL	pH = 2.0–12.0; Adsorbent dosage = 0.2 to 1 g/100 mL; Initial dye concentration = 50 to 250 mg/L	Contact time of 180 min; initial dye concentration of 50 mg/L; Adsorption temperature of 26±1°C; adsorbent dosage of 0.4 g/100L; pH of 6	64.2%	(Saad et al., 2010)

(continued)

TABLE 1.1 (Continued)
Overview of Work Done on Dye Removal from Wastewater Using Different Types of Adsorbents

S. No.	Name of pollutant	Type of adsorbent	Treated Volume (mL)	Ranges of Process parameters studied	Optimum parameters	Removal efficiency (%)/ adsorption capacity (mg/g)	References
10.	Congo Red	Bengal gram seed husk	25 mL	pH = 5.85–11.02; Adsorbent dosage = 50 to 500 mg/25 mL; Initial dye concentration = 25 to 100 mg/L	Contact time of 180 min; initial dye concentration of 50 mg/L; Adsorption temperature of 30±1°C; adsorbent dosage of 6 g/L; pH of 5.85	92%	(Somasekhara Reddy et al., 2017)
11.	Methyl blue dyes	Pumpkin Peels	V = 200 ml	pH = 3–9, Adsorbent dosage = 0.5—2 mg/L Initial dye concentration = 50–200 mg/L	Contact time of 180 min; initial dye concentration of 50 mg/L; Adsorption temperature of 50±1°C; adsorbent dosage of 0.5 g/L; pH of 5.85	-- 198.15 mg/g	(Rashid et al., 2019)

TABLE 1.2
Major Conventional Techniques for Wastewater

S. No.	Conventional Technique	Pros	Cons
1	Chemical oxidation	High output, disinfection capability, color and odor removal efficiency, combined with other process	Addition of chemical/not eco-friendly, pre-request of pretreatment, unwanted compound can be produced, generate solid waste
2	Ion exchange	Simple & straight forwarded technique, easy to apply, short process time, volatile vapor can be removed	High initial cost, large size equipment needed, frequently regeneration required, clogging of bed, routine cleaning required
3	Electrocoagulation	Use for extraction of valuable products, good over coagulation conventional technique, economic feasibility, doesn't depend upon pH	High initial cost, solid waste deposition, additional chemical required
4	Photocatalytic process	High efficiency, use to concentrate wastewater	High initial cost, moderate maintenance cost

Source: Crini and Lichtfouse (2019).

Reactive dyes connect to fiber via a covalent bond, usually cotton, but they may be in small level used on wool and nylon. This kind of dye class was first introduced commercially in 1956. A major advantage of reactive dyes over direct dyes is that they have easy chemical systems and their absorption spectra have narrower absorption bands, leading to superior overall dying capability. The main chemical training of reactive dyes is azo (which includes metallized azo), triphendioxazine, phthalocyanine, formazan, and anthraquinone. Disperse dyes are typically water-insoluble, non-ionic dyes for utilities to hydrophobic fibers from aqueous dispersion (Wu et al., 2019). They are used predominantly on polyester and to a lesser volume on nylon, cellulose, cellulose acetate, and acrylic fibers. Thermal transfer printing and dye diffusion thermal transfer (D2T2) methods for digital images represent areas of interest markets for decided on the participants of this magnificence. Direct dye water-soluble anionic dyes, when dyed from aqueous solution in the presence of electrolytes have excessive affinity for cellulosic fibers. Their most important use is the dyeing of cotton and regenerated cellulose, paper, leather-based, and, to a lesser volume, nylon. Maximum of the dyes in this class is poly-azo compounds, alongside a few stilbenes, phthalocyanines, and oxa-zines. Vat dyes are water-insoluble dyes are carried out particularly to cellulosic fibers as soluble leuco salts after reduction in an alkaline bathtub, usually with sodium hydrogen sulfite. Following exhaustion onto the fiber, the leuco forms are re-oxidized to the insoluble keto forms and after treated, typically with the aid of soaping, to redevelop the crystal structure (Mondal et al., 2018).

Numerous conventional techniques such as evaporation, precipitation, coagulation, ion exchange, solvent extraction, etc., are available to deal the waste water (Kadhom et al., 2020). Each technique has some advantages and disadvantages. In some especial cases, a single technique performs satisfactory results while in general, a combined approach is more useful.

1.3 AGRICULTURAL WASTE BIOMASS SOURCES

In India, agricultural waste materials are easily available in sufficient amounts and are renewable, inexpensive, and nontoxic. Enough of these waste materials are present to, constitute a pollution threat. Nevertheless they provides a good source of carbonaceous source for the generation of inexpensive, high-value activated carbon. The economic viability of this lignocellulosic material can be used independently or in a combined approach to other wastewater handling techniques to deal with. The important feature, making it attractive, is its eco-friendly nature for wastewater treatment. It is well known that the components present in the agricultural waste materials are cellulose, hemicelluloses, and lignin. The selection of these precious materials depends upon the activation techniques to extract them: the process affects the final outcomes in terms of their performance. In the following section some light has been through by reviewing literature, especially to agro-based waste.

1.3.1 PEANUT SHELL

Peanut shell may be a good resource because it is available in abundant amounts; however, it has low publicity for utilization as an adsorbent. In developing countries such as India, it is only used as a cattle feed ingredient. The chemical groups present mainly responsible for its activity as a good adsorbent are phenolic hydroxyl and carboxyl, which can be altered chemically to boot its performance. Therefore, peanut shell may be one of the resources to prepare activated carbon for wastewater treatment.

1.3.2 BAGASSE

In India, sugar industry produces the enormous amount of waste known as bagasse. It contains sugar cane fibrous matter (called lignocelluloses) which is obtained after crushing the sugar cane for the juice. The juice is utilized for the production of sugar crystal. After Brazil, India is the second largest sugarcane producer in the world. On the one hand, the country needs to develop a waste management strategy to handle such large bulk waste and on the other hand, India needs to of think how it can utilize the available resource. Therefore, bagasse needs special attention. The smaller amount of AC is enough for removing the color impurity from wastewater. Powder form of AC obtained from sugarcane bagasse is fine absorber for treatment of industrial and other purposes. There are many ways for converting bagasse to AC form, out of them, thermal route is the most effective one. But that requires a controlled temperature environment to avoid the loss (Swarnakar and Choubey, 2016).

1.3.3 PEAT

Peat is a material consisting of spongy material produced by the partial decomposition of organic plant material such as swamps, muskegs, bogs, fens, and moors. The formation of the peat is favored by slightly warm climatic and moist conditions; however, peat can form even in cold regions synthetically. Peat is one of the important resources to get AC. It mainly contains calcium component which might affect the formation of the AC and its end-products. But, the effect of calcium component present in peat has not been described well in the literature.

1.3.4 RICE HUSK

India is one of many countries having large consumption of rice. Most of the India's population consumes rice as a main ingredient of food. The waste material generates is rice husk, which contains approximately 32% cellulose, 21% hemicellulose, 22% lignin, and the rest ash. Due to the presence of its constituent material, it is a good source of carbon to produce activated carbon with specific pore structure and surface area (Menya et al., 2018).

1.3.5 COCONUT SHELL

Coconut shell is available in plentiful amount round the year in various tropical regions of India. In India, coconut shell is used as a fuel, which contributes significantly to the production of carbon dioxide and methane. With the advancement in technology, coconut shell is now used for the production of charcoal. The old pit method of making a charcoal has approximately yield of 30% of the dry weight of shells used. The coconut contains approximately 40% coconut husks, which further contain 30% fiber. The chemical material of coconut husks consists of mainly cellulose, lignin, pyroligneous acid, tannin, and potassium (Gratuito et al., 2007).

1.3.6 ACTIVATED CARBON

Due to enhancement of human activities' day by day, the consumption of Activated carbon (AC) in various industries, particularly to wastewater treatment processes, we are bound to increase the production of AC which grows exponentially in this era. However, the other major challenge is the reduction in the cost of AC manufacturing. To work in this direction, different works have been done and still going on specially for production of AC from natural waste along with its high physicochemical properties' development. The adsorption capability of producing AC depends on strong to its surface area, porosity and surface functional groups responsible to hold contact materials on its surface. In general, AC pores are divided into three basic types: pores having diameters less than 2 nm (called micropores), pores having diameters more than 50 nm (called macropores) and pores size between these values (called mesopores). Out of these pores, micropore AC is the most suitable of all for the adsorbent process. Activated carbon (AC) is an important carbon solid obtained from coal and agro-waste based biomass by applying only thermal or various thermochemical processes. The characteristic of the developed AC can be defined based on how it was produced, along with measurable merits such as pore morphology, high surface area, and high absorptive capacity. AC has potential use in different areas, not just in the adsorption process to treat industrial effluents. Now, the economic size of AC in the world market is approximately several billion dollars annually. AC has many applications in dealing with waste water problems and is also emerging to handle air pollution issues, and their associated health risks. Effluent discharge from the dye industries is one of the potential threats generating (Kane et al., 2016).

Before choosing the method of converting the waste biomass to AC, it is important to understand the basic components which are present. In general, most waste biomass has three components (other than few specific components in small amounts): lignin, cellulose, and hemicellulose. Chemical treatment before thermal decomposition showed a good affect in the degradation of lignin with respect to cellulose (Yu et al., 2018). Agricultural waste biomass may contain as high as 45% lignin. It is a polymer chain of aldehyde and carbonyl group and generally insoluble. During the initial phase of the thermal treatment, the various volatile gases liberated from cellulose and hemicellulose are responsible for the generation of small pores inside the solid material. The inner bonding present among the different component may be responsible for the energy required during treatment or generation of AC. In case of weak bonding, the generation cost is low. Therefor, the structure of component present in the waste biomass is essential to understand before selecting as a raw material for the production of AC.

1.3.7   PREPARATION OF AC

In broad classification, apart from acid, bases, and salt methods, the thermal decomposition method seems the low-cost and preferable choice in comparison to other, as reported by researcher (Schroder et al., 2020). The availability of the major materials such as coal, wood and shell depend upon the region and country. So, availability of other resources may be checked to generate AC from them. The amount of AC from the different waste biomass vary from one to another. The important waste biomass such as nut shell, spent grain, straw matter, olive stones and coffee grounds, peanut shell are normally used. The thermal decomposition method to get AC contains two steps. In the first step, called pyrolysis process is occurred in the presence of inert N_2 at 500–600 ^{0}C. For economic purposes, the liberated byproducts (gaseous and liquid) may be used for heating or electricity generation purpose if the process occurs at large scale. In the second step, the obtained black solid residue, called char, is further heated to high temperature at 800–1000 ^{0}C (Ukanwa et al., 2019). For heating purposes, water steam may be used to get a better pore surface area so that during adsorption, a better result may be expected. After the thermal process, the standard surface method, most preferable, BET method, is used to know about the generate pore surface area (Inthapanya et al., 2019). However, the surface area depends on the waste biomass, treatment technique and the process parameters.

1.4   ADSORPTION MODEL

The procedure of dye pollutants removal may be well defined as soon as we are able to find the interaction among AC adsorbent and adsorbate (Dye(s)) and this technique may be modeled with the assistance of numerous adsorption isotherms. The chosen isotherms are identified by means of unique constants that express the primary characteristics: out of them, the surface properties and affinity of adsorbent in the direction of the adsorbent pollutant are major concerned. Adsorption isotherm experiments are carried out by taking numerous amounts of adsorbents, starting from ppm to few grams. One-of-a-kind quantities of adsorbents are introduced in each case of the laboratory dye solution. After finding the equilibrium level that is similar to maximum extraction or decolorization, performance may be determined spectrophotometrically at specific wavelength. The quantity of dye adsorbed is calculated with the help of the graph plotted in excel or the use of any software program. The suitability of the implemented isotherms is judged from the values of the correlation coefficient. The numerical value of the R^2 is commonly used to assess the overall performance. For the adsorption isotherms, R^2 value is toward one can be considered to more in agreement with the experimental observation (Bonilla-Petriciolet et al., 2017). There are numerous adsorption isotherms available in the literature. Out of them, Freundlich, Langmuir and Temkin isotherms are the most important. The model equations and their parameters have been mentioned in Table 1.2 along with its application condition.

1.5   PROCESS PARAMETERS, DECOLORIZATION AND COD REDUCTION

The decolorization and COD reduction of dye solutions are commonly examined independently of the use of activated carbon and mixed with different strategies. The implementation of hybrid aggregate is commonly implied to enhance the dye elimination method in the case in which AC on my own capable of produce great outcomes. To reduce the cost of the overall experimentation and time, of the experiments are carried with the deliberate experimental design matrix and the parameters stages. The effects of relevant parameters (particularly; energy, temperature, time, activated carbon concentration, dye concentration, and initial pH) were investigated at the decolorization and COD discount using the fractional factorial layout a standard decolorization experimentation are happened in a batch

TABLE 1.3
Major Adsorption Isotherms Equation for the Dye Solution

S. No.	Adsorption isotherm	Equation and specific parameter or constant	Applicability condition
1.	Freundlich	$$\log\left(\frac{x}{m}\right) = \log K + \frac{1}{n}\log p$$ Where x = mass of adsorbate, m = mass of adsorbent, K and p are constants at given temperature	for multilayers adsorption on heterogeneous surface
2.	Langmuir	$$\frac{1}{q} = \frac{1}{q_0 K_L} + \frac{1}{q_0}$$ where, q_0 = saturated capacity, K_L = adsorption energy	for monolayer adsorption on homogeneous surface
3.	Temkin	$$q_e = \frac{RT}{B}\log K_T + \frac{RT}{B}\log C_e$$ Where R, B are Temkin constants, C_e is equilibrium constant.	Adsorption heat of the layer would decrease linearly with coverage due to indirect adsorbent-adsorbate interactions

Source: Suteu and Malutan, (2013).

mode as follows: particular amounts of dye solution of recognized concentration and preliminary pH are loaded into the glass beaker and maintained the favored decolorization time at the desired temperature (Mupa et al., 2017).

1.6 ADSORPTION SURFACE CHARACTERIZATION TECHNIQUE

There are various techniques available to investigate the surface properties. Out of them, the essential analytical techniques are as follows (Chand Bansal and Goyal, 2005):

- **UV vis-spectrophotometer**: It is a technique used to measure light absorbance in the ultraviolet and visible ranges of the electromagnetic spectrum. As we know, when an incident light encountered with material it can either be absorbed, reflected, or transmitted. The absorbance of this incident radiation in the UV–Vis range is responsible for atomic excitation, which leads to the transition of molecules from a low-energy ground state to an excited state. Before going to a different atomic excitation state, it must absorb enough energy through radiation for the movement of electrons to higher molecular orbits. In general, shorter band gaps usually link to absorb smaller wavelengths of light. The energy amount required for the movement of molecules to these transitions, are measured in electrochemical unit. An

UV–Vis spectrophotometer device uses the same principle to quantify the analytes present in a provided sample.

- **Atomic adsorption spectrophotometer (AAS)**: AAS able to detect elements in either liquid state or solid-state samples with the help of characteristic wavelengths of electromagnetic radiation emitted from a light source. Different elements absorb different wavelengths, and measurement of these energies' absorbance against standards will help to identify the individual element. AAS utilized the property of the different radiation wavelengths that are absorbed by different atoms. In this technique, unknown materials are first atomized so that their characteristic wavelengths are emitted and recorded. This energy supply utilized by electrons to move up one energy level in their respective atoms. When the electrons return to their original energy state, they emit energy in the form of light. This emitted light has a wavelength that is characteristic of the element. As per the emitted light wavelength and its intensity, specific elements can be detected and their concentrations measured.

- **Fourier transform infrared spectroscopy (FTIR):** It is a simple analytical technique which is mostly used to identify organic materials, but it can be used for identification of inorganic materials too. This technique utilized the amount of absorption of infrared radiation by the sample material with wavelength. Infrared absorption bands help to identify not only the molecular components but its structures as well. When we exposed a material to infrared radiation, molecules of the material get excited by absorbed of IR radiation and move into a higher vibrational state. The wavelength of light absorbed by a particular molecule is a function of the energy difference between the at-rest and excited vibrational states. The wavelengths that are absorbed by the material molecules are characteristic of its molecular structure.

1.6.1 X-Ray Spectroscopy (SEM/TEM)

In qualitative analysis, we have to look for the elements present in the sample that are needed to identify from their characteristic X-ray peaks, however, their amount is not determined. There is many commercial peak-identification software in the market which helps the identification process, but still we are not yet 100% accurate. In the SEM, if the sample is stable under high-vacuum in the electron microscope and is not susceptible to damage by exposing under the electron beam, then an accelerating voltage of 15–30 kV is recommended for SEM analysis. This is enough to generate at least one family of X-ray lines for all elements. In the TEM, the major requirement is to collect enough X-ray counts in the spectrum from a thin specimen for good analysis. This is generally accomplished at the highest accelerating voltage.

- **Thermo gravimetric analysis (TGA):** The Thermogravimetric Analyzer (TGA) uses the variation in weight of a tested sample with the progress of temperature. The TGA able to provide knowledge about chemical phenomena including chemisorption, desolvation, and dehydration, etc. In TGA, mass loss is monitored when a thermal process involves the loss of a volatile component.

- **X-ray diffraction analysis (XRD):** A main application of XRD analysis is to identify the materials based on their diffraction pattern. During the phase identification, XRD also helps to get information on how the actual structure of the material deviates from the ideal one, due to the presence of internal stresses and defects.

1.7 CONCLUSIONS

For the last few decades, waste biomass has been used as a bio-adsorbent and has a rich source of activating carbon (AC). Agricultural waste biomass have uncovered the energy encapsulate within it. Bio-adsorbent hast been financially effective, and it is conventional leading as a minimal effort and ecologically contaminant treatment strategy. There are various sources of biomass, but a few have good potential to generate AC. AC can be produced by thermal decomposition in a controlled temperature environment. Textile industries are a big source of dye solution available in the wastewater. Therefore, we need to handle the two parallel situations. One is having large sources of waste biomass and the other is colored wastewater, specially liberated from textile industries. AC is one of the better adsorbents available so far to treat the color impurity present in water bodies and industrial wastewater. There are many operating parameters which play a crucial role to achieve a good efficiency. Out of them, pH, adsorbent dose, temperature and contact time are essential affecting parameters. Various adsorption models need to fit in the experimental data so that it can be accessed which particular model can be used for best describing the experimental process. To check the surface morphology and characteristics of the produced AC, many advanced techniques such as FTIR, XRD, UV-Vis and AAS can be used. This chapter not only provides an overview regarding the different aspects for the use of waste biomass as an adsorbent in the form of AC but also provides insights on how serious is the problem of waste generation in India along with the marching solution in this direction.

REFERENCES

Bonilla-Petriciolet, A., Mendoza-Castillo, D.I., Reynel-Ávila, H.E., 2017. Adsorption isotherms in liquid phase: experimental, modeling, and interpretations, adsorption processes for water treatment and purification. https://doi.org/10.1007/978-3-319-58136-1

Boumchita, S., Lahrichi, A., Benjelloun, Y., Lairini, S., Nenov, V., Zerrouq, F., 2017. Application of peanut shell as a low-cost adsorbent for the removal of anionic dye from aqueous solutions. J. Mater. Environ. Sci. 8, 2353–2364.

Chand Bansal, R., Goyal, M., 2005. Activated Carbon Adsorption. Taylor & Francis Group, LLC.

Crini, G., Lichtfouse, E., 2019. Advantages and disadvantages of techniques used for wastewater treatment. Environ. Chem. Lett. 17, 145–155.

Gratuito, M.K.B., Panyathanmaporn, T., Chumnanklang, R.A., Sirinuntawittaya, N., Dutta, A., 2007. Production of activated carbon from coconut shell: Optimization using response surface methodology. Bioresour. Technol. 99, 4887–4895. https://doi.org/10.1016/j.biortech.2007.09.042

Hunger, K., 2002. Industrial dyes, industrial dyes. https://doi.org/10.1002/3527602011

Inthapanya, X., Wu, S., Han, Z., Zeng, G., Wu, M., Yang, C., 2019. Adsorptive removal of anionic dye using calcined oyster shells: isotherms, kinetics, and thermodynamics. Environ. Sci. Pollut. Res. 26, 5944–5954.

Kadhom, M., Albayati, N., Alalwan, H., Al-Furaiji, M., 2020. Removal of dyes by agricultural waste. Sustain. Chem. Pharm. 16, 100259. https://doi.org/10.1016/j.scp.2020.100259

Kallel, F., Chaari, F., Bouaziz, F., Bettaieb, F., Ghorbel, R., Chaabouni, S.E., 2016. Sorption and desorption characteristics for the removal of a toxic dye, methylene blue from aqueous solution by a low cost agricultural by-product. J. Mol. Liq. 219, 279–288. https://doi.org/10.1016/j.molliq.2016.03.024

Kane, S.N., Mishra, A., Dutta, A.K., 2016. Characterization of activated carbon from rice husk by HCl activation and its application for lead (Pb) removal in car battery wastewater. IOP Conf. Ser. Mater. Sci. Eng. 755. https://doi.org/10.1088/1742-6596/755/1/011001

Li, J., Du, Y., Deng, B., Zhu, K., Zhang, H., 2017. Activated carbon adsorptive removal of azo dye and peroxydisulfate regeneration: from a batch study to continuous column operation. Environ. Sci. Pollut. Res. 24, 4932–4941. https://doi.org/10.1007/s11356-016-8234-4

Lim, C.K., Bay, H.H., Neoh, C.H., Aris, A., Abdul Majid, Z., Ibrahim, Z., 2013. Application of zeolite-activated carbon macrocomposite for the adsorption of Acid Orange 7: Isotherm, kinetic and thermodynamic studies. Environ. Sci. Pollut. Res. 20, 7243–7255.

Mahmoodi, N.M., Mokhtari-Shourijeh, Z., Ghane-Karade, A., 2017. Synthesis of the modified nanofiber as a nanoadsorbent and its dye removal ability from water: Isotherm, kinetic and thermodynamic. Water Sci. Technol. 75, 2475–2487.

Menya, E., Olupot, P.W., Storz, H., Lubwama, M., Kiros, Y., 2018. Production and performance of activated carbon from rice husks for removal of natural organic matter from water: A review. Chem. Eng. Res. Des. 129, 271–296. https://doi.org/10.1016/j.cherd.2017.11.008

Mondal, S., Purkait, M.K., De, S., 2018. Advances in Dye Removal Technologies.

Mupa, M., Phineas, M.M., Isaac, G., 2017. Adsorption of a cationic dye by Marula (Sclerocarya birrea) fruit seed shell based biosorbent: Equilibrium and kinetic studies. African J. Biotechnol. 16, 1969–1976. https://doi.org/10.5897/ajb2016.15830

Rashid, J., Tehreem, F., Rehman, A., Kumar, R., 2019. Synthesis using natural functionalization of activated carbon from pumpkin peels for decolourization of aqueous methylene blue. Sci. Total Environ. 671, 369–376.

Saad, S.A., Isa, K.M., Bahari, R., 2010. Chemically modified sugarcane bagasse as a potentially low-cost biosorbent for dye removal. Desalination 264, 123–128.

Şayan, E., 2006. Optimization and modeling of decolorization and COD reduction of reactive dye solutions by ultrasound-assisted adsorption. Chem. Eng. J. 119, 175–181. https://doi.org/10.1016/j.cej.2006.03.025

Schroder, E., Thomauske, K., Oechsler, B., Herberger, S., Baur, S., Hornung, A., 2020. Activated Carbon from Waste Biomass, Intechopen.

Somasekhara Reddy, M.C., Nirmala, V., Ashwini, C., 2017. Bengal gram seed husk as an adsorbent for the removal of dye from aqueous solutions – Batch studies. Arab. J. Chem. 10, S2554–S2566.

Subbaiah, M.V., Kim, D.S., 2016. Adsorption of methyl orange from aqueous solution by aminated pumpkin seed powder: Kinetics, isotherms, and thermodynamic studies. Ecotoxicol. Environ. Saf. 128, 109–117.

Suteu, D., Malutan, T., 2013. Industrial cellolignin wastes as adsorbent for removal of methylene blue dye from aqueous solutions. BioResources 8, 427–446. https://doi. org/10.15376/biores.8.1.427-446

Swarnakar, A.K., Choubey, S., 2016. A short review on utilizing sugarcane bagasse (SCB) – Chhattisgarh (India) prospect. Int. Res. J. Eng. Technol. 3, 2395–56.

Tan, I.A.W., Ahmad, A.L., Hameed, B.H., 2008. Adsorption of basic dye on high-surface-area activated carbon prepared from coconut husk: Equilibrium, kinetic and thermodynamic studies. J. Hazard. Mater. 154, 337–346.

Ukanwa, K.S., Patchigolla, K., Sakrabani, R., Anthony, E., Mandavgane, S., 2019. A review of chemicals to produce activated carbon from agricultural waste biomass. Sustain. 11, 1–35.

Wu, H., Chen, R., Du, H., Zhang, J., Shi, L., Qin, Y., Yue, L., Wang, J., 2019. Synthesis of activated carbon from peanut shell as dye adsorbents for wastewater treatment. Adsorpt. Sci. Technol. 37, 34–48.

Yu, J., Zhang, X., Wang, D., Li, P., 2018. Adsorption of methyl orange dye onto biochar adsorbent prepared from chicken manure. Water Sci. Technol. 77, 1303–1312.

2 Agricultural Waste Biomass Utilization in Waste Water Treatment

Amit Kumar Tiwari, Dan Bahadur Pal and Nirupama Prasad
Department of Chemical Engineering, Birla Institute of Technology, Mesra Ranchi – 835215, Jharkhand, India

CONTENTS

DOI: 10.1201/9781003196358-2

2.1 INTRODUCTION

To prevent and control pollution of surface water and underground water resources, governments of many developed and developing countries have established various firm rules and regulations to maintain the quality of available water resources. Basically, prevention and control of water pollution is somehow different from the quality control of water. The quality standards of water may be different according to its intended use; quality standards of water and limits of pollutants can be prescribed in relation to the paramount use such as drinking, cooking, washing, bathing, and agricultural work. In some cases these limits and standards are not followed correctly, if they are not closely related with human consumption. Governments regulate the quality of water by the implementation of general rules, which will helps to avoid decadence of water quality and to maintain its natural state. Besides, the assurance of the purity and quality of drinking water is a duty of the department of water supply or public utility section, because of public health concerns. The prevention and control of pollution in freshwater resources is also an important duty of governments, whether these resources are used for the domestic/ drinking purpose, industrial purposes, field irrigation and stock watering, or for recreational uses. Water pollution originates from various sources, which can be easily classified into two main groups – 'point sources' and 'diffuse sources'. Point sources can be identified with pollution in certain sources such as industry outlets and municipal sewers, domestic drains and waste-water treatment plants, water wells and waste dumping sites. Entry of these point sources into water bodies, surface water, or underground water sources may pollute entire water supplies, therefore the level of pollution in point sources should be checked accurately. 'Point' sources can affect the water quality directly and indirectly, indirect effects may be due to deposition of or pollutants dispersion on the ground or only due to leakage. The first category of sources may create a 'diffuse' effect on the water quality through surface runoff or the percolation process, which may create difficulties in the tracing of the ultimate origin of pollution. The discrete origins of the second category of pollution sources are difficult to find out with accuracy; these discrete origins may be the runoff of agricultural soils containing high amount of fertilizers and pesticides or the urban water runoff etc. The reduction in quality of water due to biological and chemical contamination is a major problem in India also; more than 70% surface-water resources are suffering from this problem. The percentage of pollution of groundwater reservoirs is in increasing trend; because of they are also contaminated by organic, toxic, inorganic and biological contaminants. In several cases, due to health hazards these sources are considered as unsafe for direct consumption by human and animals; these resources are even not utilized for other activities like industrial and irrigation necessities. This indicates that the reduced quality of water can play a key role in water scarcity by limiting its availability for humans and also for the ecosystem. India will have a population of around 1.5 billion by 2040, replacing China, at present the most populous country in the world (Parikh, 2004). According to the data given in Census (2001), the population of India has grown continuously since 1951 (361 million to 1 billion), so that now every sixth person in the world is Indian. Currently the population is around 1.34 billion; according to data available in government website, population density of India was 419.80 people/km^2 in 2020. According to United Nation's projections (www.macrotrends.net, 2021), this will increase around 423.88 people/km^2 in 2021. As India has only 2.4% land of the total world's land with high population density, Indians feel that their resources and environment suffers from a high pressures.

The water pollution level of the country can be assumed by the level of water quality available throughout the India. According to a water quality monitoring organization (CPCB), the quality of water is judged by the level of BOD in water; and the presence of total coliform and faecal coliform in water showed that there is a gradual deterioration in the quality of water (CPCB, 2009). Various water samples were analyses between 1995 to 2009 and it was found that the around 57–69% samples were not able to pass the quality test because the BOD level was below 3 mg/litre; in the year 2007 the number of low-quality samples increased to 69%. During those 15 years it was also noticed that the BOD level of the 17 to 28% samples was between 3 and 6 mg/litre and in 1998, huge number of samples were tested with same concentration of BOD and it was noticed that the maximum samples were in this category with the concentration of 3 to 6 mg/litre. During testing and observation it was noticed that the various samples remained were intact, with prior concentration and a stable tendency was noticed regarding number of samples (%) with 3 to 6 mg/litre BOD concentration. It was also noticed that 13 to 19% samples with more than 6 mg/litre BOD was found during 1995 to 2009. The embarrassing side of this tendency is the higher percentage (19%) of samples expressing deniable BOD level, which may be due to non-compliance of the norms and standards is by charging sources or may be due to high amount of discharge. A high amount of discharge contributes higher concentration of pollutants (Rajaram and Das 2008). In India, another reason of water pollution is poor and scanty infrastructural facilities like lack of monitoring and testing stations, less frequencies of pollution monitoring, lack of trained manpower for testing etc. According to CPCB (2009), under GEMS and MINARS programmes, around 1,700 stations are monitored by CPCB. To proper monitoring it is required that the number of monitoring stations should be increased as early as possible. These numbers should be equal to developed countries (one station/1,935 km^2) which would be more effective for monitoring. The monitoring frequency in the country is not good, because it is noticed that 32% monitoring stations doing it on a monthly basis, around 29% doing on a 1/2 yearly basis and 39% doing monitoring on a quarterly basis (CPCB, 2009). These results indicate that not only is an increase in monitoring stations required but the frequency of monitoring needs to increase. The results acquired by the CPCB in between 1995 to 2009 on water quality indicate that the organic and microbial contamination in water bodies was very critical. The main reason of such type of contamination is of untreated waste-water discharge from domestic and industrial establishments in water resources. Another reason is lack of water flow system in the water bodies for the dilution of the received water. All these reasons are responsible for the increased oxygen demand and bacterial pollution. As per WHO (2007), in India due to shortage of water, lack of hygiene and sanitation around 400,000 lives are lost per year; whereas, air pollution is responsible for the death of significant amount of population (approx 0.52 million people) annually. According to Parikh (2004), around 1.5 million children (below the age of 5 years) die every year, approx 200 million people lost their work per year; apart from these loses, country loosing around Rs 366 billion every year due to poor water quality and water-borne diseases. Therefore, we can understand that socio-economical cost of water pollution is extremely high. Water is an important rechargeable resource which is essentially required for the survival of all the forms of life, economical development, food processing and for agricultural works. There is no substitute of water and it is a very special natural gift to human beings which provided by god. Different sources like surface and underground resources of water of the country play an important role in hydropower generation, irrigation, livestock production, forestry, industrial operations, recreational activities, and animal husbandry etc. Only 0.5% freshwater is available on the Earth's surface and a very small portion (around 0.1%) is covered by rivers which is an insignificant amount of the entire land surface. Around 0.01% of water comes from river channels. Although this amount might seem insubstantial, this water has great significance (Wetzel, 2001). India receives about 4 million litre precipitation per year including 3 million litre monsoon rainfall, which also includes snowfall. According to Kumar et al. (2005), Indian rainfalls depends on the local storms, shallow cyclonic depressions, cyclonic disturbances and north-east and south-west monsoons. Kumar et al. (2005) also suggested that the most of rainfall take place in between June and September due to influence of south-west monsoon and during the month of October and November is due to affect of north-east monsoon. India is enriched with a strong river system, which comprises around 20 big and main rivers and many other tributaries. Many of big and main rivers and tributaries are perennial, whereas few are seasonal. India has only 1/50th part of the total land and around 1/25th part of total water resources available in the world (Water Management Forum, 2003). Over the last few decenniads, the population of India has seen an increasing trend. The consequent rapid growth in industrialization and urbanization has led to a huge demand for freshwater (Ramakrishnaiah et al., 2009). In general, the health of human beings is mostly affected by agricultural activities such as excessive use of synthetic fertilizers and unsanitary conditions (Okeke and Igboanua, 2003). Other anthropogenic activities such as urban development, agricultural development, industrial development, and population increase have led to a deterioration in the quality of water in several parts of the world (Mian et al., 2010, Baig et al., 2009, Wang et al., 2010). While, small water resources have not given liberty to pollutants to enter; therefore, the water quality of these water resources is better than the bigger water resources (Bu et al., 2010). Globally, the protection of quality of water in different resources is enormously urgent due to seriousness of water pollution and lacking of water resources. Therefore, the CPCB has settled numerous monitoring stations on water resources throughout the country; those are managed and governed by state/union territory/central level authorities in India. Around 28 states and six Union Territories are covered by this network (CPCB Report, 2013). Due to worldwide expansion of industries and urbanization, the condition of water pollution is worsening (Jacob et al., 2018; Ibrahim et al., 2016; Naseem et al., 2018; Jiang et al., 2018; Rajeshkumar et al., 2018). Inorganic contaminants (metallic ions) such nickel ions, lead ions, and chromium ions are dissolved in water through industrial effluents (Morosanu et al., 2017). The eviction of these toxic metallic ions from the wastewater and drinking water is very essential because they damage human organs when they are ingested in the body through food and water as a non-biodegradable pollutant (Jaishankar et al., 2014).

2.2   WATER QUALITY

The quality of water is completely depends on its physical, chemical, and microbiological properties, on the basis of the quality water is used for a particular purpose such as irrigation, industrial, or drinking purposes. For each of the mentioned uses, water having different types of physical, chemical and microbiological standards. These required standards are set and monitored by the state/central/local governments. In general it is found that the higher standards are mandatory for drinking water and swimming pool water, whereas the water that is used for agricultural and industrial purposes may have comparatively low standards. The quality of water is widely affected by a various natural events and human activities. Climatic, geological, and hydrological factors are the most important natural factors that have a great influence in the quality and quantity of available water. The ecosystem is consistent with the quality of naturally produced water; any prominent change in the quality of water usually will not be deteriorative to the ecosystem. The term quality is used to denote the required properties of the products intended for a particular purpose; when this term is used for water, it speaks out about the quality of water required for the various purposes like for drinking, irrigation or industrial processes. Each intended use of water requires some definite properties like microbial, chemical and physical attributes; for examples, for drinking water prescribed level of TDS, pH, toxic content, and temperature and pH for aquatic life etc. The terms 'water quality' can also be defined by the range of parameters that are responsible for limits the use of water. The quality and composition of surface water and sub-terrestrial water is subjected to various natural factors such as the topographical, hydrological, biological, geological, and meteorological conditions of the basins, which always vary with weather and water levels. Human interference also has serious effects on the quality of water and quality of water resources. The result of hydrological changes such as making dams, wetlands draining, and change in water flow are the some common examples of these serious effects. More noticeable activities of pollution are domestic discharge, effluent from industries, intentional or unintentional discharge of urban wastewaters into water recourses, and unlimited application of chemicals and fertilizers in the agricultural activities. Pollution of water resources by human faeces is also a reason for water pollution, which has bad impacts on water quality; this type of pollution may be due to lack of community facilities for collection, disposal and treatment of waste facilities or lacking of on-site facilities like latrines. Faecal pollution creates intestinal diseases in humans and it is a big problem of developing countries, whereas eutrophication and heavy organic load could be a bigger affair in developed countries. The eutrophication may not be only from point sources, but it may also be from diffusible sources such as runoff from over fertilized agricultural soils (with high nutrient content like nitrogen and phosphorus) or livestock feedlots.

The excellence of water resources is a bigger concern and can be described by expressions such as water quality, the biological load and composition found in the water bodies, chemical and physical properties of the particulates present and hydrology and dimensions of water bodies. Water pollution has very harmful effects on human, animal, and living water organisms, and these effects may:

i.   be dangerous to human well-being;
ii. be harmful to organisms and animals
iii. create obstacles to aquatic activities; or
iv. be destructive of water qualities.

Several types of water bodies are found in nature, including lakes, reservoirs, surface water and groundwater resources; all are unified with each other by the hydrological rotation with many intermediate water bodies (both natural and artificial).

2.3   NATURAL PROCESSES AFFECTING WATER QUALITY

Over the last few decades, water quality has been degrading significantly, deprivation in water quality mainly resulting from the human activities. However, certain natural phenomena are also responsible for the degradation of water quality. Natural phenomena such as torrential rainfall and hurricanes increases suspended solids in the rivers and lakes due to excessive erosion and landslides. Some lakes bring water to the surface with little or no dissolved oxygen during seasonal overturn. These types of seasonal overturn can happen occasionally or frequently. In some areas due to permanent natural conditions, water quality is degraded to such an extent that it cannot be used for drinking neither for irrigation. Salinization of surface water, salt content in ground water, carbonated (hardness) of ground water, and particular ions (e.g. fluoride) and toxic elements (like selenium and arsenic) in ground water are a few common examples of natural conditions which affect the water quality. Thus, treatment of water is very essential so that it can be used in certain industries. The nature and chemical property of freshwater bodies alters due to various natural processes which are occurring inside it. These natural processes may be physical, biological, chemical, or hydrological. To a great extent water quality is affected by environmental factors. These environmental factors are mostly brought by the under certain climatic conditions, geological and geophysical changes. The few vital factors related to environmental situations are stated below:

2.3.1 DISTANCE FROM OCEANS

Oceans are rich in various different types of ions such as Na^+, Cl_2^-, Mg_2^+, SO_4, etc.

2.3.2 CLIMATE AND VEGETATION

Climate and vegetation have great effect on the amount of dissolved material in the water system. This is mainly through the evaporation and evapo-transpiration. At the particular location, regulation of erosion and mineral weathering is depends mainly on their climate and vegetation.

2.3.3 ROCK COMPOSITION (LITHOLOGY)

Composition of rocks varies according to the natural weather conditions. It is found that the vulnerability of rocks to weathering, e.g., for granite it is 01; 12 for limestone and for rock salt its 80.

2.3.4 TERRESTRIAL VEGETATION

The content of organic matter, nitrogenous compounds, and carbon in a water system depends to a great extent on the production of terrestrial plants and their tissue degradation in soil.

2.3.5 AQUATIC VEGETATION

Water vegetation has various significant consequences on the alchemy of the lake and river water system. The composition (nitrogenous, phosphorous nutrients etc.) and quality (pH, carbonates, DO, oxidative and reductive chemicals) of the water system depends on the intensification, demise and decay of aquatic vegetation and algae to the great extent.

2.4 WATER POLLUTION

Water is a true gift given to all living creatures by the nature. Water is a renewable resource and is essential for sustaining life. We cannot imagine our life without water; thus it is very important to maintain the water quality. Water is extensively used in process industries and it plays important role in economic development. Another advantage of a water system is that it can be easily manageable. Water systems can be diverted, transported, stored, and recycled. Water systems (surface and ground water) are essential to sectors such as irrigation, hydropower generation, process industries, livestock production, fisheries, forestry, and many more.

In the ecosystem, only 0.5% of the Earth's surface is covered by fresh water with the total volume of 2.84×10^5 km³; whereas rivers shares about 0.1% of the Earth's surface, carrying just 0.01% of the earth water. Despite this limited amount, this river water is essential for the planet and it has a special respect and significance (Wetzel, 2001). Around 4,000 km³ rainfall is recorded per year in India, including snowfall and annual monsoon rainfall (3,000 km³) in total precipitation. Indian rainfall is usually based on the south-west and north-east monsoons. Several other significant factors like such as storms and shallow cyclonic depressions and disturbances also effects the rainfall (Kumar et al., 2005). But rainfall from June to September in India (excluding Tamil Nadu) is mostly reliant on the south-west monsoon. Whereas, rainfall (October to November) in Tamil Nadu is reliant on the north-east monsoon (Kumar et al., 2005). Fortunately, India is blessed with more than 20 important rivers and a number of tributaries. Among these, few rivers are seasonal and few are perennial. Even though India just has geographical area of about 3.29 million km², which is just around 2.4% of the world's land area, but it supports 15% of the world's total population. In addition, according to Water Management Forum, 2003, overall India supports about 16.6% of world's population with 4% of world's water resources and 2% of world's land. Over the past few decades, demand for fresh water increased tremendously due to increasing population and industries (Ramakrishnaiah et al., 2009). The health of human beings and animals is in danger due to agricultural developments and especially excessive use of fertilizers in agricultural lands (to improve productivity) and also due to unsanitary conditions (Okeke and Igboanua, 2003). In many countries, water quality has degraded significantly due to various anthropogenic activities such as urbanization, use of fertilizers in agriculture, industrialization and increasing population (Baig et al., 2009, Mian et al., 2010, Wang et al., 2010). In addition, undersupplied water resources have gradually controlling water pollution and improving water quality (Bu et al., 2010). Government and researchers have focused attention for saving water quality and reducing water pollution.

Water is polluted by point and non-point sources. When the cause of water pollution is directly identified then it is called as point source such as municipal wastewater and effluent of industries directly discharging to the river. Non-point sources of pollution means ground water contamination due to different unrecognizable sources such as percolating through agricultural field and urban wastes. Sometime it is seen that pollutant recharged at one place and its effects are observed more than 100 and even 1,000 miles away. For example, radioactive waste that contaminated seawater can spread to nearby countries and is called as transboundary pollution.

2.5. MAJOR SOURCES OF WATER POLLUTION

Some of the most important sources responsible for water pollution are discussed below.

2.5.1 URBANIZATION

Urbanization causes increase in phosphorus content in the natural water bodies situated in urban areas (Paul and Meyer, 2001). After agriculture, urbanization is the second major cause of stream impairment. This is because of the increasing runoff from the urbanized surfaces and increasing imperviousness. Another reason is its municipal wastewater and industrial effluent discharges which increases nutrients content in urban streams.

2.5.2 SEWERAGE AND OTHER OD WASTES

Solid waste management is not effective as large amount of non-biodegradable and organic wastes are created every day. As an end result, most of the junk is not disposed scientifically in India. This increases the burden on the environment by further penetrating into the groundwater. Sometimes, sewerage can act as a fertilizer as it has nutrients such as N and P, which are required by plant life and natural world for their growth. Also, synthetic fertilizers that are utilized by the farmers to improve soil nutrient may vent into river or sea which provide fertilizing effect to the sewerage. The massive growth of algae or plankton in oceans, lakes and rivers (known as algal bloom) is mainly because of the presence of sewerage and fertilizers in water systems. This will further result in the reduction of the dissolve oxygen (DO) content which kills other aquatic life like fish, tortoise etc.

2.5.3 INDUSTRIAL EFFLUENT AND WASTES

Industrial effluents are other major reason for water pollution. Most of the commercial manufacturing units are situated at riverside. Industries like steel manufacturing units and paper manufacturing units requires huge amount of water for their manufacturing and produces large amount of effluents. Industrial effluents contain large amounts of unwanted components such as acids, alkalies, dyes, and other chemicals. These effluents are discharged into rivers. Different chemical industries discharge different type harmful compounds into the water bodies such as effluents of Aluminium industries discharge large amounts of fluoride; effluents of fertilizer industries discharge large amount of ammonia; effluents of steel industries discharge cyanide etc. Chromium salts which are used for the production of $Na_2Cr_2O_7$ and other allied chromium compounds, which act as toxicant in water bodies. Discharge of such effluents pollutes water bodies and affects human health and that of other organisms consuming it.

2.5.4 AGRO-CHEMICAL WASTES

Farmers uses water and electricity for irrigation. In agricultural segment, electricity and water are subsidized, which leads to wastage of precious water. For irrigation purposes farmers do not use optimized practices such as drip and sprinkler irrigation; instead they prefer to use traditional and wasteful methods like flood irrigation. The Indian government provided water at subsidized rate but has not taken necessary action to encourage them for judicious use of water. Breaches and seepage cause losses of water, which results in waterlogging and salinity. Agriculture sectors extensively uses fertilizers and pesticides (such as DDT, Aldrin, Dieldrin, Malathion, Hexachloro Benzene etc.) to increase the productivity of the crops. Improper disposal of these agro-chemical wastes contributes a huge amount of pollutants into the water system as well to the soils. Pesticides pollute the water bodies through surface runoff, spray drifting, precipitation washing, and application of pesticides by aspersing or by spraying it directly in low lying areas. Most of the pesticides are non-degradable in nature and remain in the soil, water, and air longer as residues. These agro-chemical wastes are consumed by humans and animals through food, water and air, which further leads to bio-magnification.

2.5.5 NUTRIENT ENRICHMENT

Nutrient in water systems can pollute them and it comes from various sources but can be broadly alienated into natural and manmade types. Natural sources have little contribution to water pollution as natural systems establish balance between the generation and utilization of nutrients over the specified period of time. Manmade processes have major contribution to water pollution. Anthropogenic sources of nutrients are the nutrients that comes from the agriculture, domestic and industrial wastes. Amount of nutrients in river and seas are significantly correlated with the disturbance gradients and land used by mankind. Nitrogen and phosphorus fertilizers used in the agricultural sector is the main source of nitrogen and phosphorus in the river water. Total nitrogen flux in rivers situated nearby North Atlantic Ocean (temperate-zone) is mainly corresponding to the net human generated input of nitrogen in their watersheds (Howarth et al., 1996; Goolsby and Battaglin, 2001). Total N and NO_3^- amounts in rivers mainly correspond to the human population density (Howarth et al., 1996). Nutrient enrichment in aquatic systems through anthropogenic sources can be further divided into point and non-point sources (Carpenter et al., 1998). Point anthropogenic sources are quite easier to supervise and control. On the other hand non-point anthropogenic sources like fertilizers, cattle and urban runoff demonstrates more spatial and temporal unevenness. These non-point sources are the main source of water in the United States. In response to the various acts related with clean water act, point source inputs should be strongly regulated (Carpenter et al., 1998).

2.5.6   THERMAL POLLUTION

Quality of water and aquatic biota are significantly affected by the change in temperature. Change in temperature in water system is called as thermal pollution and is mostly caused by the human activities. Industrialization is the major cause of thermal pollution. Industries such nuclear and hydropower plants, refineries, mineral industries, etc. use equipment which release high-intensity thermal radiation. This thermal radiation changes the biological and physic-chemical characteristics of the receiving water systems. Rise in water temperature is responsible for reduction of the amount of oxygen in water, which affects the reproductive cycle of the aquatic life.

2.5.7   OIL SPILLAGE

Oil spillage into the water bodies is another of water pollution, which hampers the aquatic life. Oil discharge into the sea comes through the oil exploration from offshore. Oil discharges into the river by the accident or leakage of petroleum products such as petrol, diesel, kerosene etc. from the cargo tankers which carry them. When this oil enters into the water system, it spreads on the water surface and forms a thin layer of water-in-oil emulsion. This thin layer of oil cuts the supply of oxygen inside the water system. Thus, aquatic life doesn't get sufficient oxygen for to sustain itself. Therefore, it is very important to control the oil spillage into the water system.

2.5.8   DISRUPTION OF SEDIMENTS

Water systems can also be polluted by the discharge of sediments into it. During the constriction works near rivers or seas, sediments (from soil, rock etc.) flows in the water system which increases the water turbidity. These extra sediments can sometimes block the gills of fish, suffocating them. Also, fabrication of dams for generation of power from hydro or water reservoirs can help to minimize the flow of sediments. This reduction in flow of sediments can affect the creation of beaches. Further, it enhances the coastal erosion and also reduces the supply of nutrients from rivers to sea. This leads to significant reduction in coastal fish stocks.

2.5.9 ACID RAIN POLLUTION

When SO_2 and Nare O_2 emitted to the atmosphere through either natural or man-made sources it reacted with the H_2 and O_2 present in the atmosphere and forms H_2SO_4 and HNO_3 in the air. These chemical substances precipitated in the form of snow and rain etc. and fall on the earth surface. This acid rain can alter plant's surrounding pH level and can even harm or kill the plant. Acid rain is a one of the major sources of water pollution. When, acid rain reaches the ground, flow into the water system that carries harmful acids. Further, when acid rain collected in water bodies it lowers the pH level of aquatic environment which affects the water lives.

2.5.10 RADIOACTIVE WASTE

Radioactive materials cause radioactive pollution in the water systems. Small dose of radioactive materials can stimulate the short-term metabolism and big dose can gradually injury the microbes initiating inherent mutation. Radioactive pollution arises in the water system from the various sources such as radioactive sediment, effluent discharge from nuclear plants, radioactive minerals utilization, application of radioisotopes in the field of medicine and research.

2.5.11 CLIMATE CHANGE

Another source of water pollution is global warming, which has a significant influence on the water quality. The global warming leads to increased evaporation, changed geographical conditions, moisture content of soil and increased intensification of droughts and floods, these leads to have serious impact on water resources. Using climate models for future projection of India, it is found out that monsoon rainfall in most of the part will increase with the increase in greenhouse gases and sulphate aerosols. In dry and semi-dry areas (North-West region), a small degree of climate change can lead to huge impact on the water resources. This will further affect drinking water, restricted water supply for hydropower generation and causing agricultural land degradation. Apart the monsoon rain water, India using perennial rivers (originates from Hindu kush and Himalaya) for water requirement, these rivers are depending on the glacial melt. Melting season of glacial coincides with the summer monsoons which eventually contribute to disasters like floods in the water catchments situated at Himalayan regions. Increasing temperature also raises the snowline which reduces the capacity of natural basins. This increases the flash flooding risk during wet season. Rising temperature further increases excessive richness of nutrients in wetlands and fresh water resources (CPCB Report, 2013).

2.6 CATEGORIES OF WATER POLLUTANTS

Water-contaminating agents or pollutants are of different types, and can be grouped as follows:

i. Organic and biotic pollutants
ii. Inorganic or abiotic pollutants
iii. Radiogenic pollutants
iv. Suspended material (Dissolved solids)
v. Pathogenic microbes
vi. Nutritional and agricultural pollutants
vii. Heat as pollutants.

The brief information about these pollutants is given below:

2.6.1 ORGANIC AND BIOTIC POLLUTANTS

Biotic pollutants are the components rich in C, H, O, N and S; these pollutants are also termed as 'decayed organic matters'. Organic compounds are usually generates from various sources such as sewerage, chemicals, urban drains, industrial effluents and decay of materials like forest plants, agricultural waste, birds, fish, microbes etc. Organic compounds may be of different types, best examples are $C_{18}H_{34}O_2$, $C_{16}H_{32}O_2$, $C_{12}H_{23}ClO$, C_2H_5OH and various other organic acids (Ekevwe et al., 2018).

2.6.2 INORGANIC OR ABIOTIC POLLUTANTS

Most of the developing countries like us are very much concerned about pollution of drinking water through injurious chemical contaminants such as NO_2, NH_4NO_3 and heavy metals (Hg, Cd, As, Cr, Tl, and Pb). Whereas in case of river water, developing countries are suffering from high levels concentration of inorganic and abiotic contaminants such as inorganic nitrogenous pollutants such as NO_2, NO_2^-, NH_4^+, and inorganic PO_4^{3-}. This type of pollution is due to mixing of contaminated water from agricultural lands, mixing of municipal sewerage or industrial water etc. into river water. Inorganic contaminants can create many health-related problems not only in humans but also in animal and aquatic lives. For example, nitrites are carcinogenic compounds and can create cancer in different parts of human body such as the oesophagus, stomach, and liver; it is also responsible for high concentration of NH_4^+ in the human body (Shah et al., 2000).

2.6.3 RADIOGENIC POLLUTANTS

Natural radioactive stuff is generated from the Earth's crust and then contaminating the surface water. A good amount of radioactive stuff is anthropogenic, emitted from the different activities and sources such as nuclear power plants, weapons testing, manufacturing of radioactive equipments and use of radioactive goods. Generally, aluminium, thorium and uranium along with the naturally occurring uranium, radium and the radioactive gas radon (radionuclides) series appears in drinking water. These pollutants are very much harmful for human health, they can cause spooky effects on human health. Radium and uranium are responsible for bone, apart from these effects uranium has toxic effect on kidney also (Alireza et al. 2010).

2.6.4 SUSPENDED MATERIAL (DISSOLVED SOLIDS)

Suspended or dissolved solids are the pollutants from the materials present in the water as suspensions, the effluent of municipalities and industries, sewerage and waste treatment plants are rich sources of these materials. There are three major types of suspended or dissolved solids are found in waste water: (a) organic content (b) sand and other solid material, and (c) dissolved solids (Bhateria and Jain, 2016).

2.6.5 PATHOGENIC ORGANISMS

Pathogenic organisms are very small living organisms which includes pillows, bacteria, moulds, viruses and various parasites that can cause diseases in human and animals. Viruses and bacteria are commonly found in wastewater; viruses (Norwalk, Hepatitis etc.), bacteria like E. Coli and Salmonella and a very common mould (Candida) are the important pathogenic organisms commonly present in water. These pathogenic microorganisms can create various diseases like cholera, diarrhoea, and other gastrointestinal problems. Some of the microbes are responsible for food poisoning e.g. salmonella bacteria, E. coli (Ksoll et al., 2006).

2.6.6 NUTRIENTS AND AGRICULTURAL POLLUTANTS

Environmental problems are also created by human activities, such as the use of chemical and fertilizers in agricultural fields. Some fertilizers contain significant amount of heavy metals like Cd and Cr along with high concentration of radionuclide. Inorganic fertilizers contain salts of NO_3^-, P, K, and NH_4^+. Industrially produced fertilizers contains excessive amount of different heavy metals like Cr, Cd, As, Cu, Hg, Pb, and Ni etc. When these fertilizers are applied in fields in excessive amount to get maximum crop production, required amount these metals are absorbed by the plants and rest amount is deposited in the soil. The plants those absorbs excessive fertilizers from the soil; they can enter in the food chain that can pollute the water (Savci, 2012). The soils with high concentration of these metals can also pollute the water.

2.6.7 THERMAL POLLUTION

The change in temperature in the water bodies is an example of thermal pollution of water. This type of pollution is normally generated by thermal power plants; the discharge of these power plants is raising the temperature of water resources up to 10 °C higher than the normal temperature. Therefore, it has very dangerous effects on the water animals and other aquatic lives due to excessive heat. It also plays an important role in the global warming, the use of renewable energy sources (solar power, wind power etc.) may be the good solution to minimize the problem. Pollution by temperature can be decreases by increasing the plantation and by reduction of CO_2 emissions in environment (Langford, 1990).

2.7 CONTROL MEASURES FOR WATER POLLUTION

Everyone understands the importance of clean water. Yet, we are unintentionally doing many things which we do contribute to water pollution in different ways. Water quality management is a key challenge in India due to uneven geographical distribution of water sources and temporal and spatial variation of rainfall. Over the past few years, water resources are persistently in drought conditions, due to the progress of commercialization and industrialization. Furthermore, lack of awareness and inefficient practices and embarrassment of other conditions led to water pollution. Polluted water adversely affects the all life forms. Rather than to discharging sewerage and industrial effluents into water bodies directly, it is better to treat them before discharging (CPCB Report, 2013). Some of the important controlling actions are given below:

i. Implementation of the National River Action Plan and Ganga Action Plan for the trapping and diversion of river water. Further, to addressing the problems and treatment of sewerage and industrial effluents before discharge it into the rivers or water bodies.

ii. In most of the country parts, due to inadequate sanitation facilities domestic waste water is hardly treated. This domestic waste water contains large amount of organic polluted with them. Further, this polluted waste water reaches to ground water and near vicinity of human habitation from where it is further consumed by human. Therefore, considerable investments are being done for the installation of water treatment plant.

iii. With the rapid progress of commercialization and industrialization, the requirement of water increased tremendously. It is estimated a rise of approximately 191.000 million m^3 (18%) of the total requirement in 2025 (CPCB Report, 2013). The effluents discharge from the various process industries are highly toxic and contains large amount of organic wastes. These toxic wastes into the surface and groundwater and pollute them. This polluted water is then used by irrigation and domestic purposes. Therefore, various regulations are enforced regarding liberation of industrial effluents and maximum value of the toxic chemical concentration level to the permissible limit value. Yet, more incentives are required for the promotion of recycling and reuse of waste water.

iv. The agriculture sector requires large amount of water for the growth of the crops. Water and electricity are subsidized in this sector for political reasons. This leads to wasteful flood irrigation instead of the adoption of irrigation methods such as sprinkling method and drip method. Therefore, advanced agricultural practices, cropping patterns and optimized irrigation processes should be encouraged.

v.   The Water Act, 1974 was revised in 1988 to preserve the quality of water. The objective of this Act is to for the prevention and control of pollution in water by stopping and managing the water pollution for national aquatic resources. The Water Cess Act was also passed in 1977 for the same purpose. This Act implemented levy and collects cess on water consumption by mankind and industrialists.

vi. For aquatic resources, several monitoring station were established across the country by the CPCB (Central Pollution Control Board). At the monitoring stations water quality is monitored and managed. According to the report of CPCB, (2013), network of these monitoring stations covers 28 states along with six Union territories. Therefore, water quality management is a pre-requisite for the maintenance and restoration of water systems.

vii. Washing cloths and laundry near bank should be prohibited.

viii. Effluent treatment plant (EPT) must be installed in the industry to control the water pollution at the source.

ix. Sewerage Treatment Plants (STPs) must be installed in every town and cities to clean sewerage effluent before discharge.

x.   Fertilizers, herbicides and pesticides must be properly used for crops growth and adoption of organic farming should be encouraged. In riparian zone, cropping practice should be prohibited in order to protect the riparian plant life (vegetation) growing there.

xi. Religious rituals that can pollute river water should be banned.

xii. In order to preserve the water table, rainwater harvesting should be accomplished.

xiii. There should be a proper awareness programme to educate people about the importance of water and how they can preserve its quality.

xiv. Polluter pays principle should be there to impose penalties to the polluters. This will prevent mankind from polluting.

xv. Coastal plant life (vegetation) helps keep water fresh and clean in rivers and the ocean etc. due to its several functions. Therefore, to stop people from the use of forests of the riparian zones for creation of roads, leisure and seeing the sights, agricultural activity, quarrying and clay withdrawal and sand mining etc., the people of the society should take part in a regulatory role.

2.8   EFFECTS OF WATER POLLUTION

According to WHO (2007), India losing around 400,000 lives due to lack of hygiene, sanitation, and water; whereas 520,000 deaths are reported due to air pollution annually. Environmental factors are responsible for the death at early age (60 years) in India. The socioeconomic value of water pollution is too high because around 1.5 million children of below five years die per year due to water diseases and approximately 200 million days of work are lost annually. Due to water diseases about Rs 3.66 billion are lost every year in the country (Parikh 2004). Similar results were also observed by McKenzie and Ray (2004) due to effects of pollution in water; they also observed in a study that India loses millions of days per year and human disabilities occurs per year due to water borne diseases, poor sanitation, water quality, and hygiene. Vast tracts of Indian groundwater resources are polluted with arsenic (As) and fluoride (F⁻). Orissa and Rajasthan are two states where fluoride is a major problem; apart from these states, around150 districts of 17 states in India having more or less same problem. Higher amount of fluoride in potable water creates 'Fluorosis' in human and animals, which leads to several bad effects such as weakening of bones, weakening of teeth, and lack of blood. The arsenic is a poisonous and carcinogenic compound found in the groundwater of Gangatic regions and causing health problems. Around 35 million to 70 million people in Bihar, West Bengal, and Bangladesh are affected by arsenic. As estimated by Murty and Kumar (2004), the industrial water pollution is cost decreased and they observed that in India it is around 2.5% of total industrial GDP; Parikh (2004) stated that the cost of rectification is very low as compared to damage cost. It can be easily understood by these estimates and adverse effects, that the abatement of pollution is the only way to maintain desirable social health and economic status.

2.9   ROLE OF AGRICULTURAL WASTE BIOMASS IN TREATMENTS OF POLLUTED WATER

In recent years, the increase in waste materials disposal have been of concern due to the large amount of solid waste produced from the agricultural industry worldwide. Intensive research is continuously undertaken to provide possible alternatives for recycling agricultural wastes. These agricultural wastes have been investigated for various purposes: for example, as bio-sorbents for the removal of heavy metal contaminants from the wastewater. Low-cost agriculture wastes either without or with processing have high surface areas with encouraging microporous characteristics and surface chemical nature to be used as adsorbents for heavy metals. They have been known as potential low- cost bio-sorbents. In general, a low-cost bio-sorbent is defined as a by-product that is available abundantly in nature with a reasonable adsorption potential. Many agricultural wastes have revealed promising adsorption ability; this includes olive stones, orange wastes, grape stalk waste, papaya wood, broad bean, peas, medlar peels and peels of lemons, oranges, grapefruits, apples etc, apple seeds, core of apples, skin of grapes, different products of coconut waste such as shell powder, copra meals coconut husk etc. and neem bark. The agricultural wastes are mainly composed lignocellulosic components which refer to lignin, cellulose and hemicelluloses. They possess numerous functional groups i.e.; carboxylic acids, phenolic, carboxyl and hydroxyl that act as the precursor for adsorption, in addition, their spongy and buoyancy properties have· also enabled the adsorption process to occur easily. The agriculture wastes are viable bio-sorbents because they are abundantly available, low-cost and environmental friendly which generate very low greenhouse emissions.

Currently, it is noticed that the peoples are taking interest in the low-cost, commercial and easily available agricultural waste materials for the commercial purpose as absorbent for heavy metals. Adsorption technologies have various advantages, such as its efficiency to reduce heavy metal ions even at very low concentration and its inexpensive material requirement. Different agricultural waste materials has been utilized to develop metal adsorbents to remove Pb_2^+ ions from wastewater and also reviewed for their technical feasibility. Generally the metal binding and metal removing capacity of these materials depends upon the dose of sorbent, optimum pH of substrate, substrate temperature, and contact time etc. The activated carbon is widely used for the treatment of wastewater, but due to its high cost, its commercial use is very limited. The application of agricultural waste and by-products as an adsorbent material for the purification of water contaminated by heavy metals has become more popular since last few decade because they are biodegradable, less costly, efficient, abundant, and easily available. From the available and reviewed literature it is noticed that more attention was given to inexpensive waste materials such as agricultural and forest waste and it was found that these materials could be the alternative adsorbents they have binding capacity to remove metal ions from wastewater. As per Njoroge (2007), the major environmental contaminants are wastes from various sources, which may contain heavy metals and regular organic contaminants such as harmful chemicals, PCB, furans, dioxins etc. These wastes may also contain toxic and infectious components or radioactive materials (Njoroge, 2007).

Various other methods such as filtration, coagulation, precipitation, ozonation, ion exchange, RO, microfiltration and advanced oxidation processes are also commercially used for the removal of organic contaminants from the wastewater. But due to very high application cost, the use of these processes have been limited for this purpose; even then, reverse osmosis (RO) and ions exchange are more satisfying methods because of their best capacity of pollutants removal. The process of adsorption is commonly used for the treatment of wastewaters released by different industries and that are contaminated by inorganic and organic pollutants. At present, agricultural produces such as cereal crops, fruit and vegetable crops, by-products of agro-produce and waste are the useful waste materials, requiring reasonable disposal. If these wastes will be burned, then they will produces more carbon dioxide and creates environmental pollution. Therefore, it is needed that the agricultural waste should be get converted into useful, optimistic and valuable products. Use of these wastes as an ion exchange or adsorbent material could be a reasonable and meaningful way, which helps to remove toxic elements from the wastewaters. In between 2000 to 2010, several agricultural waste materials has been utilized to develop metal adsorbents, few selected work is given in Table 2.1 as examples of utilization of agricultural waste biomass as adsorbents.

TABLE 2.1
Researches on Utilization of Agricultural Waste Biomass as Adsorbents

Agricultural wastes biomass (as adsorbents)	Year wise references
Rice husk ash	Feng et al. (2004)
Husk of Black gram	Saeed et al. (2005)
Cereal chaff	Han et al. (2005)
Maize bran	Singh et al. (2006)
Rice husk	Zulkali et al. (2006)
Capsicum annuum seeds	Özcan et al. (2007)
Arca shell	Dahiya et al. (2008)
Apricot stone	Kazemipour et al. (2008)
Antep pistachio	Yetilmezsoy and Demirel (2008)
Almond	Kazemipour et al. (2008)
Rice stem	Gupta et al. (2009a, 2009b)
Almond	Pehlivan et al. (2009)
Bamboo dust	Kannan and Veemaraj (2009)
Banana peels	Gupta et al. (2009a, 2009b)
Ash of Rice husk	Naiya et al. (2009)
Citrus peels	Schiewer and Balaria (2009)
Coffee residue	Boudrahem et al. (2009)
Orange peel	Lugo-Lugo et al. (2009)
Activated charcoals (bamboo)	Lalhruaitluanga et al. (2010)
Banana peels	Anwar et al. (2010)
Corncobs native	Tana et al. (2010)
Charcoals raw (bamboo)	Lalhruaitluanga et al. (2010)
Corncobs chemically modified	Tana et al. (2010)

2.10 CONCLUSION

Water is one of the most importance substances on earth. All life forms need water to survive. Apart from survival, water is now required for numbers of activities in our everyday life from cooking the food to cleaning the things. However, it adversely affects all life forms when it gets polluted. It is important therefore, to save water from its depletion and pollution. There is a need of taking early action before it adversely affects our lives. Generally, water pollution is induced by humans. Major sources of water pollution are industrialization, urbanization, deforestation, unorganized agricultural practices, and social and religious practices. Thus, it is important to be well organized and responsible enough to save, protect, and preserve water. There are varieties of treatment processes have been developed over the years which control water pollution to a large extent. However, most of these techniques are complicated and require high cost and energy consumption. Removal of heavy metals from the aqueous solutions using various adsorbents has proven to be a very promising technique. Activated carbon is the most popular adsorbent, but is very expensive and requires regeneration process to clean. Over the past few years, adsorbent materials derived from agricultural wastes have been used for the effective removal of heavy metals from the wastewater. These agricultural wastes are abundantly available in nature and are of low cost. The adsorption efficiency of adsorbent not only depends on the adsorbents properties, but also on various adsorption parameters such as pH, temperature, concentration, contact time, particle size of adsorbent, etc. From the past studies, it is observed that the some of the agricultural waste such as citrus peels, cocoa shells, leaf powder A. indica (neem), leaves peepul, leaves, Casuarina glauca tree, maize bran, mango peel waste, orange peel, palm shell, onion skins, ponkan peel, rice husk ash, rose petals, seaweed Ascophyllum nodosum, Senecio anteuphorbium, sugarcane bagasse, valonia tannin resin, album leaves and wheat bran have better adsorption properties than the activated carbon. Agricultural waste biomass is widely utilized for the removal of heavy metals and other impurities from waste water, more research can be conducted for refining of these adsorbents and their improved properties.

ACKNOWLEDGEMENT

We are really thankful for the valuable help taken from the scholars and researchers whose research articles are cited and written as references. We are also grateful to many other peoples those are authors, editors or publishers of the books, chapters and articles, from which related literature and evidences are cited. We would also swell with pride to thank my parent institute (Birla Institute of Technology, Mesra, Ranchi) who provide us a healthy and peaceful environment for such academic activities. Last but not least we would like to mention our special thanks to our family members for their constant support.

REFERENCES

<a href='www.macrotrends.net/countries/IND/india/population-density'>India Population Density 1950-2021</a>. www.macrotrends.net.

Alireza, B., Mohammadi, S., Mowlavi, A., Parvaresh, P. (2010) Measurement of heavy radioactive pollution: radon and radium in drinking water samples in Mashhad. Int J Curr Res. 10: 54–58.

Anwar J., Umer S., Waheed-uz-Zaman, Salman M., Dar A., Anwar S. (2010) Removal of Pb(II) and Cd(II) from water by adsorption on peels of banana. Bioresour Technol. 101(6): 1752–1755.

Baig, J. A., Kazi, T. G., Arain,M. B., Afridi, H. I., Kandhro, G. A., Sarfraz, R. A., Jamali, M. K., Shah, A. Q. (2009). Evaluation of arsenic and other physico-chemical parameters of surface and ground water of Jamshoro, Pakistan. Journal of Hazardous Materials. 166: 662–669.

Bhateria, R., Jain, D. (2016). Water quality assessment of lake water: a review. Sustain. Water Resour. Manag. 2: 161–173. https://doi.org/10.1007/s40899-015-0014-7

Boudrahem, F., Aissani-Benissad, F., Aıt-Amar, H. (2009) Batch sorption dynamics and equilibrium for the removal of Pb(II) ions from aqueous phase using activated carbon developed from coffee residue activated with zinc chloride. J Environ Manag. 90: 3031–3039.

Bu, H., Tan, X., Li, S., Zhang, Q. (2010). Water quality assessment of the Jinshui River (China) using multivariate statistical techniques. Environ Earth Sci. 60: 1631–1639.

Carpenter, S. R., Caraco, N. F., Correll, D. L., Howarth, R. W., Sharpley, A. N., & Smith, V. H. (1998). Nonpoint pollution of surface waters with phosphorus and nitrogen. Ecol App. 8(3): 559–568.

Census of India 2001, Series-1 India, Paper-1 of 2001.

Central Pollution Control Board [CPCB] (2009), 'Status of Water Quality in India—2009', Monitoring of Indian Aquatic Resources Series, MINARS/ /2009–10, New Delhi.

CPCB Report. (2013). Status of Water Quality in India, 2011. Monitoring of Indian National Aquatic Resources, Series: MINARS/35/2013-14: 1–212.

Dahiya, S., Tripathi, R. M . A., Hegde, G. (2008) Biosorption of heavy metals and radio-nuclide from aqueous solutions by pretreated arca shell biomass. J Hazard Mater 150: 376–386.

Ekevwe, A. E., Isaacl, A., Baertholomew, G., and Aroh, O. (2018) Review of organic pollutants in wastewater along the Course of River Gwagwarwa and River Rafin Malam in Kano State-Nigeria. Journal of Biotechnology and Bioengineering, 2: 36–39.

Feng, Q., Lin, Q., Gong, F., Sugita, S., Shoya, M .(2004) Adsorption of Pb(II) and mercury by rice husk ash. J Colloid Interface Sci. 278: 1–8.

Goolsby, D., Battaglin, W. (2001). Long-term changes in concentrations and flux of nitrogen in the Mississippi River Basin, USA, Hydrological Processes, 15: 1209–1226.

Gupta, S., Kumar, D., Gaur, J. P. (2009b) Kinetic and isotherm modeling of Pb(II) sorption onto some waste plant materials. Chem Eng J 148: 226–233.

Gupta, V. K., Carrott, P. J. M., Carrott, M .M .L., Suhas, R. (2009a) Low-cost adsorbents: growing approach to wastewater treatment—a review. Crit Rev Env Sci Tech. 39: 783–842.

Han, R., Zhang, J., Zou, W. J., Liu, S. H. (2005) Equilibrium biosorption isotherm for Pb(II) ion on chaff. J Hazard Mater. 125: 266–271.

Howarth, R. W., Billen, G., Swaney, D., Townsend, A., Jaworski, N., Lajtha, K., Downing, J. A., Elmgren, R., Caraco, N., Jordan, T., Berendse, F., Freney, J., Kudeyarov, V., Murdoch, P., and Zhu, Z. L. (1996). Regional nitrogen budgets and riverine N&P fluxes for the drainages to the North Atlantic Ocean: natural and human influences. Biogeochemistry, 35: 75–139.

Ibrahim, W. M.; Hassan, A.F., Azab, Y. A. (2016). Biosorption of toxic heavy metals from aqueous solution by Ulva lactuca activated carbon. Egypt. J. Basic Appl. Sci. 3: 241–249.

Jacob, J. M., Karthik, C., Saratale, R. G., Kumar, S. S., Prabakar, D., Kadirvelu, K., Pugazhendhi, A. (2018) Biological approaches to tackle heavy metal pollution: A survey of literature. J. Environ. Manag. 217: 56–70.

Jaishankar, M., Tseten, T., Anbalagan, N., Mathew, B. B., Beeregowda, K. N. (2014) Toxicity, mechanism and health effects of some heavy metals. Interdiscip. Toxicol. 7: 60–72.

Jiang, R., Wang, M., Chen, W., Li, X. (2018). Ecological risk evaluation of combined pollution of herbicide siduron and heavy metals in soils. Sci. Total Environ. 626, 1047–1056.

Kannan, A., Thambidurai, S. (2007) Removal of Pb (II) from aqueous solution using palmyra palm fruit seed carbon. Electron J Environ Agric Food Chem. 6: 1803–1819.

Kannan, N., Veemaraj, T. E. (2009) Removal of Pb (II) ions by adsorption onto bamboo dust and commercial activated carbons—a comparative study. J Chem. 6: 247–256.

Kazemipour, M., Ansari, M., Tajrobehkar, S., Majdzadeh, M., Kermani, H. R. (2008) Removal of Pb(II), cadmium, zinc, and copper from industrial wastewater by carbon developed from walnut, hazelnut, almond, pistachio shell, and apricot stone. J Hazard Mater B. 150: 322–327.

Ksoll, W. B., Ishii, S., Sadowsky, M. J., Hicks, R. E. (2007).Presence and Sources of Faecal Coliform Bacteria in Epilithic Periphyton Communities of Lake Superior. Appl. Environ. Microbiol. 73: 3771–3778.

Kumar, V., Singh, P., Jain, S. K. (2005) Rainfall trends over Himachal Pradesh, Western Himalaya, India. In: Development of Hydro Power Projects – A Prospective Challenge (Conf. Shimla, 20–22 April, 2005)

Lalhruaitluanga, H., Jayaram, K., Prasad, M. N. V., Kumar, K. K. (2010) Pb(II) adsorption from aqueous solutions by raw and activated charcoals of Melocanna baccifera Roxburgh (bamboo)- a comparative study. J Hazard Mater. 175: 311–318.

Langford, T. E. L. (1990). Ecological Effects of Thermal Discharges. London and New York: Elsevier Applied Science Publishers Ltd.

Lugo-Lugo, V., Hernandez-Lopez, S., Barrera-Dıaz, C., Urena-Nunez, F., Bilyeu, B. (2009) A comparative study of natural, formaldehyde-treated and copolymer-grafted orange peel for Pb(II) adsorption under batch and continuous mode. J Hazard Mater. 161: 1255–1264.

McKenzie, D. Ray, I. (2004), 'Household Water Delivery Options in Urban and Rural India', Working Paper No. 224, Stanford Centre for International Development, Stanford University.

Mian, I. A., Begum, S., Riaz, M., Ridealgh, M., McClean, C. J. and Cresser, M. S. (2010). Spatial and temporal trends in nitrate concentrations in the River Derwent, North Yorkshire, and its need for NVZ status. Science of the Total Environment.408: 702–712.

Morosanu, I., Teodosiu, C., Paduraru, C., Ibanescu, D., Tofan, L. (2017). Biosorption of lead ions from aqueous effluents by rapeseed biomass. New Biotechnol. 39: 110–124.

Murty, M.N., Kumar, S. (2004). Environmental and economic accounting for industry. New Delhi: Oxford University Press.

Naiya, T. K., Bhattacharya, A. K., Mandal, S., Das, S. K. (2009) The sorption of Pb(II) ions on rice husk ash. J Hazard Mater. 163: 1254–1264.

Naseem, K., Farooqi, Z. H., Ur Rehman, M. Z., Ur Rehman, M. A., Begum, R., Huma, R., Shahbaz, A., Najeeb, J., Irfan, A. A. (2018). Systematic study for removal of heavy metals from aqueous media using Sorghum bicolor: An efficient biosorbent. Water Sci. Technol. 77: 2355–2368.

Njoroge, G.K. (2007) A pilot study report environmental pollution and impact to public health Nairobi, Kenya. www.korogocho.org/english/index.php?option=com_docman.

Özcan, A.S., Özcan, A., Tunali, S., Akar, T., Kiran, I., Gedikbey, T. (2007) Adsorption potential of Pb(II) ions from aqueous solutions onto Capsicum annuum seeds. Sep Sci Technol. 42: 137–151.

Okeke, C. O., Igboanua, A. H. (2003). Characteristics and quality assessment of surface water and groundwater resources of Akwa Town, Southeast, Nigeria. J. Niger. Assoc. Hydrol. Geol. 14: 71–77.

Parikh, J. (2004), 'Environmentally Sustainable Development in India', available at http://scid.stanford.edu/events/ India2004/JParikh.pdf

Paul, M., Meyer, J. L. (2001). Streams in the Urban Landscape. Annual Review of Ecology and Systematics. 32: 333–365. 10.1146/annurev.ecolsys.32.081501.114040.

Pehlivan, E., Altun, T., Cetin, S., Bhanger, M. I. (2009) Pb(II) sorption by waste biomass of hazelnut and almond shell. J Hazard Mater 167:1203–1208.

Rajaram, T., Das, A. (2008), Water pollution by industrial effluent in India: discharge scenarios and case for participatory ecosystem specific local regulation. Futures, 40: 56–69.

Rajeshkumar, S., Liu, Y., Zhang, X., Ravikumar, B., Bai, G., Li, X. (2018). Studies on seasonal pollution of heavy metals in water, sediment, fish and oyster from the Meiliang Bay of Taihu Lake in China. Chemosphere, 191: 626–638.

Ramakrishnaiah, C. R., Sadashivalah, C., Ranganna, G. (2009). Assessment of water quality index for groundwater in Tumkur Taluk, Karnataka State. Indian J. Chem. 6: 523–530.

Saeed, A., Iqbal, M., Akhtar, W. (2005) Removal and recovery of Pb(II) from single and multimetal (Cd, Cu, Ni, Zn) solutions by crop milling waste (black gram husk). J Hazard Mater. 117: 65–73.

Savci, S. (2012) An agricultural pollutant: Chemical fertilizer. International Journal of Environmental Science and Development, 3: 73.

Schiewer S, Balaria A (2009) Biosorption of Pb(II) by original and protonated citrus peels: equilibrium, kinetics, and mechanism. Chem Eng J 146: 211–219

Shah, T. M., Molden, D., Sakthivadivel, R., Seckler, D. (2000). The global groundwater situation: overview of opportunities and challenges. IWMI Books, Reports H025885, International Water Management Institute.

Singh, K. K., Talat, M., Hasan, S. H. (2006) Removal of Pb(II) from aqueous solutions by agricultural waste maize bran. Bioresour Technol. 97: 2124–2130.

Tana, G., Yuan, H., Yong, L., Dan, X. (2010) Removal of Pb(II) from aqueous solution with native and chemically modified corncobs. J Hazard Mater. 174: 740–745.

Wang, X., Han, J., Xu, L., Zhang, Q. (2010). Spatial and seasonal variations of the contamination within water body of the Grand Canal, China. Environmental Pollution. 158: 1513–1520.

Water Management Forum. (2003). Inter-basin Transfer of Water in India-Prospects and Problems. The Institution of Engineers (India), New Delhi.

Wetzel, G. W. (2001) Limnology: Lake and River Ecosystems. Academic Press, New York. pp. 15–42.

World Health Organization [WHO] (2007) Guidelines for drinking-water quality, Incorporation First Addendum, Volume 1, Recommendations, 3rd edition, WHO, Geneva.

Yetilmezsoy, K., Demirel, S. (2008) Artificial neural network (ANN) approach for modeling of Pb(II) adsorption from aqueous solution by Antep pistachio (Pistacia vera L.) shells. J Hazard Mater. 153: 1288–1300.

Zulkali, M. M. D., Ahmad, A.L., Norulakmal, N. H., Sharifah, N. S. (2006) Comparative studies of Oryza sativa L. husk and chitosan as Pb(II) adsorbent. J Chem Technol Biotechnol. 81: 1324–1327.

3 Phytochemical Extraction from Waste Biomass

Bidyut Mazumdar, Amit Keshav and Alok Sharma

Department of Chemical Engineering, National Institute of Technology, Raipur, India

CONTENTS

DOI: 10.1201/9781003196358-3

It is a well-known fact that plants can produce their own food. Since, these chemicals are derived from plants they are called as Phytochemicals. They are also referred to as secondary metabolites or natural products or bioactive compounds. They offer defense in plants by preventing the insect attack and grazing animals. Some of the phytochemicals provide color, aroma and flavor to fruits, vegetables and other plant parts. The agricultural waste and food processing waste mainly contribute to the waste biomass. Besides being economical such type of waste is a potential source for the recovery of phytochemicals. Different type of plant parts contributes to the waste include fruits, leaves, stems, peels, arils, seeds etc. The food waste can also contribute to the generation of energy to a greater extent owing to the use of technologies like anaerobic digestion, fermentation, pyrolysis, gasification, incineration etc. (Pham et al., 2015; Jeevahan et al., 2018)

3.1 PHYTOCHEMICAL SOURCES

Phenolic compounds are mainly found in peels and skin (Pande and Akoh, 2010). The pulp of jamun fruit was found to be a rich source of phenolic compounds (Prakash Maran and Manikandan, 2012). Hesperidin and eriocitrin were extracted from albedo and flavedo of lemon residue (Peiró et al., 2019). The pomace of sour cherry was used to recover the phenolic compounds (Okur et al., 2019). The waste leaves of prickly water lily were also reported to have a high anthocyanin content (Wuet al., 2020). Tomato processing produces waste biomass in form of tomato pomace, which was reported to be a good source of carotenoids (Silva et al., 2019). The orange peel, specifically the flavedo contains a high number of flavonoids and terpenes such as D-limonene (Negro, Ruggeri and Fino, 2018; Victor et al., 2020). Sunflower seed cake is considered as waste after the oil extraction, phenolic acids and flavonoids are present in a good amount (Zardo et al., 2019). The lutein yield was 29.70 µg/g as reported for the carotenoid content in kinnow peel (Saini, Panesar and Bera, 2020). Subproduct of pine and eucalyptus forest like branches, wood, bark and leaves were used to recover the phenolic compounds (Xavier, Freire and González-Álvarez, 2019). The pistachio processing waste includes hulls and shells, which was found to exhibit polyphenolic content (Zalazar-García, Feresin and Rodriguez, 2020). The extract of walnut residual biomass is comprised of phenolic acids like gallic acid, ferulic acid and ellagic acid (Fernández-Agulló et al., 2020; Özbek et al., 2020). The press cake of bilberry fruit juice is a good source of flavonol and anthocyanin (Varo et al., 2019). The tomato processing produces tomato peel as waste biomass, which was observed to a potential source of polyphenolic compounds (specially chlorogenic acid) (Ninčević Grassino et al., 2020). The extraction of pectin and phenolic compounds was reported from the waste peel of mangoes (Rojas et al., 2020). The stems and leaves of Ginja cherry are considered as agricultural waste; it possesses good polyphenolic profile (Demiray et al., 2011). The peels of red apple also contribute to the waste biomass produced during apple processing, and were reported to contain polyphenolic phytochemicals like quercetin, catechin etc. (Blidi et al., 2015). Polyphenols were also recovered from the bio-oil produced from agricultural waste biomass, including palm fruit bunch and sugarcane bagasse (Mantilla, Manrique and Gauthier-Maradei, 2015). The cold-pressed pomace produced after chokeberry processing has anthocyanin content 456.7 mg/L (Roda-Serrat et al., 2020). Some cruciferous vegetables are known to synthesize phytochemicals containing sulfur such as glucosinolates (Cieślik et al., 2007). Aliphatic glucosinolates were found in leaves and florets of broccoli and cauliflower whereas indolyl glucosinolates were present in roots (Mirou et al., 2009). Coumarins are reported to exist in fruits and vegetables of Rutaceae and Umbelliferae families, which include parsnips, carrot, celery etc. (Ostertag et al., 2002). The occurrence of betacyanin and betaxanthin in the beetroot peels was reported by Sharma, Mazumdar and Keshav (2020a).

Terpenoids	Polyphenols	Alkaloids and nitrogenous
❖ Monoterpenoids	❖ **Phenolic acids**	metabolites
❖ Diterpenoids	❖ **Flavonoids**	❖ Indole alkaloids
❖ Triterpenoids	❖ **Anthocyanidin**	❖ Isoquinoline alkaloids
❖ Sesquiterpenoids	❖ **Proanthocyanidin**	❖ Steroidal alkaloids
❖ Polyterpenoids	❖ **Coumarins**	❖ Tropane alkaloids
❖ Carotenoids	❖ **Xanthones**	❖ Pyridine alkaloids
❖ Xanthophylls	❖ **Stillbenoids**	❖ Pyrrolizidine alkaloids
	❖ **Lignans**	

Betalains	Capsaicinoids	**Allium compounds**
❖ **Betacyanin**	❖ Capsaicin	❖ **Methiin**
❖ **Betaxanthin**	❖ Dihydrocapsaicin	❖ **Propiin**
❖ **Indicaxanthin**	❖ Nonivamide	❖ **Isoalliin**
❖ **Vulgaxanthin**		

FIGURE 3.1   Broad classification of phytochemicals divided in subgroups.

3.2   CLASSIFICATION

The phytochemicals are a diverse group of natural products, which has been classified into broad groups. Figure 3.1 shows the different groups of phytochemicals covered in this chapter. The phytochemicals exist as clusters in plants. A single plant species can contain various phytochemicals.

3.2.1   TERPENOIDS

Terpenes are the largest group of phytochemicals which exist in plants. Basically, they are polymers of isoprene units (C_5H_8) and hence have the molecular formula $(C5H8)_n$. Terpenes are modified by addition or removal of methyl groups at different positions producing Terpenoids. As shown in Figure 3.1, terpenoids are subdivided into monoterpenoids, diterpenoids, triterpenoids, sesquiterpenoids. They also have special class of carotenoids and xanthophylls. Terpenoids have a wide variety of applications in flavorings, therapeutics and fragrances. The section gives a brief introduction about terpenes and terpenoids.

3.2.1.1   Monoterpenoids

When a terpenoid contains two isoprene units along with ten carbon atoms. A monoterpenoids is denoted by the molecular formula $C_{10}H_{16}$. They are known to naturally exist in essential oils of plants and their parts. On the basis of structure, they are divided as acyclic, monocyclic and bicyclic terpenoids. These compounds have a wide application in pharmaceutical industry because their strong color and aroma. Most of the monoterpenoids have bioactive properties and their mixture is also used in cosmetics and perfumes. Citral, pinene, camphor, limonene and menthol are some common examples of monoterpenoids.

3.2.1.2 Diterpenoids

Diterpenoids are denoted by the molecular formula $C_{20}H_{32}$ with four isoprene units. They are known to exhibit different bioactive properties such as antifungal, antimicrobial, anti-inflammatory and anticancer. Nicaeenin F and Nicaeenin G are derived from latex of *Euphoria nicaeensis* have been reported to show anticancer properties (Krstić et al., 2018).

3.2.1.3 Triterpenoids

Triterpenoids are the secondary metabolites which consist 30 carbon atoms and six isoprene units. They are oxidized to form different aldehydes, alcohols and carboxylic acids. They also undergo glycosylation to produce saponins, as the end product. Centelloids are the pentacyclic triterpenoids produced in the leaves and roots of *Centella asiatica* (also known as Asian pennywort). It is known to show various pharmacological activities (James and Dubery, 2009).

3.2.1.4 Sesquiterpenoids

If the terpenoids consist of three isoprene units, they are called as Sesquiterpenoids. Structurally, they can exist in linear, cyclic, bicyclic and tricyclic units. They also exist in form of lactone ring. They have a good antimicrobial and anti-insecticidal properties. Santhemoidin A is produced by *Tarchonanthus camphoratus* (also known as Camphor bush) Trypanosoma parasites (Kimani et al., 2018).

3.2.1.5 Carotenoids

Carotenoids are the tetra-terpenoids produced by some plant species such as *Basella rubra, Moringa oleifera, Spinacia oleracea etc.* β-carotene is an important carotenoid because it is the precursor of retinol and abscisic acid. They play a vital role in the protection of photosynthetic organisms by preventing the damage of photosynthetic apparatus. They impart primary red and orange color to fruits and vegetables (Liang, Zhu and Jiang, 2018).

3.2.1.6 Xanthophylls

The carotenes which contain hydroxyl and keto groups, undergo oxygenation to produce xanthophylls. Lutein, zeaxanthin and violaxanthin are commonly found xanthophylls in nature. They are responsible for the yellow colour in fruits and vegetables. Tomato and chickpea varieties are the common sources of xanthophylls (Liang, Zhu and Jiang, 2018).

3.2.2 POLYPHENOLS

Polyphenols contribute a high proportion of phytochemicals to be found the biomass. Their structure consists an aromatic ring with hydroxyl groups attached to it. Polyphenols are further sub categorized into phenolic acids, benzoic acids, flavones, flavonones, flavonols, isoflavones, anthocyanidin, proanthocyanidin, phenylacetic acid, coumarins, stilbenes, xanthones, lignans, cinnamic acid, acetophenone, tannins etc. Polyphenols are known to show preventive action against reactive oxygen species and nitrogen species. Also, they have shown the preventive biological activity against UV light, parasites, plant pathogens, etc (Manach et al., 2004; Brglez Mojzer et al., 2016). We have briefly described about the different subcategories in the below section.

3.2.2.1 Phenolic Acids

Structurally, when the phenols carry a carboxylic acid as functional group it is termed as phenolic acid. They play a vital role in plant growth and reproduction. They are synthesized in response to various biotic and abiotic stress to the plants. The phenolic acids are further subdivided into two categories: hydroxybenzoic acids and hydroxycinnamic acids. Hydroxycinnamic acids are produced as simple esters along with hydroxy carboxylic acid and glucose (Ghasemzadeh and Ghasemzadeh, 2011). Ferulic acid, caffeic acid and sinapic acids are the common examples of hydroxycinnamic acids (Alam et al., 2016). Hydroxybenzoic acids are the phenolic derivatives of benzoic acids. Salicin and salicylic acid are the two commonly known benzoic acids (Juurlink et al., 2014). The phenolic acids have a wide application in pharmaceuticals, foods, cosmetics etc.

3.2.2.2 Flavonoids

Flavonoids are considered as the most important phytochemicals for the colouration and flavoring in plant species. They are also known to participate in UV light filtration, floral pigmentation and symbiotic nitrogen fixation. They exist as aglycones, glycosides and methylated derivatives (Kumar and Pandey 2013). The flavonoids constitute a broader category of flavones, flavonones, flavonols and isoflavones. They are synthesized from the phenylalanine as the precursor by the polypropanoid pathway. The structural conformation of flavonoids consists of C6-C3-C6 along with a heterocyclic ring (contains an oxygen molecule) and two C6 aromatic rings (Ghasemzadeh and Ghasemzadeh, 2011). Flavones consist of the 2-phenylchromen-4-one as their primary skeleton. Kaempferol is a commonly known flavone which exist in nature (Kshatriya, Shaikhand and Nazeruddin, 2013). Flavonones occur as glycosides and aglycones, they are derived from flavones. Naringenin and hesperidin are the common examples of flavonones (Brodowska, 2017). Flavonols can be found as O-glycoside or as C-glycosides. Rutin, myricetin and quercetin are the example of flavonols. They have been reported for the major contribution in the screening of UV radiation (Pollastri and Tattini, 2011). Isoflavones are generally considered as phytoestrogens. They are primarily found the Fabaceae family. Genistein and daidzein are commonly occurring isoflavones (Brodowska, 2017).

3.2.2.3 Anthocyanins

The polyhydroxy and polymethoxy derivatives of flavyliyum (2-phenylbenzopyrylium cation) are known as anthocyanidins. The water-soluble acylglycosides and glycosides of anthocyanidins are termed as Anthocyanin. The oligomeric and polymeric flavan-3-ols are commonly known as Proanthocyanidin (Wu et al., 2020). They are primarily found in berry fruits, providing them the rich and dark blue-purple colour. The supplementation of these phytochemicals is reported to prevent various cardiovascular diseases in humans. Blueberries, raspberries, black currants, cabbage etc. are some common sources of anthocyanins and proanthocyanidin (Kruger et al., 2014).

3.2.2.4 Coumarins

Coumarins belong to the family of benzopyrones existing in nature. The pleasant smell of sweet grass is due to the presence of coumarins (Matos et al., 2015). They have several pharmacological applications due to their ability to inhibit selective enzymes. Their potential of inhibiting selected targets has led them to be used in the treatment of Alzheimer's and Parkinson's disease. They are found in cinnamon tree, liquorice and some berries (Stefanachi et al., 2018).

3.2.2.5 Xanthones

Xanthones are belongs to the class of oxygenated heterocyclic compounds. The biological activity of xanthone is associated with its tricyclic structure existing in nature (Mazimba et al., 2013). These include antihypertensive, antiproliferative, cytotoxic, antibacterial, cardioprotective, immunosuppression etc. (Negi et al., 2013). Mangostin is well known xanthone, which is found in Garcinia fruits (El-Kenawy, Hassan and Osman, 2018). Mangiferin is one more commonly existing xanthone in mango plant varieties (Imran et al., 2017).

3.2.2.6 Stillbenoids

A stilbene consists of 1, 2-diphenylethylene as the primary skeleton (Chou, Ho and Pan, 2018). The hydroxylated derivatives of stilbenes are known as stillbenoids, belonging to the phenylpropanoids family (Sirerol et al., 2016) Their biological activity is still completely unknown but they are reported to show anticancer, anti-inflammatory and several other antioxidant activities. A stilbene undergoes isomerization, glycosylation, methoxylation and oligomerization to synthesize stillbenoid derivatives (Chong, Poutaraud and Hugueney, 2009). Resveratrol, Pterostillbene, piceatannol and pinosylvin are some example of stillbenoids. Resveratrol is a primary stillbenoid found in grape varieties and it is believed to be associated with anti-aging properties. However, this fact is still unproven.

3.2.2.7 Lignans

Lignans are called as diphenolic compounds because they consist of two propyl-benzene units attached together by a β, β' bond. They are the precursor to several phytoestrogens. It was found to exist in a number of nuts (almond, cashew, peanut, walnut etc.), pulses and lentils (mung bean and white bean), and in soy and soy products. Secoisolariciresinol is a major lignan found in sesame seed and flax seeds. Matairesinol, lariciresinol and pinoresinol are the other lignans present in nature (Rodríguez-García et al., 2019).

3.2.3 ALKALOIDS

The naturally occurring organic compounds which contain a nitrogen atom (amino groups) on a cyclic ring and showing any biological activity are termed as Alkaloids. Due to the presence of the nitrogenous group, they behave like alkalis. Generally, they are colorless and odorless, but some alkaloids with yellowish appearance have also been found (Kurek, 2019). The alkaloids contain single amine group has been categorized as primary, secondary, tertiary or quaternary amines. On the contrary, the alkaloids having multiple amine groups are termed as amides having the neutral character has also been found in the nature. They are primarily found in the bark, leaves, stems and branches of plants. They are widely present in the Berberidaceae, Buxaceae, Chenopodiaceae, Euphorbiaceae, Fabaceae, Loganiaceae, Magnoliaceae and several other plant families (Kukula-Koch and Widelski, 2017). Since they are water soluble at low pH and fat soluble at neutral pH and high pH, their use in pharmacology and drugs is very popular (El-Sayed and Verpoorte, 2007). Alkaloids are categorized on the basis of their structure. Two broad categories of alkaloids are indole alkaloids and isoquinoline alkaloids, each having more than 4,000 compounds. Some other alkaloid groups existing in nature are steroidal alkaloids, tropane alkaloids, pyridine and pyrrolizidine alkaloids.

3.2.3.1 Indole Alkaloids

This group of alkaloids have a benzene ring conjugated with a nitrogenous pyrrole ring. The bioactive properties of indole alkaloids are due to the presence of pyrrole ring with nitrogen (El-Sayed-Verpoorte2007). The common examples are reserpine and vincristine found in *Rouvolfia serpentina* and *Catharanthus roseus* (Sagi et al., 2016). Some of the indole alkaloids have been as anti-depressants since a long time (Hamid, Ramli and Yusoff, 2017).

3.2.3.2 Isoquinoline Alkaloids

Isoquinoline alkaloids are the derivatives of tyrosine and phenylalanine. The basic structural skeleton consists of isoquinoline or tetrahydroisoquinoline ring. They are structurally dissimilar groups. They are further categorized on the basis of degree of oxygenation and intramolecular rearrangements (Kukula-Koch and Widelski, 2017). Aporphine, protopine, morphinan, benzylisoquinoline, protoberberine and emetine alkaloids (Leitao da-Cunha et al., 2005). Chelodine, copsitine, berberine, sanguinarine and chelerthrine are some commonly known isoquinoline alkaloids.

3.2.3.3 Steroidal Alkaloids

The structural conformation of steroidal alkaloids consists of a steroid molecule fused with one or two nitrogen. *Solanum* and *Veratrum* alkaloids belong to the distinct class of steroidal alkaloids, which exist in nature. They are reported to possess antiparasitic and hypotensive properties (Kukula-Koch and Widelski, 2017).

3.2.3.4 Tropane Alkaloids

Tropane alkaloids are a unique class of alkaloids which contain tropane ring in their structural skeleton. Also, they are the esters of either 3β-tropanole (pseudotoprine) or 3α-tropanole (tropine) (Grynkiewicz and Gadzikowska, 2008). They are basically subdivided as scopolamine and hyoscyamine existing in Solanaceae family, cocoa alkaloids e.g., cocaine found in *Erythoxylum cocoa* and polyhydroxylated nortropane alkaloids such as calystegines present in Convulvulaceae and Brassicaceae family (Dräger, 2004). Though this class of alkaloids have a wide application in pharmaceuticals but due to their use in drug addiction, most of them are banned by the authorities (Kohnen-Johannsen and Kayser, 2019).

3.2.3.5 Pyridine Alkaloids

These alkaloids consist of a pyridine ring in their structural conformation. Nicotine and anabasine are the commonly found pyridine alkaloids.

3.2.3.6 Pyrrolizidine Alkaloids

Pyrrolizidine alkaloids are the group of alkaloids which is composed of either saturated or unsaturated units of necine. On the basis of presence of necine units, they are classified as retronecine, otonecine, platynecine and heliotridine (Xu et al., 2019). They are also known as the derivatives of ornithine. They are well known for their anti-microbial, anti-inflammatory, anti-cancer, anticholinesterase and cytotoxic properties (Moreira et al., 2018).

3.2.4 Capsaicinoids

Capsaicinoids are the phytochemicals responsible for the pungency in chilli and pepper varieties (Sarpras et al., 2016). The measurement of pungency is such fruits is done on the basis of Scoville heat units (Scoville, 1912). The vanillyl amine synthesized from phenylpropanoid pathway and branched fatty acid synthesized via amino acid catabolism are condensed to produce capsaicinoids (Castro-Concha et al., 2016). Capsaicin is the naturally existing phytochemical in chilli pepper which gives the burning sensation. They are synthesized in the central epidermis layer of the chilli pepper. Dihydrocapsaicin is the structural analogue of capsaicin, also existing in chilli peppers. They are synthesized by the condensation of fatty acids and vanillylamine. They help to reduce the body fat by the energy expenditure and heat sensation in the body (Kumar, Bhatt and Kumar, 2018).

3.2.5 Betalains

Betalains are one the popular and unique subclass of phytochemicals existing in the Caryophyllales family. They contain nitrogen in their structure and are water soluble in nature. Based on their structure, they are further classified as betacyanin and betaxanthin. Betacyanin are responsible for red-violet pigmentation while the yellow pigmentation is due to the presence of betaxanthin (Azeredo, 2009). Betalains are structural analogues of anthocyanins, both of these phytochemicals never exist together in any of the species (Stafford, 1994). The common sources of betalains are beetroot, *Opuntia* fruits, Pitaya fruit, *Amaranthus* species and *Basella rubra* fruits (Gengatharan, Dykes and Choo, 2015). Chemically, they are immonium derivatives of betalamic acid, coloration of betalains is due to the resonance of double bonds (Mabry, 2001). Betalains have gained popularity as food additives and colorants since they are derived from natural sources.

3.2.5.1 Betacyanin

Betacyanins have polysaccharide or acyl groups in their structure as 5-*O*- or 6-*O*-glucosides (Delgado-Vargas, Jimenez and Paredes-Lopez, 2010). Betacyanin are further subdivided into betanin, amaranthin, gomphrenin and bougainvellein. Among these betanin is the most explored phytochemical from beetroot species (Stintzing and Carle, 2004). Due to the presence of *cyclo*-DOPA in their structure, they tend to show two absorption spectra. One in the range of 270–280 nm and the other at 540 nm (Belhadj et al., 2017).

3.2.5.2 Betaxanthin

Betaxanthin possess amines and amino acid groups in their structure (Delgado-Vargas, Jimenez and Paredes-Lopez, 2010). On the basis of the presence of amine or amino group they are classified amine-derived group and amino-acid derived group. Indicaxanthin, vulgaxanthin humilixanthin etc. are some naturally existing betaxanthins (Stintzing, Schieber and Carle, 2002). They show the absorption spectra in the range of 460 to 480 nm wavelength (Slimen, Najar and Abderrabba, 2017).

3.2.6 Allium Compounds

Allium compounds are widely found in plants of *Allium* species like onion, garlic and shallots. When the alkyl cysteine sulphoxide is cleaved by the enzyme alliinase, the pungent odor and taste of allium compound is generated. In case of garlic, alliin is the main ACSO, which produces allicin and thiosulfinate compounds. However, in onions, isoalliin is the ACSO, it produces thiosulfinates, cepaenes and zweibelanes (Bianchini and Vainio, 2001). They are known to possess various biological activities such anticarcinogenic, antibiotic, anticoagulant, antihyperlipidemic, antihypertensive, hepatotoxic, insecticidal etc. (Block, 2005).

3.3 DIFFERENT TYPES OF EXTRACTION TECHNIQUES

The efficiency of extraction of phytochemicals from different sources primarily depends on the extraction technique (Azmir et al., 2013). This is the main reason for the proper application and optimization of extraction techniques. However, the phytochemical also depends on raw materials, solvent chosen and process variables of extraction (Tiwari, 2015). The extraction techniques have been categorized into two groups i.e., conventional and non-conventional techniques. Both conventional and non-conventional techniques have been used for the extraction of phytochemicals. Soxhlet extraction, maceration, hydro distillation are some popular conventional techniques of extraction. These techniques require large quantity of solvents and sample, time, energy. The non-conventional techniques used for extraction are microwave assisted extraction (MAE), ultrasound assisted extraction (UAE), supercritical fluid extraction (SFE) etc. (Rashed et al., 2016). The non-conventional techniques are also referred as modern techniques or novel techniques because of the novel and modern fundamentals being used for the equipment and technique. In contrast to the conventional techniques, non-conventional techniques require less sample quantity and solvent volume, time and energy. They are also considered as clean or green techniques because of this reason. Table 1 illustrates different extraction techniques used for the recovery of phytochemicals from different sources. This section briefly explains about the techniques so far used for phytochemical extraction.

3.3.1 CONVENTIONAL EXTRACTION

3.3.1.1 Soxhlet Extraction

A unique apparatus is required for the extraction by Soxhlet technique, also named as Soxhlet apparatus. The large amount of dried and powdered sample is placed inside a thimble and the

3.3.1.2 Maceration

The principle of extraction by maceration lies in the size reduction. The size of sample is reduced by grinding due to continuous stirring. The smaller particles have more surface area to be introduced to the solvent. This technique also facilitates extraction by enhancing the diffusion and by replacing the concentrated sample from the sample surface. It has been used to recover the essential oils and phytochemicals since a long time (Azmir et al., 2013). The technique is also known as cold maceration. In some instances, the temperature has also been applied to improve the efficiency of extraction. But the yield of the extract and loss of desired compounds was observed as the limitation (Wu et al., 2015; Deng et al., 2017).

3.3.1.3 Hydrodistillation

The extraction of volatile organic compounds was primarily done by hydrodistillation by using distilled water. This technique does not require the use of organic solvents and requires 6–8 h of extraction time. It works on the principle of hydrolysis, decomposition by heat and hydrodiffusion. The degradation of thermolabile compounds during extraction is the major disadvantage of hydrodistillation (Wu et al., 2015). Though, this technique offers certain advantages wherein both volatile and non-volatile compounds can be extracted separately. The non-volatile component is extracted in the water which is in contact with the plant matrix. The volatile component is separated by the using azeotropic distillation. The energy consumption is high with prolonged extraction time (Da Porto and Decorti, 2009; Zhao and Zhang, 2014). The technique is also known as steam distillation.

3.3.1.4   Percolation

Percolation is a better technique as compared to maceration. It is a continuous process where the solvent is replaced by fresh solvent at regular intervals. Distilled water as well as organic solvents can be used for the extraction. It is mainly performed at room temperature; heat is applied to achieve better extraction at some instances. The necessity of large solvent volume and long extraction time are the limitations of this technique (Zhang, Lin and Ye, 2018).

3.3.1.5   Decoction

Decoction has been used in many the recovery of various herbal extracts for a long time. Water soluble impurities also get extracted while preparing the decoction of plant sample. This technique is not suited for the thermolabile and volatile compounds. The dissolution of some bioactive compounds was reported, which is comparable to maceration. The herbal extraction has also resulted in the interaction of different bioactive compounds (Zhang, Lin and Ye, 2018).

3.3.1.6   Reflux Extraction

It is a type of solid-liquid extraction method. The extraction is carried out by continuous evaporation and condensation of the solvent at constant temperature (Chua, Latiff and Mohamad, 2016). The reflux extraction consumes less time and solvent amount when compared to percolation and maceration. It also has better efficiency than the latter two techniques (Arias et al., 2009).

The polarity of desired compounds and choice of solvent are the major factors affecting the efficiency of conventional techniques. Sometimes different solvents are required to isolate and identify number of compounds from a single source. An ideal solvent offers low boiling point, low toxicity, dissociation of complex extracts and better mass transfer. Extraction time and temperature are also responsible to efficient and safe extraction of compounds (da Silva, Rocha-Santos and Duarte, 2016). Due to the above-mentioned disadvantages of conventional techniques, eco-friendly extraction techniques with better sustainability and efficiency must be developed.

3.3.2   Non-conventional Techniques

Non-conventional techniques of extraction are also known as novel extraction techniques. They are believed to overcome the constraints observed during the conventional extraction techniques (Putnik et al., 2018). The use of green solvents has enhanced the extraction efficiency of non-conventional techniques.

TABLE 3.1
Different Techniques of Phytochemical Extraction and Their Respective Sources

S. No.	Technique	Source	Phytochemical	Reference
1	UAE	Olive fruits	Phenolic compounds	Deng et al., 2017
2	UAE, MAE, SE, HD, CM	*Lonicera macranthoides*	Volatile compounds	Wu et al., 2015
3	UAE, HD	Mentha spicata	Essential Oils	Da Porto and Decorti, 2009
4	SFE, SE, HD	Eucalyptus leaves	Essential Oils	Zhao and Zhang 2014
5	Percolation	*Undaria pinnatifida* seaweed	Xanthophylls	Zhang et al., 2014
6	Reflux extraction	*Andrographis paniculata*	Terpenoid	Chua et al., 2016
7	MAE	Pomegranate peels	Phenolic compounds	Kaderides 2019
8	UAE	*Coccinia indica fruits*	Phytochemicals	Sharma 2020
9	MAE, UAE, SE	*Amaranthus tricolor* leaves	Betalains	Sharma 2021
10	UAE, MAE	Beetroot peels	Betacyanin	Sharma 2020
11	MAE	Potato peels	Antioxidants	Singh 2014
12	UAE, PEF, HVED	*Vitis vinifera* vine shoots	Polyphenols	Rajha et al., 2015
13	PEF	*Agaricus bisporus* mushroom	Polyphenols	Parniakov et al., 2014
14	SFE	Red pepper	Oleoresin	Silva and Martinez 2014
15	HVED	Rape seeds	Isothiocyanates	Barba et al., 2015
16	HPP Extraction		Flavonoids	Escobedo-Avellaneda et al., 2011
17	UAE, HPP Extraction	Chilean papaya seeds	Antioxidant compounds	Briones-Labarca et al., 2015
18	UAE, HPP extraction and PEF	Grape by-products	Anthocyanin	Corrales et al., 2008
19	Maceration	Apple waste peels	Polyphenols	Blidi et al., 2015
20	SFE	Tomato leaf waste	Phenolic compounds and flavonoids	Arab et al., 2019

UAE: Ultrasound assisted extraction, MAE: Microwave assisted extraction, SE: Soxhlet extraction, HD: Hydro distillation, CM: Cold maceration, SFE: Super critical fluid extraction, HPP: High pressure processing, PEF: Pulse electric field, HVED: High voltage electrical discharge thimble is placed inside the extraction unit of Soxhlet apparatus. Large amount of solvent is heated in a round bottom flask and is continuously refluxed back by a siphon tube after condensation in the condenser. The process is repeated continuously until the completion of extraction, which can take hours. It is studied extensively among the researchers and has been optimized for various plant samples. The requirement of large solvent volume, time and energy has been the major limitation of this technique. Also, very low amount of extract is recovered after the extraction and most of the phytochemicals are degraded due to prolonged exposure to temperature.

These techniques do not require high temperature and pressure; also, they can be operated in the absence of light and oxygen. They do not require large amount of solvent and longer extraction time (Fomo, Madzimbamuto and Ojumu, 2020). Since phytochemicals are now used by a number of processing industries such as refineries, food, pharmaceuticals, textiles, etc. The chemical industries need to use safe and sustainable techniques of extraction from fermentation broths or biomass or plant samples. Nowadays microwave assisted extraction (MAE), ultrasound assisted extraction (UAE), pulse electric field extraction (PEF), supercritical fluid extraction (SFE) and pressurized liquid extraction (PLE) are the novel extraction methods being used. These techniques have shown to improve the yield of extraction and have minimal effect on the compound of interest. Some of these techniques as described in the below section with their principle and application on different sources.

3.3.2.1 Microwave Assisted Extraction

The application of microwaves for the extraction of phytochemicals from different sources has become popular over the time (Kaderides et al., 2019). Microwaves are the electromagnetic radiation with the frequency ranging from 1GHz to 1,000 GHz. But the frequency range used in microwave used for domestic and industrial application ranges from 915 MHz to 2,450 MHz. Higher heat and mass transfer rates are achieved due to the heating effects of microwaves (Kaderides et al., 2019). Microwaves can penetrate into the matrix of sample and interact with polar moieties. This ability of microwaves can cause either direct heating or bulk heating of the solvent initially and the sample eventually (Azmir et al., 2013). Due to this direct and bulk heating, the requirement of sample and solvent is greatly reduced, especially in case of large-scale extraction. The heating potential of microwaves depends on the dielectric properties of the solvent and samples. During the extraction, the target phytochemical is forced to move into the solvent due to the direct and bulk heating by microwaves (Singh et al., 2014). A closed or an open system can be used for MAE, wherein the pressure can be increased or decreased above the atmospheric pressure depending on the requirement of process. MAE can be performed either in an open vessel system or closed vessel system. Of the two, the open vessel system is safer and is more suitable for the extraction of thermolabile compounds. However, the closed system is more efficient because the boiling point of solvent is reduced due to the increase in pressure inside the system (Vinatoru, Mason and Calinescu, 2017). Figure 3.2 shows a MAE instrument with its different parts. Process variables such as microwave power, irradiation time, moisture content, particle size of sample, sample to solvent ratio, temperature, etc. affect the MAE. Above all, the choice of solvent still remains a crucial factor. The solvent selection depends on the dielectric constant, dissipation factor and solubility of the solvent selected for the extraction. The polar solvents such as water have high dielectric constant and tend to absorb more microwave energy (Ajila et al., 2011). MAE has several advantages like low thermal gradients, smaller size of the apparatus, increased yield of extraction and enhanced heating (Marić et al., 2018). Non

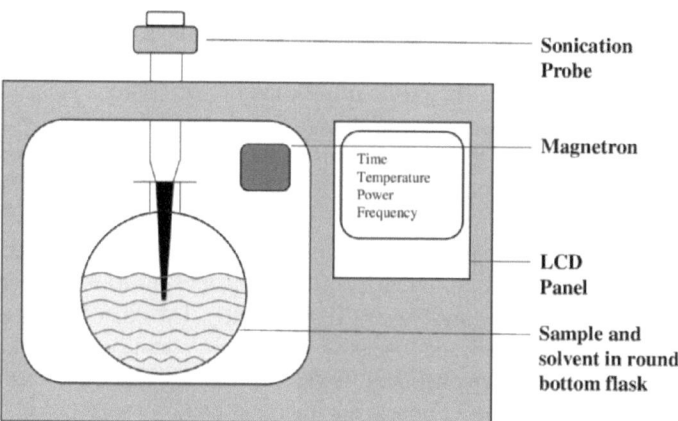

FIGURE 3.2   MAE and UAE instrument with different parts.

polar solvent is not preferred in the MAE due to their poor ability to absorb energy. But if the desired compound is soluble in non-polar solvents, certain modifiers are added to enhance the extraction. Also, some pretreatment techniques must be adopted to facilitate the extraction using non-polar solvents. The dependency on extraction temperature and solvent selection are major disadvantages of this technique (Adetunji et al., 2017). MAE can be better alternative for conventional techniques because of its reproducibility, stability and better efficiency.

3.3.2.2   Ultrasound Assisted Extraction

UAE has been widely used as an alternative to conventional techniques, offering clean, green and economic processing (Chemat et al., 2017). The frequency of ultrasonic waves ranges from 20 kHz to 100 MHz. These waves can penetrate any object by producing compression and expansion. In case of liquid medium, it causes the formation, growth and collapse of bubbles due to the acoustics, the effect is known as cavitation. It has been widely used to improve the food pro-cessing (Li and Sun, 2002; Tao, Zhang and Sun, 2014; Marić et al., 2018). The cavitation effect produced by the acoustic waves is the main driving force for enhanced UAE. The mass transfer between sample and solvent is increased because of the high shear forces produced by cavitation (Chemat et al., 2017). When the cavitation bubbles are collapsed, a turbulence is generated causing the collision and agitation of sample particles in the solvent (Putnik et al., 2018). Meanwhile, the explosion of cavitation bubble at the interface of solid and liquid causes breakdown of sample particles and erosion of surface by shockwaves and microjets produced by ultrasonic waves (Tiwari, 2015). Basically, the two types of UAE equipment used are ultrasonic bath and ultrasound probe. The ultrasonic bath is assembled with ultrasonic transducers at the bottom and on the side walls of tank. In some ultrasonic baths, a transducer array box is installed which can be placed in any direction and position to facilitate the propagation of waves in the medium. The intensity of sound waves is low since it is distributed from the large surfaces (Wen et al., 2020). The probe-ultrasound device (Figure 2) is assembled with probe having an indigenous transducer. The probe is directly submerged in the medium, which lowers the loss of ultrasonic energy. Since, the probes have a smaller surface area they tend to produce high intensity of ultrasounds. The design of reactor and the shape of probe must be considered properly while choosing the UAE device because they can affect the ultrasonic intensity (Azmir et al., 2013). The process variables such as intensity, frequency and ultrasonic power influence the extraction by ultrasonic waves (Tiwari, 2015). The other factors affecting the performance of UAE are the viscosity of solvent, stability and solubility of both solvent and target phytochemical. Extraction time, temperature, sample-to-solvent ratio and solvent concentration are the other factors which can play a vital role in efficient UAE (Sharma, Mazumdar and Keshav, 2020b). For large-scale application of UAE, it offers the safety, economical processing, eco-friendliness and sustainability after proper intensification and optimization of the process (Wen et al., 2020).

3.3.2.3 Pulse Electric Field Extraction

The pulse electric field extraction involves the disintegration of electroporation of cellular membrane by electricity pulse to enhance the extraction. The range of pulse electric field ranges from 100 V/cm to 80 kV/cm. The separation of molecules takes place on the basis of electric charge, when the electric current passed through the sample. Pores are formed on the membrane due to this repulsion which increases the permeability (Rajha et al., 2015). PEF has been widely studied for its application in food preservation and extraction of phytochemicals from plants and their biomass. The PEF causes softening of tissues and disruption of membranes, which facilitates the release of cytosolic compounds, providing an economic and sustainable perspective (Roselló-Soto et al., 2016). A PEF extraction unit (Figure 3.3) is assembled with a pulse generator and a high voltage generator, a product holding and treatment chamber and a set of controlling and monitoring devices (Soliva-Fortuny et al., 2009). The factors influencing PEF extraction are field intensity, energy, temperature, pulse number and sample properties (Azmir et al., 2013). It is considered as green technology because of the use of alternative green solvents such as water, ethanol, methanol and various vegetable oils; the energy consumption is less and the extract of more purity is obtained (Parniakov et al., 2014). This technique can be effectively used for the recovery and protection of phytochemicals.

3.3.2.4 Supercritical Fluid Extraction

In supercritical fluid extraction, the alteration in temperature and pressure enables the transformation of liquid and gaseous phase into a supercritical fluid. A critical temperature is the temperature at which the gas gets converted into the liquid. Similarly, a critical pressure is the pressure at which the liquid gets

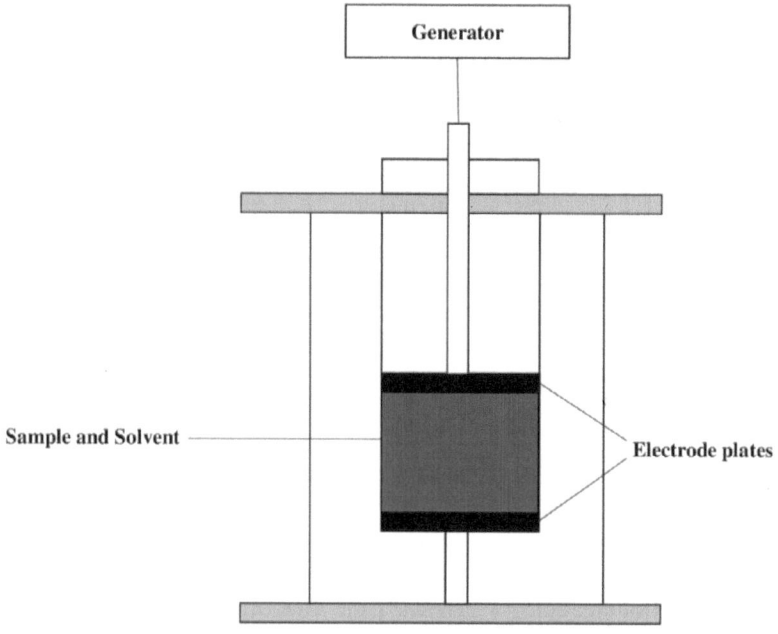

FIGURE 3.3 Experimental set-up of PEF extraction unit.

converted into the gaseous phase (Soquetta, Terra and Bastos, 2018). The main transport process is the convection in supercritical solvent phase (Silva and Martínez, 2014). The sample requirement is less and the extraction is selective, fast and does not need further cleaning (Oroian and Escriche, 2015). SFE can also be coupled with any analytical technique like Gas chromatography-mass spectroscopy for further characterization. This technique two steps, firstly the phytochemical present in the sample gets solubilized and gets separated in the supercritical solvent. In the next step, the solvent containing target compound is passed through a packed bed. Due to the increase in temperature and decrease in pressure the extract becomes solvent free when it leaves the extraction assembly (da Silva, Rocha-Santos and Duarte, 2016). It is used for the extraction of non-polar phytochemicals like carotenoids. The extraction of polar phytochemicals can be done after adding some modifiers like methanol, water and ethanol (Herrero et al., 2013). Supercritical fluid has better transport properties such as low viscosity, low surface tension, more spreadable in the plant matrix. These features of supercritical fluid favor the penetration solvent into the sample and improves extraction efficiency (Pouliot, Conway and Leclerc, 2014). The extraction is also affected by the solubility of extracts which is influenced by the density of the solvent. The supercritical fluids commonly used in SFE are CO_2, ethene, ethane, propane, LPG, methanol, water, n-butane, n-pentene, nitrous oxide (da Silva, Rocha-Santos and Duarte, 2016).

3.3.2.5 Pressurized Liquid Extraction

PLE is a non-conventional technique in which solutes are transported over a solid matrix. At high temperature and pressure liquid solvents are used, the surface tension of solvent is decreased which facilitates the solvent transport. The sample matrix gets disrupted and the mass transfer of solvent is improved (Barbosa et al., 2019). The selection of solvent depends on the solubility of compound of interest. The pressurized liquids are very flexible in nature because of their physicochemical properties such as viscosity, density, diffusivity and dielectric constants. These properties can be altered by changing the pressure and temperature of the extraction unit (Pronyk and Mazza, 2009). This technique offers green extraction and sustainability compared to the conventional techniques. Since this method involves frequent variation in temperature and pressure, it requires a highly sophisticated extraction system as shown in Figure 4.

3.3.2.6 High Voltage Electrical Discharge Extraction

HVED extraction involves the formation of a channel of plasma by high voltage electrical discharge between two electrodes submerged in extraction solvent (Boussetta and Vorobiev, 2014; Rajha et al., 2015). The electric field intensity creates a flood of electrons, which starts the flow of positive streamer for negatively charged electrode. Due to this mechanism, several other phenomena such as turbulence, cavitation bubble and shock waves occur, causing causes the cellular damage. The intracellular compounds are released into the solvent due to this damaged (Rajha et al., 2015). It was reported that the extraction parameters must be optimized for individual product because efficiency is affected by the nature of raw materials (Barba, Boussetta and Vorobiev, 2015). This technique has given better results as compared to the conventional techniques because it increased the yield and extraction kinetics. Since the process involves application of high voltage electrical discharge, it produces free radicals or chemical electrolysis can also take place. It can hinder with the phytochemicals being extracted and diminish their properties.

3.3.2.7 High Pressure Processing Base Extraction

It is also known as high hydrostatic processing, or HPP. It was developed to process food and make it microbiologically safe without losing its nutritional content, sensory and physicochemical properties, (Escobedo-Avellaneda et al., 2011). The range of pressure involved in this technique ranges from 100 to 1000 MPa (Briones-Labarca et al., 2015). It is considered as a green technique because it does not produce any waste and only consumes electricity (Andrés et al., 2016). The mass transfer is improved and the phytochemicals are diffused based on phase transitions (Oroian and Escriche, 2015). The basic mechanism involved in HPP is the deprotonation of charged molecules, breaks the salt bridges and hydrophobic bonds, resulting in the protein denaturation. The amount of solvent entering in the cell increases and extraction yield is also increased (Briones-Labarca et al., 2015).

The non-conventional techniques can also be used in combination to improve extraction by saving resources and energy. The ultrasound is combined with microwave, heat reflux and supercritical fluid extraction (Yang and Wei, 2015). The combination of HPP and PEF has also been reported for the anthocyanin extraction from grapes (Corrales et al., 2008). The extraction of bioactive using mango peels was done by supercritical fluid extraction and pressurized liquid extraction in combination (Garcia-Mendoza et al., 2015). There many other examples of extraction being done using the combination of techniques.

3.4 PHYTOCHEMICAL EXTRACTION FROM DIFFERENT SOURCES

The biomass also contributes to a large volume of wastes including the solid and liquid. This biomass contains a variety of ingredients which can be converted into value added products or can be used in production of such products. Generally, this kind of waste biomass does not a good economic value but their further processing can improve the economic importance (Stintzing, Schieber and Carle, 2002). Most of the fruits and vegetables are used for the edible part and the non-edible part such as peels, seeds, skins etc. are left over.

The extraction of the phytochemicals depends on the nature of source. It may vary based on the suitability and selection of plant material or biomass. The by-products of fruit and vegetable processing and other perishable sources create challenges in extraction of phytochemicals. The basic steps involved in recovery of a desired compound includes selection of sample and its cleaning or washing to eliminate physical impurities. It is followed by the removal or separation of peel or skin or pericarp and seed from main fruit or vegetable. The main fruits and vegetables sample is now subjected to drying to remove the moisture content to obtain the solid part. Size reduction of the dried sample is done to increase the surface area and facilitate the extraction. The selected extraction technique is then applied to recover the compound of interest from the sample. The extract is filtered and the solvent is removed by evaporation or lyophilization. Evaporation or lyophilization is performed under the influence of vacuum to prevent the loss of phytochemicals. The extracts are further stored under optimum conditions depending on the nature of extracted phytochemicals.

The phytochemicals have been extracted from various sources which are perishable in nature and can end up as waste biomass. Carotenoids such as lycopene and β-carotene has been extracted from tomato paste waste, skins and seeds by SFE. β-carotene was also extracted from apricot waste and press cake of carrot processing by SFE. Polyphenols like resveratrol and naringenin were extracted by SFE from the waste biomass of grape processing. Epicatechin, catechin and gallic acid were also extracted from the pomace of white grapes (Wijngaard et al., 2012). Maceration technique was applied for the extraction of polyphenols like catechin and quercetin from the apple peel wastes. A comparative study was done on the aqueous mixture of ethanol, glycerol and butanediol

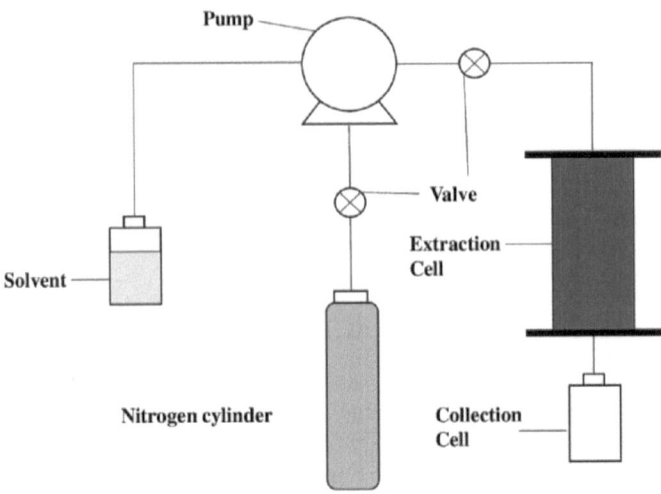

FIGURE 3.4   PLE extraction unit with different parts.

showing the optimum results obtained from ethanolic extract (Blidi et al., 2015). Carbon dioxide was used as supercritical fluid for the phytochemical extraction from tomato leaf waste. The recovery was reported to be efficient for phenolic (240 mg GAE/g), flavonoids (184 mg QE/g) and phylloquinone (29 µg/g) (Arab et al., 2019). UAE of eggplant field waste was performed for the recovery of anthocyanin, polyphenols and steroidal alkaloids (Mauro et al., 2020). Microwave extraction of antioxidant compounds was performed from the peanut skins and the obtained phenolic content was 143.6 mg GAE/g. The LC-MS-MS analysis of the extract confirmed the presence of resveratrol in the extracts (Ballard et al., 2010). The recovery of antioxidants from mango peel was done by maceration technique (Ajila et al., 2007). An ultrasonic bath was used for the UAE of the extraction of mangiferin from the mango peels (Kulkarni and Rathod, 2014). The UAE of jabutica peels was done for the extraction and determination of anthocyanin and phenolic compounds (Rodrigues et al., 2015). A green and solvent free technique (integrated with MAE and UAE) was used for the extraction of pectin, polyphenols and essential oils from waste orange peels (Boukroufa et al., 2015). The β-carotene was extracted from the carrot peels by microwave assisted extraction (Chumnanpaisont, Niamnuy and Devahastin, 2014). The leaves of plants contribute in a high proportion to the biomass waste. They are highly perishable in nature and does not have a long shelf life. Hence, the perishable and unsalable leaves can be a potential source for phytochemical extraction. SFE and Soxhlet extraction was applied for the extraction of flavonoids from the spearmint leaves. Epicatechin, catechin, naringenin, luteolin, myricetin, rutin and apigenin were identified by the HPLC analysis of spearmint leaf extract (Bimakr et al., 2011). SFE, MAE, UAE and PLE has been used to determine the bioactive compounds in the olive leaves (Rahmanian, Jafari and Wani, 2015). Rosemary leaves was subjected to the MAE and HPLC-ESI-QTOF-MS was used for the characterization of phytochemicals i.e., flavonoids, phenolic acids, terpenoids (Borrás-Linares et al., 2014). The antioxidant and anti-inflammatory activity of the phytochemicals from *Trichodesma khasianum* leaves was determined after the maceration extraction (Chen et al., 2020). Pressurized hot water extraction which is a type of PLE was used for the recovery of phytochemicals and glycosidic compounds from the stevia leaves (Kovačević et al., 2018). The optimization of MAE of Pitacia leaves was compared with UAE and Soxhlet extraction and the latter proved to be more efficient (Dahmoune et al., 2014). The recovery of phytochemicals from the perishable Moringa leaves was done by UAE technique. Flavonoids like quercetin and kaempferol were detected by HPLC-DAD in their glycosidic form (Pollini et al., 2020). The leaves of *Cassia auriculata* was subjected for the extraction of phytochemicals by UAE technique (Sharmila et al., 2016).

3.5 ISOLATION, PURIFICATION AND CHARACTERIZATION OF PHYTOCHEMICALS FROM PLANT BIOMASS

In the recent years, the techniques involved in isolation, purification and characterization of phytochemicals from the plant biomass has become more advanced (Altemimi et al., 2015). It is much easier to perform the complex and advanced biological assays with the advent of novel techniques. These techniques also provide better accuracy for the isolation, purification and characterization. The selection of appropriate method for the downstream processing of phytochemicals focusses on retaining its biological activities such as antimicrobial, antioxidant, cytotoxic and other properties (Mulinacci et al., 2004). Since the in vivo models require animal experimentation they are time consuming, expensive and ethical clearance of experimentation. However, the in vitro models do not have such constraints and can be performed conveniently. It becomes impossible to select and finalize a particular protocol for the isolation and characterization of a specific phytochemical due to certain factors involved. It can be due to the origin of plant, geographical conditions, genetics etc. The physicochemical properties and complexity of phytochemicals also influence the selection process (Sarajlija and Novotni, 2012). The identification, selection and collection of plant and biomass sample are considered as major step which affect the isolation and characterization of phytochemicals. The ethno-botanical information of the target compound is also a necessary step. This is followed by using different types of solvents for the isolation and purification of phytochemicals. Chromatographic techniques such as column chromatography are widely used for the separation, isolation and purification of phytochemicals. The purification process of phytochemical is accelerated by the application of advanced equipment such as HPLC. Many spectroscopic techniques like Fourier transform infrared (FTIR), UV-visible, mass spectroscopy (MS) and nuclear magnetic resonance (NMR) are used for the identification of phytochemicals (Popova, Hall and Kubátová, 2009).

3.5.1 Phytochemical Purification from Extract

As mentioned earlier, various chromatographic techniques such as Thin-layer chromatography (TLC) and Column chromatography have been used for separation and purification of phytochemicals. Column chromatography is still preferred because of its economy, availability of stationary phase and convenience. Cellulose, polyamide, alumina and silica are some efficient and widely used stationary phases in column chromatography. The separation becomes difficult due to the complex nature of phytochemicals. Hence, multiple mobile phases (solvents) with increasing polarity are used for proper separation. The fractions obtained after column chromatography is most often analyzed by TLC technique. TLC and Silica gel-based chromatography can be coupled with analytical techniques for better separation and characterization (Cotas et al., 2014).

3.5.2 Structural Elucidation of Phytochemicals

The data obtained from various techniques, including FTIR, UV-visible, MS and NMR are used to determine the structure of the compound. Spectroscopic techniques work on the principle of absorption of electromagnetic radiation. When the electromagnetic waves pass through the organic molecule, it absorbs the radiation. A spectrum is generated by recording the amount of electromagnetic radiation being absorbed the target compound or molecule. The bonds involved in conformation of molecular structure are bound to a specific spectrum. The structural elucidation of target compound is done the basis of the spectra obtained. For the precise structural elucidation, the data from different regions such as IR, UV, electron beam and radiofrequency are used by the researchers (Popova, Hall and Kubátová, 2009).

3.5.3 UV-visible Spectroscopy for Phytochemical Identification

The qualitative analysis performed for the determination of group of phytochemicals is done by UV-visible spectroscopy for both pure and crude extracts. The aromatic compounds are potentially good chromatophores under the influence of UV light, which property is utilized for the quantitative analysis. The complexes of iron with anthocyanins, phenols and other antioxidants can be easily determined by UV-visible spectroscopy. This technique is also used in the determination of phenolic content, flavonoid content, anthocyanin content and antioxidant activity at wavelength range of 500–600 nm, 400–600 nm, 500–700 nm and 500–550 nm, respectively. It is economical and less time consuming as compared to other techniques (Luque et al., 2006).

3.5.4 Infrared Spectroscopy of Phytochemicals

As mentioned earlier in this chapter that the phytochemicals and its group have the ability of absorbing the light. But they also have the potential to transmit the light which is not absorbed. When the molecule is exposed to the infrared radiation, they tend to undergo vibrational changes due to the absorption of infrared radiation. This technique is also termed as vibrational spectroscopy. The bonds existing in the compounds have different vibrational frequencies. These bonds can be C-O, C=O, C-C, C=C, C≡C, N-H and O-H depending on the phytochemical group. The band of frequency absorption in the spectrum is used for the detection of bonds present in the target compound (Cuadrado et al., 2006). The determination of structure and identification of chemicals in any phytochemical is done by FTIR, which is a high-resolution analytical technique. It provides a nondestructive, robust and rapid examination of crude plant extracts.

3.5.5 NUCLEAR MAGNETIC RESONANCE SPECTROSCOPY

The NMR spectroscopy is based on the magnetic properties of nucleus contained in the atom. The nuclei of carbon atom, proton, hydrogen atom and isotope of carbon atom are involved in NMR spectroscopy. The difference between the magnetic intensity of nuclei is measured and recorded. This data obtained is used by the researchers to determine the precise location and position of a particular nuclei in the atom. The atoms present in the neighboring sites can also be detected by this technique. So, the number of atoms present can be eventually determined. NMR is followed by the various techniques like TLC, column chromatography, LC and GC for the identification of conformation of a molecule (Altemimi et al., 2017).

3.5.6 MASS SPECTROSCOPY OF PHYTOCHEMICALS

In mass spectroscopy, charged ions of high energy are produced when the bombardment between organic molecule and laser or electron takes place. A graph is plotted between relative abundance of ion fragments and their charge/mass ratio which is known as mass spectrum. The precise molecular weight of the identified compound is determined by mass spectroscopy. Also, the location of ion fragments and the molecular formula is determined by this technique (Altemimi et al., 2017). The identification and characterization of target compound is simultaneously done by UV-visible, FTIR, MS and NMR. The target compound might be in hydrolyzed and the characterization might show its derivatives instead. By applying the tandem MS, the structural elucidation of the target compound can be done. A rapid, robust and precise data is obtained for the identification of desired compound, if HPLC is coupled with MS even if the standards are not available (Altemimi et al., 2017). Nowadays most of the highly sophisticated chromatographic techniques such as GC and LC are coupled with MS for the simultaneous separation and detection of phytochemicals. Both electron spray ionization and chemical ionization are applied in the MS but the latter is more preferred due to its high ionization efficiency for phytochemicals.

3.6 CONCLUSION

The waste biomass contains a high proportion of plants and their parts, which are highly perishable in nature. This biomass is a good and economical source of phytochemicals existing in nature. These phytochemicals have a wide range of application is pharmaceutical industry, food industry, dyeing and textile industry. The recovery of phytochemicals was being done by applying the conventional techniques like maceration, hydro-distillation and Soxhlet extraction. But due the limitation of time, cost and labor their alternative technique must be adopted for efficient and better extraction. The non-conventional techniques (MAE, UAE, SFE, PLE, PEF and HVED extraction) of extraction offer convenience, robustness, safety and environment friendly approach for the phytochemical extraction. This chapter has displayed that different biomass waste generated from fruits and vegetables viz. skins, peels, seeds, perishable leaves were used for phytochemical extraction. The separation and characterization of phytochemical has been improved because of the introduction of novel techniques. Hence, this chapter clearly describes the importance of phytochemical extraction from biomass in terms of environment and economy.

REFERENCES

Adetunji, L.R., Adekunle, A., Orsat, V. and Raghavan, V., 2017. Advances in the pectin production process using novel extraction techniques: A review. Food Hydrocolloids, 62, pp.239–250.

Ajila, C.M., Brar, S.K., Verma, M., Tyagi, R.D. and Valéro, J.R., 2011. Solid-state fermentation of apple pomace using *Phanerocheate chrysosporium*–Liberation and extraction of phenolic antioxidants. Food Chemistry, 126(3), pp.1071–1080.

Ajila, C.M., Naidu, K.A., Bhat, S.G. and Rao, U.P., 2007. Bioactive compounds and antioxidant potential of mango peel extract. Food Chemistry, 105(3), pp.982–988.

Alam, M.A., Subhan, N., Hossain, H., Hossain, M., Reza, H.M., Rahman, M.M. and Ullah, M.O., 2016. Hydroxycinnamic acid derivatives: a potential class of natural compounds for the management of lipid metabolism and obesity. Nutrition & Metabolism, 13(1), p.27.

Altemimi, A., Lakhssassi, N., Baharlouei, A., Watson, D.G. and Lightfoot, D.A., 2017. Phytochemicals: Extraction, isolation, and identification of bioactive compounds from plant extracts. Plants, 6(4), p.42.

Altemimi, A., Lightfoot, D.A., Kinsel, M. and Watson, D.G., 2015. Employing response surface methodology for the optimization of ultrasound assisted extraction of lutein and β-carotene from spinach. Molecules, 20(4), pp.6611–6625.

Andrés, V., Mateo-Vivaracho, L., Guillamón, E., Villanueva, M.J. and Tenorio, M.D., 2016. High hydrostatic pressure treatment and storage of soy-smoothies: Colour, bioactive compounds and antioxidant capacity. LWT-Food Science and Technology, 69, pp.123–130.

Arab, M., Bahramian, B., Schindeler, A., Valtchev, P., Dehghani, F. and McConchie, R., 2019. Extraction of phytochemicals from tomato leaf waste using subcritical carbon dioxide. Innovative Food Science & Emerging Technologies, 57, p.102204.

Arias, M., Penichet, I., Ysambertt, F., Bauza, R., Zougagh, M. and Ríos, Á., 2009. Fast supercritical fluid extraction of low-and high-density polyethylene additives: Comparison with conventional reflux and automatic Soxhlet extraction. The Journal of Supercritical Fluids, 50(1), pp.22–28.

Azeredo, H.M., 2009. Betalains: properties, sources, applications, and stability–a review. International Journal of Food Science & Technology, 44(12), pp.2365–2376.

Azmir, J., Zaidul, I.S.M., Rahman, M.M., Sharif, K.M., Mohamed, A., Sahena, F., Jahurul, M.H.A., Ghafoor, K., Norulaini, N.A.N. and Omar, A.K.M., 2013. Techniques for extraction of bioactive compounds from plant materials: A Review. Journal of Food Engineering, 117(4), pp.426–436.

Ballard, T.S., Mallikarjunan, P., Zhou, K. and O'Keefe, S., 2010. Microwave-assisted extraction of phenolic antioxidant compounds from peanut skins. Food Chemistry, 120(4), pp.1185–1192.

Barba, F.J., Boussetta, N. and Vorobiev, E., 2015. Emerging technologies for the recovery of isothiocyanates, protein and phenolic compounds from rapeseed and rapeseed press-cake: effect of high voltage electrical discharges. Innovative Food Science & Emerging Technologies, 31, pp.67–72.

Barbosa, A.M., Santos, K.S., Borges, G.R., Muniz, A.V., Mendonça, F.M., Pinheiro, M.S., Franceschi, E., Dariva, C. and Padilha, F.F., 2019. Separation of antibacterial biocompounds from *Hancornia speciosa* leaves by a sequential process of pressurized liquid extraction. Separation and Purification Technology, 222, pp.390–395.

Belhadj, S.I., Najar, T. and Abderrabba, M., 2017. Chemical and antioxidant properties of betalains. Journal of Agricultural and Food Chemistry, 65(4), pp.675–689.

Bianchini, F. and Vainio, H., 2001. Allium vegetables and organosulfur compounds: do they help prevent cancer? Environmental Health Perspectives, 109(9), pp.893–902.

Bimakr, M., Rahman, R.A., Taip, F.S., Ganjloo, A., Salleh, L.M., Selamat, J., Hamid, A. and Zaidul, I.S.M., 2011. Comparison of different extraction methods for the extraction of major bioactive flavonoid compounds from spearmint (*Mentha spicata L.*) leaves. Food and Bioproducts Processing, 89(1), pp.67–72.

Blidi, S., Bikaki, M., Grigorakis, S., Loupassaki, S. and Makris, D.P., 2015. A comparative evaluation of bio-solvents for the efficient extraction of polyphenolic phytochemicals: apple waste peels as a case study. Waste and Biomass Valorization, 6(6), pp.1125–1133.

Block, E., 2004, April. Biological activity of allium compounds: recent results. In IV International Symposium on Edible Alliaceae, 688 (pp. 41–58).

Borrás-Linares, I., Stojanović, Z., Quirantes-Piné, R., Arráez-Román, D., Švarc-Gajić, J., Fernández-Gutiérrez, A. and Segura-Carretero, A., 2014. Rosmarinus officinalis leaves as a natural source of bioactive compounds. International Journal of Molecular Sciences, 15(11), pp.20585–20606.

Boukroufa, M., Boutekedjiret, C., Petigny, L., Rakotomanomana, N. and Chemat, F., 2015. Bio-refinery of orange peels waste: A new concept based on integrated green and solvent free extraction processes using ultrasound and microwave techniques to obtain essential oil, polyphenols and pectin. Ultrasonics Sonochemistry, 24, pp.72–79.

Brglez Mojzer, E., Knez Hrnčič, M., Škerget, M., Knez, Ž. and Bren, U., 2016. Polyphenols: Extraction methods, antioxidative action, bioavailability and anticarcinogenic effects. Molecules, 21(7), p.901.

Briones-Labarca, V., Plaza-Morales, M., Giovagnoli-Vicuna, C. and Jamett, F., 2015. High hydrostatic pressure and ultrasound extractions of antioxidant compounds, sulforaphane and fatty acids from Chilean papaya (*Vasconcellea pubescens*) seeds: Effects of extraction conditions and methods. LWT-Food Science and Technology, 60(1), pp.525–534.

Brodowska, K.M., 2017. Natural flavonoids: classification, potential role, and application of flavonoid analogues. European Journal of Biological Research, 7(2), pp.108–123.

Boussetta, N. and Vorobiev, E., 2014. Extraction of valuable biocompounds assisted by high voltage electrical discharges: A review. Comptes Rendus Chimie, 17(3), pp.197–203.

Castro-Concha, L.A., Baas-Espinola, F.M., Ancona-Escalante, W.R., Vázquez-Flota, F.A. and Miranda-Ham, M.L., 2016. Phenylalanine biosynthesis and its relationship to accumulation of capsaicinoids during Capsicum chinense fruit development. Biologia Plantarum, 60(3), pp.579–584.

Chemat, F., Rombaut, N., Sicaire, A.G., Meullemiestre, A., Fabiano-Tixier, A.S. and Abert-Vian, M., 2017. Ultrasound assisted extraction of food and natural products. Mechanisms, techniques, combinations, protocols and applications. A review. Ultrasonics Sonochemistry, 34, pp.540–560.

Chen, S.Y., Wang, G.Y., Lin, J.H. and Yen, G.C., 2020. Antioxidant and anti-inflammatory activities and bioactive compounds of the leaves of *Trichodesma khasianum clarke*. Industrial Crops and Products, 151, p.112447.

Chong, J., Poutaraud, A. and Hugueney, P., 2009. Metabolism and roles of stilbenes in plants. Plant Science, 177(3), pp.143–155.

Chou, Y.C., Ho, C.T. and Pan, M.H., 2018. Stilbenes: chemistry and molecular mechanisms of anti-obesity. Current Pharmacology Reports, 4(3), pp.202–209.

Chua, L.S., Abd Latiff, N. and Mohamad, M., 2016. Reflux extraction and cleanup process by column chromatography for high yield of andrographolide enriched extract. Journal of Applied Research on Medicinal and Aromatic Plants, 3(2), pp.64–70.

Chumnanpaisont, N., Niamnuy, C. and Devahastin, S., 2014. Mathematical model for continuous and intermittent microwave-assisted extraction of bioactive compound from plant material: Extraction of β-carotene from carrot peels. Chemical Engineering Science, 116, pp.442–451.

Cieślik, E., Leszczyńska, T., Filipiak-Florkiewicz, A., Sikora, E. and Pisulewski, P.M., 2007. Effects of some technological processes on glucosinolate contents in cruciferous vegetables. Food Chemistry, 105(3), pp.976–981.

Corrales, M., Toepfl, S., Butz, P., Knorr, D. and Tauscher, B., 2008. Extraction of anthocyanins from grape by-products assisted by ultrasonics, high hydrostatic pressure or pulsed electric fields: A comparison. Innovative Food Science & Emerging Technologies, 9(1), pp.85–91.

Cotas, J., Leandro, A., Monteiro, P., Pacheco, D., Figueirinha, A., Gonçalves, A.M., da Silva, G.J. and Pereira, L., 2020. Seaweed phenolics: From extraction to applications. Marine drugs, 18(8), p.384.

Cuadrado, M.U., De Castro, M.L., Juan, P.P. and Gómez-Nieto, M.A., 2005. Comparison and joint use of near infrared spectroscopy and Fourier transform mid infrared spectroscopy for the determination of wine parameters. Talanta, 66(1), pp.218–224.

da-Cunha, E.V.L., Fechine, I.M., Guedes, D.N., Barbosa-Filho, J.M. and da Silva, M.S., 2005. Protoberberine alkaloids. The Alkaloids: Chemistry and Biology, 62, pp.1–75.

da Porto, C. and Decorti, D., 2009. Ultrasound-assisted extraction coupled with under vacuum distillation of flavour compounds from spearmint (carvone-rich) plants: comparison with conventional hydrodistillation. Ultrasonics Sonochemistry, 16(6), pp.795–799.

Dahmoune, F., Spigno, G., Moussi, K., Remini, H., Cherbal, A. and Madani, K., 2014. Pistacia lentiscus leaves as a source of phenolic compounds: Microwave-assisted extraction optimized and compared with ultrasound-assisted and conventional solvent extraction. Industrial Crops and Products, 61, pp.31–40.

Delgado-Vargas, F., Jiménez, A.R. and Paredes-López, O., 2010. Natural pigments: carotenoids, anthocyanins, and betalains—characteristics, biosynthesis, processing, and stability. Critical Reviews in Food Science and Nutrition, 40(3), pp.173–289.

Demiray, S., Piccirillo, C., Rodrigues, C.L., Pintado, M.E. and Castro, P.M.L., 2011. Extraction of valuable compounds from Ginja cherry by-products: effect of the solvent and antioxidant properties. Waste and Biomass Valorization, 2(4), p.365.

Deng, J., Xu, Z., Xiang, C., Liu, J., Zhou, L., Li, T., Yang, Z. and Ding, C., 2017. Comparative evaluation of maceration and ultrasonic-assisted extraction of phenolic compounds from fresh olives. Ultrasonics Sonochemistry, 37, pp.328–334.

Dräger, B., 2004. Chemistry and biology of calystegines. Natural Product Reports, 21(2), pp.211–223.

El-Sayed, M. and Verpoorte, R., 2007. Catharanthus terpenoid indole alkaloids: biosynthesis and regulation. Phytochemistry Reviews, 6(2–3), pp.277–305.

Escobedo-Avellaneda, Z., Moure, M.P., Chotyakul, N., Torres, J.A., Welti-Chanes, J. and Lamela, C.P., 2011. Benefits and limitations of food processing by high-pressure technologies: effects on functional compounds and abiotic contaminants. CyTA-Journal of Food, 9(4), pp.351–364.

Fernández-Agulló, A., Freire, M.S., Ramírez-López, C., Fernández-Moya, J. and González-Álvarez, J., 2020. Valorization of residual walnut biomass from forest management and wood processing for the production of bioactive compounds. Biomass Conversion and Biorefinery, pp.1–10.

Fomo, G., Madzimbamuto, T.N. and Ojumu, T.V., 2020. Applications of nonconventional green extraction technologies in process industries: Challenges, limitations and perspectives. Sustainability, 12(13), p.5244.

Garcia-Mendoza, M.P., Paula, J.T., Paviani, L.C., Cabral, F.A. and Martinez-Correa, H.A., 2015. Extracts from mango peel by-product obtained by supercritical CO_2 and pressurized solvent processes. LWT-Food Science and Technology, 62(1), pp.131–137.

Gengatharan, A., Dykes, G.A. and Choo, W.S., 2015. Betalains: Natural plant pigments with potential application in functional foods. LWT-Food Science and Technology, 64(2), pp.645–649.

Ghasemzadeh, A. and Ghasemzadeh, N., 2011. Flavonoids and phenolic acids: Role and biochemical activity in plants and human. Journal of medicinal plants research, 5(31), pp.6697–6703.

Grassino, A.N., Djaković, S., Bosiljkov, T., Halambek, J., Zorić, Z., Dragović-Uzelac, V., Petrović, M. and Brnčić, S.R., 2020. Valorisation of tomato peel waste as a sustainable source for pectin, polyphenols and fatty acids recovery using sequential extraction. Waste and Biomass Valorization, 11(9), pp.4593–4611.

Grynkiewicz, G. and Gadzikowska, M., 2008 Tropane alkaloids as medicinally useful natural products and their synthetic derivatives as new drugs. Pharmacological Reports, 60(4), pp. 439–463.

Hamid, H.A., Ramli, A.N. and Yusoff, M.M., 2017. Indole alkaloids from plants as potential leads for antidepressant drugs: A mini review. Frontiers in Pharmacology, 8, p.96.

Herrero, M., Castro-Puyana, M., Mendiola, J.A. and Ibañez, E., 2013. Compressed fluids for the extraction of bioactive compounds. TrAC Trends in Analytical Chemistry, 43, pp.67–83.

Imran, M., Arshad, M.S., Butt, M.S., Kwon, J.H., Arshad, M.U. and Sultan, M.T., 2017. Mangiferin: a natural miracle bioactive compound against lifestyle related disorders. Lipids in Health and Disease, 16(1), p.84.

James, J.T. and Dubery, I.A., 2009. Pentacyclic triterpenoids from the medicinal herb, *Centella asiatica (L.)* Urban. Molecules, 14(10), pp.3922–3941.

Jeevahan, J., Anderson, A., Sriram, V., Durairaj, R.B., Britto Joseph, G. and Mageshwaran, G., 2018. Waste into energy conversion technologies and conversion of food wastes into the potential products: a review. International Journal of Ambient Energy, pp.1–19.

Juurlink, B.H., Azouz, H.J., Aldalati, A.M., AlTinawi, B.M. and Ganguly, P., 2014. Hydroxybenzoic acid isomers and the cardiovascular system. Nutrition journal, 13(1), p.63.

Kaderides, K., Papaoikonomou, L., Serafim, M. and Goula, A.M., 2019. Microwave-assisted extraction of phenolics from pomegranate peels: Optimization, kinetics, and comparison with ultrasounds extraction. Chemical Engineering and Processing-Process Intensification, 137, pp.1–11.

Kimani, N.M., Matasyoh, J.C., Kaiser, M., Brun, R. and Schmidt, T.J., 2018. Antiprotozoal sesquiterpene lactones and other constituents from *Tarchonanthus camphoratus and Schkuhria pinnata*. Journal of Natural Products, 81(1), pp.124–130.

Kohnen-Johannsen, K.L. and Kayser, O., 2019. Tropane alkaloids: Chemistry, pharmacology, biosynthesis and production. Molecules, 24(4), p.796.

Kovačević, D.B., Barba, F.J., Granato, D., Galanakis, C.M., Herceg, Z., Dragović-Uzelac, V. and Putnik, P., 2018. Pressurized hot water extraction (PHWE) for the green recovery of bioactive compounds and steviol glycosides from *Stevia rebaudiana Bertoni* leaves. Food chemistry, 254, pp.150–157.

Krstić, G., Jadranin, M., Todorović, N.M., Pešić, M., Stanković, T., Aljančić, I.S. and Tešević, V.V., 2018. Jatrophane diterpenoids with multidrug-resistance modulating activity from the latex of Euphorbia nicaeensis. Phytochemistry, 148, pp.104–112.

Kruger, M.J., Davies, N., Myburgh, K.H. and Lecour, S., 2014. Proanthocyanidins, anthocyanins and cardiovascular diseases. Food Research International, 59, pp.41–52.

Kshatriya, R., Shaikh, Y. and Nazeruddin, G., 2013. Synthesis of flavone skeleton by different methods. Orient J Chem, 29, pp.1475–1487.

Kukula-Koch, W.A. and Widelski, J., 2017. Alkaloids. In Pharmacognosy (pp. 163–198). Academic Press.

Kulkarni, V.M. and Rathod, V.K., 2014. Mapping of an ultrasonic bath for ultrasound assisted extraction of mangiferin from Mangifera indica leaves. Ultrasonics sonochemistry, 21(2), pp.606–611.

Kumar, V., Bhatt, V. and Kumar, N., 2018. Amides From Plants: Structures and Biological Importance. In Studies in Natural Products Chemistry (Vol. 56, pp. 287–333). Elsevier.

Kumar, S. and Pandey, A.K., 2013. Chemistry and biological activities of flavonoids: An overview. The Scientific World Journal.

Kurek, J., 2019. Introductory Chapter: Alkaloids-Their Importance in Nature and for Human Life. In Alkaloids-Their Importance in Nature and Human Life. IntechOpen.

Li, B. and Sun, D.W., 2002. Effect of power ultrasound on freezing rate during immersion freezing of potatoes. Journal of Food Engineering, 55(3), pp.277–282.

Liang, M.H., Zhu, J. and Jiang, J.G., 2018. Carotenoids biosynthesis and cleavage related genes from bacteria to plants. Critical Reviews in Food Science and Nutrition, 58(14), pp.2314–2333.

Mabry, T.J., 2001. Selected topics from forty years of natural products research: betalains to flavonoids, antiviral proteins, and neurotoxic nonprotein amino acids. Journal of Natural Products, 64(12), pp.1596–1604.

Manach, C., Scalbert, A., Morand, C., Rémésy, C. and Jiménez, L., 2004. Polyphenols: food sources and bioavailability. The American Journal of Clinical Nutrition, 79(5), pp.727–747.

Mantilla, S.V., Manrique, A.M. and Gauthier-Maradei, P., 2015. Methodology for extraction of phenolic compounds of bio-oil from agricultural biomass wastes. Waste and Biomass Valorization, 6(3), pp.371–383.

Maran, J.P. and Manikandan, S., 2012. Response surface modeling and optimization of process parameters for aqueous extraction of pigments from prickly pear (*Opuntia ficus-indica*) fruit. Dyes and Pigments, 95(3), pp.465–472.

Marić, M., Grassino, A.N., Zhu, Z., Barba, F.J., Brnčić, M. and Brnčić, S.R., 2018. An overview of the traditional and innovative approaches for pectin extraction from plant food wastes and by-products: Ultrasound-, microwaves-, and enzyme-assisted extraction. Trends in Food Science & Technology, 76, pp.28–37.

Matos, M.J., Santana, L., Uriarte, E., Abreu, O.A., Molina, E. and Yordi, E.G., 2015. Coumarins—an important class of phytochemicals. Phytochemicals-Isolation, Characterisation and Role in Human Health, pp.113–140.

Mauro, R.P., Agnello, M., Rizzo, V., Graziani, G., Fogliano, V., Leonardi, C. and Giuffrida, F., 2020. Recovery of eggplant field waste as a source of phytochemicals. Scientia Horticulturae, 261, p.109023.

Mazimba, O., Nana, F., Kuete, V. and Singh, G.S., 2013. Xanthones and anthranoids from the medicinal plants of Africa. In Medicinal Plant Research in Africa (pp. 393–434). Elsevier.

Moreira, R., Pereira, D.M., Valentão, P. and Andrade, P.B., 2018. Pyrrolizidine alkaloids: chemistry, pharmacology, toxicology and food safety. International Journal of Molecular Sciences, 19(6), p.1668.

Mulinacci, N., Prucher, D., Peruzzi, M., Romani, A., Pinelli, P., Giaccherini, C. and Vincieri, F.F., 2004. Commercial and laboratory extracts from artichoke leaves: estimation of caffeoyl esters and flavonoidic compounds content. Journal of Pharmaceutical and Biomedical Analysis, 34(2), pp.349–357.

Negi, J.S., Bisht, V.K., Singh, P., Rawat, M.S.M. and Joshi, G.P., 2013. Naturally occurring xanthones: chemistry and biology. J Appl Chem, 2013(1), pp.1–9.

Negro, V., Ruggeri, B. and Fino, D., 2018. Recovery of energy from orange peels through anaerobic digestion and pyrolysis processes after D-limonene extraction. Waste and Biomass Valorization, 9(8), pp.1331–1337.

Okur, İ., Baltacıoğlu, C., Ağçam, E., Baltacıoğlu, H. and Alpas, H., 2019. Evaluation of the Effect of Different Extraction Techniques on Sour Cherry Pomace Phenolic Content and Antioxidant Activity and Determination of Phenolic Compounds by FTIR and HPLC. Waste and Biomass Valorization, 10(12), pp.3545–3555.

Omirou, M.D., Papadopoulou, K.K., Papastylianou, I., Constantinou, M., Karpouzas, D.G., Asimakopoulos, I. and Ehaliotis, C., 2009. Impact of nitrogen and sulfur fertilization on the composition of glucosinolates in relation to sulfur assimilation in different plant organs of broccoli. Journal of Agricultural and Food Chemistry, 57(20), pp.9408–9417.

Oroian, M. and Escriche, I., 2015. Antioxidants: Characterization, natural sources, extraction and analysis. Food Research International, 74, pp.10–36.

Ostertag, E., Becker, T., Ammon, J., Bauer-Aymanns, H. and Schrenk, D., 2002. Effects of storage conditions on furocoumarin levels in intact, chopped, or homogenized parsnips. Journal of Agricultural and Food Chemistry, 50(9), pp.2565–2570.

Özbek, H.N., Halahlih, F., Göğüş, F., Yanık, D.K. and Azaizeh, H., 2020. Pistachio (*Pistacia vera L.*) Hull as a potential source of phenolic compounds: Evaluation of ethanol–water binary solvent extraction on antioxidant activity and phenolic content of pistachio hull extracts. Waste and Biomass Valorization, 11(5), pp.2101–2110.

Parniakov, O., Lebovka, N.I., Van Hecke, E. and Vorobiev, E., 2014. Pulsed electric field assisted pressure extraction and solvent extraction from mushroom (*Agaricus bisporus*). Food and Bioprocess Technology, 7(1), pp.174–183.

Peiró, S., Luengo, E., Segovia, F., Raso, J. and Almajano, M.P., 2019. Improving polyphenol extraction from lemon residues by pulsed electric fields. Waste and Biomass Valorization, 10(4), pp.889–897.

Pham, T.P.T., Kaushik, R., Parshetti, G.K., Mahmood, R. and Balasubramanian, R., 2015. Food waste-to-energy conversion technologies: Current status and future directions. Waste Management, 38, pp.399–408.

Pollastri, S. and Tattini, M., 2011. Flavonols: old compounds for old roles. Annals of Botany, 108(7), pp.1225–1233.

Pollini, L., Tringaniello, C., Ianni, F., Blasi, F., Manes, J. and Cossignani, L., 2020. Impact of Ultrasound Extraction Parameters on the Antioxidant Properties of *Moringa oleifera* Leaves. Antioxidants, 9(4), p.277.

Popova, I.E., Hall, C. and Kubátová, A., 2009. Determination of lignans in flaxseed using liquid chromatography with time-of-flight mass spectrometry. Journal of Chromatography A, 1216(2), pp.217–229.

Pouliot, Y., Conway, V. and Leclerc, P.L., 2014. Separation and concentration technologies in food processing. Food Processing: Principles and Applications, pp.33–60.

Pronyk, C. and Mazza, G., 2009. Design and scale-up of pressurized fluid extractors for food and bioproducts. Journal of Food Engineering, 95(2), pp.215–226.

Putnik, P., Lorenzo, J.M., Barba, F.J., Roohinejad, S., Režek Jambrak, A., Granato, D., Montesano, D. and Bursać Kovačević, D., 2018. Novel food processing and extraction technologies of high-added value compounds from plant materials. Foods, 7(7), p.106.

Rahmanian, N., Jafari, S.M. and Wani, T.A., 2015. Bioactive profile, dehydration, extraction and application of the bioactive components of olive leaves. Trends in Food Science & Technology, 42(2), pp.150–172.

Rajha, H.N., Boussetta, N., Louka, N., Maroun, R.G. and Vorobiev, E., 2015. Effect of alternative physical pretreatments (pulsed electric field, high voltage electrical discharges and ultrasound) on the dead-end ultrafiltration of vine-shoot extracts. Separation and Purification Technology, 146, pp.243–251.

Rashed, M.M., Tong, Q., Abdelhai, M.H., Gasmalla, M.A., Ndayishimiye, J.B., Chen, L. and Ren, F., 2016. Effect of ultrasonic treatment on total phenolic extraction from *Lavandula pubescens* and its application in palm olein oil industry. Ultrasonics Sonochemistry, 29, pp.39–47.

Roda-Serrat, M.C., Andrade, T.A., Rindom, J., Lund, P.B., Norddahl, B. and Errico, M., 2020. Optimization of the recovery of anthocyanins from chokeberry juice pomace by homogenization in acidified water. Waste and Biomass Valorization, pp.1–13.

Rodrigues, S., Fernandes, F.A., de Brito, E.S., Sousa, A.D. and Narain, N., 2015. Ultrasound extraction of phenolics and anthocyanins from jabuticaba peel. Industrial Crops and Products, 69, pp.400–407.

Rodríguez-García, C., Sánchez-Quesada, C., Toledo, E., Delgado-Rodríguez, M. and Gaforio, J.J., 2019. Naturally lignan-rich foods: a dietary tool for health promotion? Molecules, 24(5), p.917.

Rojas, R., Alvarez-Pérez, O.B., Contreras-Esquivel, J.C., Vicente, A., Flores, A., Sandoval, J. and Aguilar, C.N., 2020. Valorisation of mango peels: Extraction of pectin and antioxidant and antifungal polyphenols. Waste and Biomass Valorization, 11(1), pp.89–98.

Roselló-Soto, E., Parniakov, O., Deng, Q., Patras, A., Koubaa, M., Grimi, N., Boussetta, N., Tiwari, B.K., Vorobiev, E., Lebovka, N. and Barba, F.J., 2016. Application of non-conventional extraction methods: Toward a sustainable and green production of valuable compounds from mushrooms. Food Engineering Reviews, 8(2), pp.214–234.

Sagi, S., Avula, B., Wang, Y.H. and Khan, I.A., 2016. Quantification and characterization of alkaloids from roots of Rauwolfia serpentina using ultra-high performance liquid chromatography-photo diode array-mass spectrometry. Analytical and Bioanalytical Chemistry, 408(1), pp.177–190.

Saini, A., Panesar, P.S. and Bera, M.B., 2020. Valuation of *Citrus reticulata* (kinnow) peel for the extraction of lutein using ultrasonication technique. Biomass Conversion and Biorefinery, pp.1–9.

Sarajlija, H., Čukelj, N., Mršić, G.N.D., Brnčić, M. and Ćurić, D., 2012. Preparation of flaxseed for lignan determination by gas chromatography-mass spectrometry method. Czech Journal of Food Sciences, 30(1), pp.45–52.

Sarpras, M., Ahmad, I., Rawoof, A. and Ramchiary, N., 2019. Comparative analysis of developmental changes of fruit metabolites, antioxidant activities and mineral elements content in Bhut jolokia and other Capsicum species. LWT – Food Science and Technology, 105, pp.363–370.

Scoville, W.L., 1912. Note on capsicums. Journal of the American Pharmaceutical Association, 1(5), pp.453–454.

Sharma, A., Mazumdar, B. and Keshav, A. 2020a. Green Extraction of Betacyanin from Beetroot Peel using Microwave and Ultrasound Technology. Journal of Indian Chemical Society, 97, p. 2017

Sharma, A., Mazumdar, B. and Keshav, A. 2020b. Extraction and Phytochemical Analysis of *Coccinia indica* fruit using UV-VIS and FTIR Spectroscopy. Advances in Biomedical Engineering, pp. 1–6.

Sharma, A., Mazumdar, B. and Keshav, A., 2021. Valorization of unsalable Amaranthus tricolour leaves by microwave-assisted extraction of betacyanin and betaxanthin. Biomass Conversion and Biorefinery, pp. 1–17.

Sharmila, G., Nikitha, V.S., Ilaiyarasi, S., Dhivya, K., Rajasekar, V., Kumar, N.M., Muthukumaran, K. and Muthukumaran, C., 2016. Ultrasound assisted extraction of total phenolics from *Cassia auriculata* leaves and evaluation of its antioxidant activities. Industrial Crops and Products, 84, pp.13–21.

Silva, L.P.S. and Martínez, J., 2014. Mathematical modeling of mass transfer in supercritical fluid extraction of oleoresin from red pepper. Journal of Food Engineering, 133, pp.30–39.

Da Silva, R.P., Rocha-Santos, T.A. and Duarte, A.C., 2016. Supercritical fluid extraction of bioactive compounds. TrAC Trends in Analytical Chemistry, 76, pp.40–51.

Silva, Y.P., Ferreira, T.A., Celli, G.B. and Brooks, M.S., 2019. Optimization of lycopene extraction from tomato processing waste using an eco-friendly ethyl lactate–ethyl acetate solvent: a green valorization approach. Waste and Biomass Valorization, 10(10), pp.2851–2861.

Singh, A., Nair, G.R., Liplap, P., Gariepy, Y., Orsat, V. and Raghavan, V., 2014. Effect of dielectric properties of a solvent-water mixture used in microwave-assisted extraction of antioxidants from potato peels. Antioxidants, 3(1), pp.99–113.

Sirerol, J.A., Rodríguez, M.L., Mena, S., Asensi, M.A., Estrela, J.M. and Ortega, A.L., 2016. Role of natural stilbenes in the prevention of cancer. Oxidative Medicine and Cellular Longevity, 2016.

Soliva-Fortuny, R., Balasa, A., Knorr, D. and Martín-Belloso, O., 2009. Effects of pulsed electric fields on bioactive compounds in foods: a review. Trends in Food Science & Technology, 20(11–12), pp.544–556.

Soquetta, M.B., Terra, L.D.M. and Bastos, C.P., 2018. Green technologies for the extraction of bioactive compounds in fruits and vegetables. CyTA-Journal of Food, 16(1), pp.400–412.

Stafford, H.A., 1994. Anthocyanins and betalains: evolution of the mutually exclusive pathways. Plant Science, 101(2), pp.91–98.

Stefanachi, A., Leonetti, F., Pisani, L., Catto, M. and Carotti, A., 2018. Coumarin: A natural, privileged and versatile scaffold for bioactive compounds. Molecules, 23(2), p.250.

Stintzing, F.C. and Carle, R., 2004. Functional properties of anthocyanins and betalains in plants, food, and in human nutrition. Trends in Food Science & Technology, 15(1), pp.19–38.

Stintzing, F.C., Schieber, A. and Carle, R., 2002. Identification of betalains from yellow beet (*Beta vulgaris L.*) and cactus pear [*Opuntia ficus-indica (L.) Mill.*] by high-performance liquid chromatography– electrospray ionization mass spectrometry. Journal of Agricultural and Food Chemistry, 50(8), pp.2302–2307.

Tao, Y., Zhang, Z. and Sun, D.W., 2014. Experimental and modeling studies of ultrasound-assisted release of phenolics from oak chips into model wine. Ultrasonics Sonochemistry, 21(5), pp.1839–1848.

Tiwari, B.K., 2015. Ultrasound: A clean, green extraction technology. TrAC Trends in Analytical Chemistry, 71, pp.100–109.

Varo, M.A., Jacotet-Navarro, M., Serratosa, M.P., Mérida, J., Fabiano-Tixier, A.S., Bily, A. and Chemat, F., 2019. Green ultrasound-assisted extraction of antioxidant phenolic compounds determined by high performance liquid chromatography from bilberry (*Vaccinium myrtillus L.*) juice by-products. Waste and Biomass Valorization, 10(7), pp.1945–1955.

Victor, M.M., David, J.M., Cortez, M.V., Leite, J.L. and da Silva, G.S., 2020. A High-Yield Process for Extraction of Hesperidin from Orange (*Citrus sinensis L. osbeck*) Peels Waste, and Its Transformation to Diosmetin, A Valuable and Bioactive Flavonoid. Waste and Biomass Valorization, pp.1–8.

Vinatoru, M., Mason, T.J. and Calinescu, I., 2017. Ultrasonically assisted extraction (UAE) and microwave assisted extraction (MAE) of functional compounds from plant materials. TrAC Trends in Analytical Chemistry, 97, pp.159–178.

Wen, L., Zhang, Z., Sun, D.W., Sivagnanam, S.P. and Tiwari, B.K., 2020. Combination of emerging technologies for the extraction of bioactive compounds. Critical Reviews in Food Science and Nutrition, 60(11), pp.1826–1841.

Wijngaard, H., Hossain, M.B., Rai, D.K. and Brunton, N., 2012. Techniques to extract bioactive compounds from food by-products of plant origin. Food Research International, 46(2), pp.505–513.

Wu, C., Wang, F., Liu, J., Zou, Y. and Chen, X., 2015. A comparison of volatile fractions obtained from Lonicera macranthoides via different extraction processes: ultrasound, microwave, Soxhlet extraction, hydrodistillation, and cold maceration. Integrative Medicine Research, 4(3), pp.171–177.

Wu, C.Y., Wang, H., Fan, X.H., Yue, W. and Wu, Q.N., 2020. Waste *Euryale ferox salisb.* leaves as a potential source of anthocyanins: extraction optimization, identification and antioxidant activities evaluation. Waste and Biomass Valorization, 11(8), pp.4327–4340.

Xavier, L., Freire, M.S. and González-Álvarez, J., 2019. Modeling and optimizing the solid–liquid extraction of phenolic compounds from lignocellulosic subproducts. Biomass Conversion and Biorefinery, 9(4), pp.737–747.

Xu, J., Wang, W., Yang, X., Xiong, A., Yang, L. and Wang, Z., 2019. Pyrrolizidine alkaloids: an update on their metabolism and hepatotoxicity mechanism. Liver Research, 3(3–4), pp.176–184.

Yang, Y.C. and Wei, M.C., 2015. Kinetic and characterization studies for three bioactive compounds extracted from *Rabdosia rubescens* using ultrasound. Food and Bioproducts Processing, 94, pp.101–113.

Zalazar-García, D., Feresin, G.E. and Rodriguez, R., 2020. Optimal operational variables of phenolic compound extractions from pistachio industry waste (*Pistacia vera var. Kerman*) using the response surface method. Biomass Conversion and Biorefinery, pp.1–10.

Zardo, I., de Espíndola Sobczyk, A., Marczak, L.D.F. and Sarkis, J., 2019. Optimization of ultrasound assisted extraction of phenolic compounds from sunflower seed cake using response surface methodology. Waste and Biomass Valorization, 10(1), pp.33–44.

Zhang, Q.W., Lin, L.G. and Ye, W.C., 2018. Techniques for extraction and isolation of natural products: a comprehensive review. Chinese Medicine, 13(1), p.20.

Zhang, Z., Pang, X., Xuewu, D., Ji, Z. and Jiang, Y., 2005. Role of peroxidase in anthocyanin degradation in litchi fruit pericarp. Food Chemistry, 90(1–2), pp.47–52.

Zhao, S. and Zhang, D., 2014. Supercritical CO_2 extraction of Eucalyptus leaves oil and comparison with Soxhlet extraction and hydro-distillation methods. Separation and Purification Technology, 133, pp.443–451.

4 Biomass (Agricultural Waste) as Sustainable Reinforcement in Polymer Composite

Nirupama, Dan Bahadur Pal and Amit Kumar Tiwari
Department of Chemical Engineering, Birla Institute of Technology Mesra, Ranchi (JH)-India

CONTENTS

4.1 INTRODUCTION

The term 'composite' is defined as a heterogeneous material made by the combination of two or more chemically different components. Two or more materials combined in such a way that it utilizes desirable properties of one material and suppress the undesirable properties as far can be practicable. Compared to their individual components, composite materials have better properties, such as high

DOI: 10.1201/9781003196358-4

specific strength, stiffness, and fatigue characteristics. Composite materials often lead to both cheaper and better solutions. They are formed by embedding discontinuous phases (reinforcement) in a continuous phase (matrix). The reinforcements are usually harder, stronger, and stiffer than the matrix phase. Reinforcement is a load-carrying component and provides strength and rigidity while the matrix is a load transferring medium that helps to maintain the reinforcement position and orientation and protect them from the external impact due to elevated temperature, humidity, etc. The performance of the composite materials is not only determined by the properties of their constituents but also by their distribution and the interaction between them. Bavan and Kumar (2010) reported intrinsic properties of the reinforcement, which includes fibre geometry, fibre orientation, fibre content, and packing arrangement, also affect the performance of the composite materials.

In the past few years, polymeric matrixes in composite materials have gained considerable attention in various fields of applications. This is due to advantages offered by polymeric matrix over other conventional materials. These advantages include ease in processing, excellent hydrophobicity, corrosion resistance, low water absorption, low density, and low-cost characteristics. During the 1960s and early 1970s, the use of polymer composites was boosted extensively because of the development of advanced synthetic fibres like boron, carbon, and aramid fibres. These synthetic fibres have extremely high modulus as reported by Chawla (1987). Despite the good mechanical properties, these synthetic fibres reinforced polymer composites failing their charm. This is because of the drawbacks of synthetic fibres such as recycling, reusability, and biodegradability after their designed service life. Synthetic fibres such as glass and carbon fibre can cause various health issues like an irritation to the skin, eye, and respiratory tract. Long-term exposure in synthetic fibres can even cause cancer and lung scarring. In addition to that these synthetic fibres are quite expensive. To overcome these problems, 'Bio-composites' are proposed as a key concept in the field of polymer science in order to preserve our environment. The 'Biocomposites' are defined as an environmentally conscious composite materials or ecologically orientated composite materials which are socially acceptable with minimum environmental impact.

The concept of utilizing agricultural waste and green materials has developed considerable interest among the scientific researchers and the industrialist. With the attentiveness of protecting the environment, sincere efforts have been made in the search of the material which will utilize bio-based resources and should be bio-degradable (Vigneshwaran et al., 2020). In most developing countries like India, agriculture supports a major role in the contribution of their economy and employment rate. Thousands of tons of different crops are produced each year which leaves tons of wastes. Now, these agricultural wastes are utilized in making fibre reinforced composite for commercial use and increase employment in the rural sector. Natural fibres obtained from agricultural wastes such as hemp, kenaf, jute, ramie, cotton, abaca, and coir fibres found as a promising alternative to these synthetic fibres.

Natural fibres are a bio-based fibre which can be obtained either from the plant (cellulosic) or from animal (protein). In this chapter, only plant-based fibres are considered. Plant-based fibre such as hemp, Kenaf, sisal, coir, jute, ramie, pineapple leaf, banana, bagasse, bamboo, etc. has been already used with polymer matrix. These natural fibres-based polymer composites offered innumerable advantages. The tensile strength of conventional fibres (like glass fibre, boron fibre, aramid fibre, etc.) is substantially higher, but ease in processing, biodegradability, and cost reduction aspects offer great potential in natural fibre-based composites. Other advantages are abundant, low density, low energy requirement, less abrasion to equipment, and less skin and respiratory irritation. However, certain drawbacks such as poor water resistance, hydrophilic nature of natural fibre which gives poor bonding at fibre/matrix interface with hydrophobic polymers, aggregates during processing, and low thermal degradation temperature reduce the performance of the composites. Bledzki and Gassan (1999) reviewed readily used natural fibres up until 1999 and gave many significant information regarding composite's property improvement, proper process and process condition selection and its technical applications. Later, Faruk et al. (2012) has done comprehensive literature survey and gave a sensible overview of present development and future the trend of natural fibre/polymer composites. The overview of the potential of biomass (agricultural waste) for polymer composites is presented in this chapter.

Although the concept of utilizing agricultural waste (natural fibre) as a reinforcing agent in the composite is not new, their use vanished due to the development of advanced synthetic fibres. Nearly 3000 years back in Egypt people used to make walls with clay mixed with straw. But after the development of advanced synthetic fibres during 1960s, natural fibre lost its value. Therefore, in 2006 by the request of Food and Agriculture Organization (FAO, 2006), the year 2009 was declared as the 'International Year of Natural Fibres' by the general assembly of United Nations. Their main objective was to raise the awareness of natural fibres globally. They thought this will help to sustain farmer's income. Also, encouraged appropriate government policies and efficient growing techniques for making the environment green. According to the European end of life vehicles (ELV) directive, from 2015 onwards, all new vehicles must be manufactured of 95% recyclable materials, in which 85% recoverable through reuse or mechanical recycling; 10% can be for energy recovery and only 5% can be landfilled. In Lucintel's forecasts, it was reported that market growth of natural fibre composite will be 8.2% from 2015 to 2020 at a compound annual growth rate (CAGR). The report also highlighted the major driving factor for the growth of this market in various applications, such as automotive, building & construction, and others is due to their benefits such as lightweight and environmentally sustainable.

4.2   NATURAL FIBRES

Fibre is defined as a continuous filament or discrete elongated piece, similar to a thread that has a high aspect ratio and obtained from renewable resources. Natural

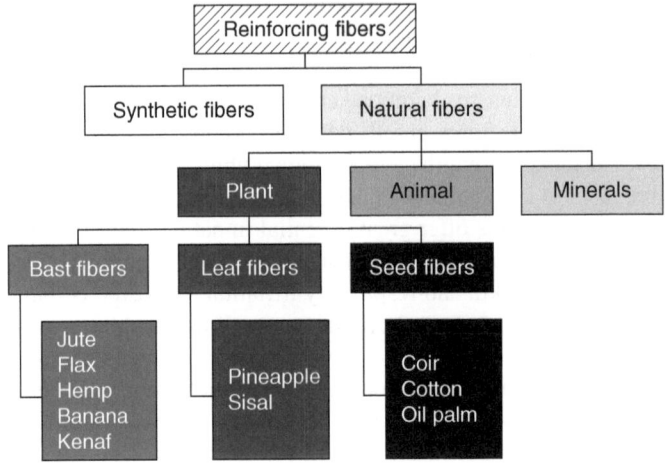

FIGURE 4.1   Classification of natural fibres.

fibres can be obtained from the minerals, animals and plants (Figure 4.1). Fibres driven from plants can be further classified as woody and non-woody fibres. In the same way, depending upon the sources non-woody fibres can be classified as into seed, leaf and bast fibres. Plant fibres (as shown in Figure 4.2) are mainly consisting of cellulose, hemicellulose, lignin, pectin, and waxy components in varying proportion. Chemical compositions of few plant based fibres are listed in the Table 4.1.

A physical typeset of the fibres is generally determined by the quantities of cellulose, hemicellulose, and lignin. Among these components, cellulose is the strongest and the stiffest and provides structural stability to the fibres. Mechanical performance of composite is main dependent on the cellulose content in fibres. Fibres with higher cellulose content provide higher mechanical properties. These celluloses are semi-crystalline polysaccharide with a large amount of hydroxyl (OH) group in it. These OH group present in cellulose makes them hydrophilic in nature as reported by Alvarez et al. (2003).

Hemicelluloses are strongly bonded to the cellulose microfibril by the hydrogen bonds have lower molecular weight compared to cellulose. Hemicelluloses are amorphous component in plant fibre. This is mainly consisting of OH and acetyl groups which make them hygroscopic in nature. Hemicellulose shows the least resistance towards biodegradability, water absorption, and thermal degradation.

Lignin act as a coupling agent in amorphous, phenylpropane polymer unit (mainly aromatic) and have the least water sorption. In plant fibre, lignin is the thermally stable component but these are also responsible for the UV- degradation (Bledzki et al. 2008).

Pectin is a heteropolysaccharide and provides flexibility to the plant. Waxy components usually found at the surface of the fibre which consists of different types of alcohols, which affects the wettability and adhesion characteristics (John et al., 2008; Saheb and jog, 1999).

FIGURE 4.2 Photographs of sources of some plant fibres.

The microstructure of naturally occurring fibre is very complex in structure. A single or elementary fibre is made up of concentric cell walls. Each concentric cell wall is different from the others. Components of each elementary fibre have different arrangement and thickness. Plant fibre is a complex structure of a hollow tube called lumen and cell walls. The lumen is accountable for the water and nutrient uptake. The plant fibre cell wall is composed of primary and secondary cell wall. Primary cell wall is comparatively thinner than the secondary cell wall. Secondary cell wall encircles inside the primary cell wall. During the initial growth of the cell wall primary cell wall forms and has a loose irregular arrangement of cellulose microfibrils. The secondary cell wall has three distant layers: S1 (outer), S2 (middle), and S3 (inner). Among the three layers, outer layer (S2) is the thickest and provides strength to the fibre. S2 contains a series of helically wound cellulose microfibrils and these cellulose microfibrils embedded in an amorphous matrix of lignin and hemicellulose (Pereira et al., 2015). The microfibrillar angle play a significant function in defining the strength of the fibre and is varies from one plant to another. The microfibrillar angle is defined as the angle between the fibre axis and the cellulose microfibrils (Yan et al., 2014).

Compared to conventional synthetic fibres (like glass or carbon fibres), plant-based fibres have gained significant focus as potential substitutes. Natural fibre-reinforced polymer composite has gained much attention in various industrial applications. This is due to the benefits offered by the natural fibres over synthetic fibres. These advantages include its biodegradability, lighter, cheaper and acceptable specific strength. Furthermore, these fibres are abundantly available from renewable sources (Joseph et al., 2002; Li et al., 2007; Eichhorn et al., 2001; Bledzki et al., 1999; Milosevic et al., 2020).

The potential of various plant fibres such as jute, hemp, banana, sisal, flax, and coir as reinforcement in polymer composites has already investigated and found applications in various fields. This is mainly due to the encouraging monetary and ecological impact, in addition to their capability to distinctively meet human desires worldwide (Holbery and Houston, 2006; Dittenber and GangaRao, 2012). Saheb and Jog, (1999) have done a comprehensive review with special focus of the effect of type of fibres, polymer matrix, fibre surface modification, and fibre–matrix interface. Bavan and Kumar (2010) reported the potential of plant fibre-based polymer composite materials in India. This is mainly due to the encouraging monetary and ecological impact, in addition to their capability to distinctively meet human desires worldwide. Faruk et al. (2014) reported that the cost of natural fibre which is one third of the cost of glass fibre or less. Ho et al. (2012) reported a review with special reference to different processing techniques and their effect on the performance of plant fibre and its composites.

From past studies, it has been understood that the properties of the natural fibre-polymer composite strongly dependents on various aspects such as fibre content, fibre aspect ratio, fibre/matrix interfacial adhesion, fibre orientation, etc. The hydrophilic nature of natural fibres makes them unsuited with the hydrophobic polymeric matrix and leads to provide poor interfacial bonding. This poor interfacial bonding causes insufficient stress distribution from matrix to fibre. Researchers have now developed different techniques to improve their bonding with the polymer matrix. There is a number of articles available that have done a comprehensive review of different chemical modifications. Chemical modification to the natural fibres helps to improve the compatibility with the polymer matrix for the development of advanced composite materials. In addition, fibre hybridization is a promising strategy to strengthen composite materials (Yousif, 2012; Ku et al., 2011; Kabir et al., 2012). Jawaid and Khail, 2011 reported a review focusing on the recent expansion of cellulosic/synthetic and cellulosic/cellulosic fibres based reinforced polymer hybrid composites. Swolfs et al., 2014 presented an outline of the mechanical performance of the hybrid fibre composites and discussed some current trends in fibre hybridization.

4.2.1 PROPERTIES AND CHARACTERISTICS OF NATURAL (PLANT) FIBRES

The characteristics values of natural and artificial fibres are summarised in Table 4.1 (Dittenber and GangaRao, 2012). Plant fibres are non-abrasive in nature. Thus, contribute to significant reductions in the equipment maintenance cost. Compared to conventional glass fibre, plant fibres offer safer working conditions. It reduces the threat of dermal and respiratory problems. Moreover, these natural fibres are

TABLE 4.1

Physical, Chemical and Mechanical Properties of Natural and Synthetic Fibres

Fibre	Cellulose (wt%)	Hemi-cellulose (wt%)	Lignin (wt%)	Waxes (wt%)	Microfibrillar angle (deg.)	Tensile strength (MPa)	Young's modulus (GPa)	Elongation at break (%)	Density [g/cm³]
Abaca	56–63	20–25	7–9	3	–	400	12	3–10	1.5
Bagasse	55.2	16.8	25.3	–	–	290	17	–	1.25
Bamboo	26–43.2	30	21–31	–	–	140–230	11–17	–	0.6–1.1
Flax	71	18.6–20.6	2.2	1.5	5–10	345–1035	27.6	2.7–3.2	1.5
Hemp	68	15	10	0.8	2–6.2	690	70	1.6	1.48
Jute	61–71	14–20	12–13	0.5	8	393–773	26.5	1.5–1.8	1.3
Kenaf	72	20.3	9	–	–	930	53	1.6	–
Sisal	65	12	9.9	2	10–22	511–635	9.4–22	2–2.5	1.5
Ramie	68.6–76.2	13–16	0.6–0.7	0.3	7.5	560	24.5	2.5	1.5
Oil palm	65	–	29	–	42–46	248	3.2	25	0.7–1.55
Pineapple	81	–	12.7	–	14	400–627	1.44	14.5	0.8–1.6
Coir	32–43	0.15–0.25	40–45	–	30–49	175	4–6	30	1.2
Curaua	73.6	9.9	7.5	–	–	500–1,150	11.8	3.7–4.3	1.4
E-glass	–	–	–	–	–	2,000–3,500	70	0.5	2.5
S-glass	–	–	–	–	–	4,570	86	2.8	2.5
Aramids	–	–	–	–	–	3,000–3,150	63–67	3.3–3.7	1.4
Carbon	–	–	–	–	–	4,000	230–240	1.4–1.8	1.4

Source: Dittenber and GangaRao (2012).

biodegradable in nature hence, reduces extra load on the environment. It also requires very little energy for production (Mohanty et al., 2002).

Apart from many advantages natural fibres have many drawbacks. First major drawback is its inconsistent, its properties vary plant to plant from same cultivation. It depends on various parameters such as a variety of the plant, the climate in which the plant is grown, the maturity of the plant, harvesting, fibre extraction methods (mechanical, chemical, or steam explosion treatment), fibre modification, etc. (Velde and Kiekens, 2001; Petroudy, 2017; John et al., 2008; John and Thomas, 2008; Godara et al., 2019). Second major drawback of natural (plant) fibres is its hydrophilic nature. Hydrophilic nature in natural fibre is basically due to the presence of a large amount of hydroxyl group in it. This makes them incompatible with hydrophobic polymer matrix. Incompatibility leads provide weak fibre/matrix interfacial adhesion, further caused insufficient stress distribution between them. This weak fibre/matrix adhesion leads to give composite with inferior properties. Hydrophilicity of natural fibre further results in high moisture uptake, causing swelling and void formation at the interface. If moisture content is not removed prior to the fabrication of the composites, this will result in a porous product. High moisture absorption can also deteriorate the performance and distort the dimensional stability of the composite (Alvarez et al., 2004). Fibre/matrix interaction can be improved by fibre surface treatment and/or by matrix modifications (Li et al., 2007; Kabir et al., 2012; John and Ananadjiwala, 2008). Third major drawback is the thermal stability of natural fibre. It usually degrades above 200°C. This restricts the selection polymer matrix (Bismarck et al., 2006; Prasad et al., 2018b).

4.3   NATURAL FIBRE-POLYMER COMPOSITE

Polymer matrix composites are the most common advanced composite. Reasons are their low cost, better strength and stiffness, simple manufacturing principles, and resistance to corrosion (Elanchezhian et al., 2018; Gowda et al., 2018). Both thermosets and thermoplastics are attractive as matrix materials. Epoxy, polyester, Polyvinyl chlorides, polypropylene, and polyethylene are the most widely used polymer matrices for natural fibre. Generally, thermoset composites show better mechanical properties than thermoplastics composites. But thermoplastics matrix composite which has benefits like recyclability, low operating cost, easy to mould complex parts, and design flexibility overshadow the use of thermoset matrix (Faruka et al., 2012; Nabi Saheb and Jog, 1999).

However, polymers are non-biodegradable materials. Therefore, in short life cycle applications, these composite materials increase the burden on the environment. The recycling cost of plastics is so high and so that no choice left other than incineration and landfilling. Due to public awareness for the environment, researchers and industrialists are now grown interested in the biodegradable polymer which degrades by the action of microorganisms and converted into CO_2, CH_4, water, inorganic compounds, or biomass.

Starch, polylactic acid (PLA), and polyhydroxyalkanoate (PHAs) are the most commonly used biodegradable polymers (Ho et al., 2011). The application of bio-composite will increase in near future in various applications. Bio-composite shows non-linear mechanical behavior. These bio-composites also have poor long term application, and can be used only for low strength application. Nanotechnology can be used to overcome these limitations, as it shows good prospects like nanotechnology-based coatings. This coating helps to improve water resistance, reduce bio-degradation and volatile organic compounds, and improve flame resistance.

4.4 CHALLENGES

4.4.1 INTERFACE

The fibre/matrix adhesion plays a deciding role in the manufactured composite properties. The major drawback faced during the fabrication of plant fibre based polymer composites is the poor fibre and matrix adhesion at the interface. This disturbs the final composite product's performance (Tserki et al., 2006). Furthermore, the incorporation of natural fibres in the polymer matrix often leads to agglomeration, which causes insufficient distribution of fibres in the matrix (Kabir et al., 2012; Prasad et al., 2015).

4.4.2 WATER ABSORPTION

Humidity aging is another major problem which hinders the application of natural fibre based composites. When these composites exposed to the atmosphere or come in contact with any aqueous media, this hampers its properties as well as dimensional stability. Azwa et al. (2013) reported that the plant fibre-polymer composites can absorb moisture up to 0.7–2% in 24 hours, 1–5% after a week, and up to 18–22% after several months. The penetration of moisture into the natural fibre-polymer composites can be mainly facilitated by three mechanisms. The first mechanism involves water molecules diffusion into the micro-gaps in the polymer chain. The second mechanism involves due to the gaps and flaws generated at the fibre/matrix interface due to the incomplete wetting and impregnation of fibres into the matrix. Water transports into the gaps and flaws through the capillary. The third mechanism involves due to micro-cracks forms in matrix during processing or due to the swelling of the fibres. Water transports through these micro-cracks (Dhakal et al., 2007). Therefore, study of the water absorption behaviour is very important. It helps to estimate the consequences that absorbed water can hamper the durability of the composite material underwater.

4.4.3 CHEMICAL MODIFICATION

The major challenge faced by natural fibre composite is to develop strong interface between the fibre and the matrix. Hydrophilic nature of natural fibre obstructs its bonding with the polymer matrix which causes poor interfacial bonding. Therefore, there is a need for fibre surface modification in order to improve its hydrophobicity. Surface treatment helps to improve its water resistance, wettability, and interfacial adhesion. Fibre surface treatment can be carried out with physical or chemical methods or using both.

4.4.3.1 Physical Treatment

Physical treatment involves stretching, calendaring, thermo-treatment, or the production of hybrid yarns. Another method used is electric discharge such as plasma, corona, laser, and γ-ray (Bledzki et al., 2008). This is an environmentally friendly method. This method improves the surface of cellulosic fibres by altering the structure and surface property, not chemical composition. This results in the improved mechanical performance of the composite.

4.4.3.1.1 Plasma Treatment

Fibre surface is modified with the use of plasma. In this method, reactive free radicals and groups are generated using different types of gases. Plasma treatment increases or decreases the surface energy and introduces cross-linking for better fibre/matrix bonding (Faruk et al., 2012). Marais et al. (2005) prepared flax fibre/polyester composite and observed some improvement in fibre/matrix adhesion and mechanical properties after helium cold plasma treatment. Martin et al. (2000) prepared sisal fibre reinforced HDPE composite and observed some improvement in mechanical properties after the plasma treatment. Seki et al. (2010) reported that the radio frequency plasma able to improve interfacial adhesion between the jute fibre and the polyester matrix. The interlaminar shear strength increased to 26.3 MPa for radio frequency plasma jute/polyester composite from 11.5 MPa for untreated jute/polyester composite.

4.4.3.1.2 Corona Treatment

In this technique, surface energy of the natural fibres is altered by surface oxidation activation. This technique can be applied to both fibres as well as to matrix in order to achieve strong interfacial adhesion between them (Faruk et al., 2012). Ragoubi et. al., (2010) prepared hemp fibre/ polypropylene composites and showed improved tensile strength after corona treatment to hemp fibre. Corona treatment improved 30% of tensile modulus. Pizzi et al. (2009) reported only 5 min corona treatment is sufficient to improve mechanical properties.

4.4.3.2 Chemical Treatment

On the other hand, chemical treatment permanently alters the nature of cell walls of the natural fibre. Chemical treatment is generally carried out using chemicals. Selected chemicals contain some functional groups that make the natural fibre hydrophobic and exposes hydroxyl group present in it. Chemical treatment also helps in removing the impurities as well as non-cellulosic part. Furthermore, these treatments make the fibre surface rougher. All these alteration makes natural fibre well-suited to hydrophobic polymer matrix and helps in generating strong fibre/matrix bonding (Akil et al., 2011).

Use of compatiblizer and coupling agent are also very common technique to improve fibre-matrix interaction. Compatibilizer is a polymeric interfacial agent. Functional groups that present in the compatibilizer, graft onto the polymer chain as well as with the hydroxyl group of natural fibre. Whereas, coupling agent is a chemical substance that reacts with both the natural fibre as well as with the polymer matrix. Coupling agent can be used prior to processing as well as during processing to build strong fibre/matrix bonding.

Properties of fibre reinforced composite is mostly influenced by the distribution of stress between fibre and the matrix. Load carrying capacity of the composite depends on how efficiently stress is transferred from matrix to the fibre. Efficient stress distribution is achieved through the strong interfacial adhesion between the fibre and the matrix. There are number of chemical modification techniques developed and are reported in the literature. Alkalization, silanization, acetylation, acrylation and acrylonitrile grafting, benzoylation, maleated coupling, benzene diazonium salt, peroxide, permanganate, and isocyanate treatment are the few chemical treatment methods applied to increase the hydrophobicity of the natural fibre. The reported trend shows that the chemical treatment is more preferred over physical treatment (Faruk et al., 2014; Li et al., 2007; Azwa et al., 2013).

4.4.3.2.1 Mercerization

Mercerization is an alkalization treatment method used to modify the natural fibre. This is one of the most common and effective modification techniques. Sodium hydroxide (NaOH) is a common chemical used in this technique. In this method, natural fibre is dipped in NaOH solution for specific period of time. Concentration of NaOH varies from 0.5 to 10 wt.% according to the requirement and type of fibre in use. This method helps to extract the non-cellulosic constituents such as hemicellulose, lignin, pectins, and waxes from the fibre. This treatment also removes impurities and some parts of amorphous cellulose. These impurities and non-cellulosic components are mainly responsible for weak fibre/matrix interfacial. Moreover, elimination of these non-cellulosic constituents improves the roughness over the fibre surface which takes additional part in establishing strong fibre-matrix bonding and resulted to provide better composite properties (Venkateshwaran et al., 2013).

4.4.3.2.2 Acetylation

Acetylation is a well-recognized etherification technique. This technique introduces plasticization to the natural fibres. In this method, acetyl group (CH_3CO) reacts with the hydroxyl group (OH) of the natural fibre, improving the fibre's hydrophobicity. Moreover, this treatment endows with a rough surface which leads to provide strong interlocking between the fibre and the matrix. Strong interfacial adhesion further leads to improve the dimensional stability of the composites (Kabir et al., 2012).

4.4.3.2.3 Acrylation

Acrylic acid (CH_2=CHCOOH) is used as a pretreatment to improve hydrophobicity of natural fibres. The Hydrophobicity of the natural fibre is improved by generating an ester linkage. Application of acrylic acid to natural fibre generates strong ester linkage by interacting carboxylic group of acrylic acid with the cellulosic OH group of natural fibre. It is observed that the acrylation helps to improve composite properties like strength, storage modulus, and thermal stability as well as to improves water resistance (Bogoeva-Gaceva et al., 2007; Zahran and Rehan, 2006; Prasad et al., 2016).

4.4.3.2.4 Benzoylation

This treatment is carried out using benzoyl chloride. This treatment helps to improve the hydrophobicity and thermal stability of the fibre. First, fibre is treated with NaOH which helps to remove non-cellulosic components and exposes OH groups of the fibre. Then fibres are immersed in the benzoyl chloride solution, where OH group of the fibres gets replaced with benzoyl chloride (Gholampour and Ozbakkaloglu, 2020). This will further lead to provide better interaction with polymer matrix, thus improving the mechanical performance of the composite (Nair et al., 2001). Alkaline benzoyl chloride solution was used for the surface modification of sisal (Joseph et al., 2002) and bamboo fibre (Kushwaha et al., 2011). Alkaline benzoyl chloride improved the tensile strength and water resistance behaviour of the composite.

4.4.3.2.5 Silanization

In this method, Silane (SiH_4) is used as a coupling agent. This method can be used to modify any or both the fibre and the matrix. Hydroxyl groups (OH) present in the natural fibre are the reason for the moisture absorption. Silane treatment helps to reduce the cellulose OH group present in it. Silane forms silanols in the presence of moisture. This silanols then react with OH group of the cellulose and chemisorbed onto the fibre surface by generating a strong stable covalent bond to the cell wall. Silane forms hydrocarbon chains and resist fibre swelling. Silanation also improve adhesion of the fibre into the polymer matrix. The silane treatment further leads to provide improved mechanical properties and minimizes the effects of moisture that can weaken the composite's properties (Xie et al., 2010).

4.4.3.2.6 Compatibilization

In the recent years, maleated coupling agent is extensively used to improve natural fibre/polymer composite properties. These coupling agents interact with both the fibre and the matrix and forms strong ester linkage between them. This will help to achieve strong linkage between the fibre and the matrix and with high performance natural fibre-polymer composite materials. (Mohanty et al., 2006; Prasad et al., 2016).

4.5 PROCESSING TECHNIQUES

A suitable processing technique must be selected to fabricate defect free composite product (Jaafar et al, 2019). Polymer composites can be processed by means of any of the following tradition processing techniques:

- Hand lay-up, extrusion,
- Injection moulding,
- Compression moulding, and
- Resin transfer moulding (RTM)
- Pultrusion

Selection of processing technique depends on the properties of raw materials, temperature, pressure, and speed of processing. Another criterion on which selection depends are manufacturing cost, targeted composite properties and their shape and size. For the small to medium-sized composites, injection and compression mouldings are preferred. These methods are simple to use and have a fast-processing cycle. On the other hand, for large-sized composite products, open moulding and autoclave processes (e.g. RTM and hand lay-up) are preferred. In addition, semi-finished products are also prepared in the form of mat, fibre yarn, slivers, and granules (Faruk et al., 2012).

Granulated composite products are manufactured by compounding the constituents. In this process, short fibre is blended with a thermoplastic matrix in the compounder machine to obtain a high degree of consistency. In order to produce a palletized feedstock, the compounded mixture can be further processed in any other thermoplastic processing techniques like injection or extruders. There are many compounding processes available that includes extrusion, kneading, and high-shear. Injection moulding is the most commonly used processing techniques for making composite products. In the case of short fibre composite, fibres are chopped as per the critical fibre length criterion. Fibre can be loaded to its full capacity, if its length maintains according to is critical length. Composite with critical fibre length provides uniform stress distribution between fibre and matrix. Uniform stress distribution resulted due to strong fibre matrix adhesion. A good-quality composite product can be obtained by the optimization of the process parameter such as melt temperature, screw and injection speeds, injection pressure, and mould temperature.

The compression moulding technique is a combination of hot press and auto-clave process. Chopped fibre, fibre mat, or stitched fibres can be used for manufacture composite products using compression moulding technique. The lay-up technique is a simple but labour-intensive process. This technique is commonly applied for thermoset plastic such as epoxy and polyester (Holbery and Houston,

2006). Bledzki et al. (2008) have investigated the effect of three different compounding processes on the performance of the polymer composite. These processes are mixer-injection moulding, mixer-compression moulding, and direct compression moulding process. Among these three processes, the mixer-injection moulding leads to provides better mechanical performance nearly 90% higher tensile strength than the other processes with a lower damping index. While the direct compression moulding process showed nearly 170% higher impact strength than other processes. In the case of mixer-injection and mixer-compression moulding, the fibre breaks into shorter length regardless of initial fibre length due to the fibre agglomeration. Whereas, in the case of the direct compression moulding process, the fibres take place one after another like the layers in the composite, with its initial higher length responsible for higher impact strength.

4.6   PROS AND CONS OF NATURAL FIBRES COMPARED TO CONVENTIONAL FIBRES

–   Natural fibre is obtained from renewable sources. Its production requires less energy and CO_2 emission compared to conventional fibres. But its quality is not consistent as it depends on the plant's species, its age, growing location, climate condition, harvesting time, and processing techniques. Other affecting parameters are temperature, moisture, and UV radiation.
–   Plant fibres have relatively low cost as it is abundantly available. But due to inconsistent in harvesting conditions and government policies its price fluctuates.
–   Plant fibres have relatively low density, which results in providing high specific strength which improves fuel efficiency and reduce emission especially in automotive applications due to weight reduction. But they have very low impact strength compared to conventional fibres.
–   Natural fibre is very friendly to work with as they are less susceptible to tool wearing and skin irritation. But it has a restricted operating temperature of 200°C after that fibre degradation start, which confines the matrix choice.
–   Natural fibres are biodegradable. Natural fibre-based polymer composite requires more fibre content for equivalent performance, which reduces more polluting polymer. But hydrophilic nature of natural fibre absorbs water and incorporates a poor adhesion interface with hydrophobic polymers.
–   Natural fibre-based composites have good electrical resistance, thermal and acoustic insulation but have poor fire resistance and lower durability.

High water absorption, poor fibre/matrix adhesion, aggregates formation during processing, low thermal degradation temperature are the major challenges for cellulosic fibre-based polymer reinforcement as these reduces mechanical properties of the composite (Dittenber and GangaRa, 2012; Ku et al., 2011). But, for most of the limitations have the remedial measures in the form of fibre's surface modification treatments which outweigh the disadvantages.

4.7 APPLICATIONS

Applications of natural fibre-polymer composite are presented in Table 4.2.

TABLE 4.2
Application of Natural Fibre-Polymer Composite

Potential application	Fibre	Examples
Automotive	Kenaf, hemp, flax, jute, abaca, wood	Carrier for door panels/inserts, backseats, headliners, dash boards, trunk trim, rear parcel shelves, instrument panels, seat cover, other interior trim, load floors, etc.
Aircraft	Hemp, flax and sisal	Interior panelling
Construction	Wood, flax, rice husk, bagasse, coir, stalk	Railing, decking, bridge, siding profiles, roofing seats, particle boards, etc.
Household products and furniture	Wood, flax, rice husk, bagasse, coir, stalk	Frames, door, furniture, fencing, Door mats, rugs, interior panel, Window frame, food tray, partition, etc.
Electrical and electronics	Kenaf and flax	Mobile cases, laptops cases, etc.
Sports	Flax, hemp and kenaf	Tennis Racket, bicycle, Frames, Snowboards

Sources: Ho et al. (2012), Lucintel (2011).

4.8 RECENT DEVELOPMENTS AND FUTURE TRENDS

In the last few decays, Natural fibre composites have experienced healthy growth. Mainly these markets are divided into a woody and non-woody fibre-based composite. North American's natural fibre composites market was the largest for wood-based composite for building and construction application whereas Europe is a leader in non-wood-based composite in the automotive sector. Government support, environmental regulations, and customer acceptance are the driving factor for natural fibre composite growth. Electrical and electronics and Sporting goods applications are new in the Natural fibre composites industries (Potluri and Krishna, 2020; Karthi et al., 2020).

Lucintel, a global management consulting and market research firm, said in their report that the growth of natural fibre composites market in 2010 achieved $2.1 billion with compound annual growth rate (CAGR) of 15% in the last five years and it is expected to reach $ 3.8 billion (10% CACR) by 2016. Natural fibre composite materials now comprise a major proportion of the market ranging from various household products to sophisticated applications. Ease in processing, biodegradability, weight and cost reduction aspects offer great potential in natural fibre-based composites. According to Markets and Markets research report on the natural fibre composites markets, in 2015 natural fibre composite market was of worth USD 3.36 billion which is projected to reach USD 6.5 billion in 2021. They expected to register a CAGR of 11.68% between 2016 and 2021. In their research report, they considered 2014 as the history year and 2015 as the base year for estimating market size of natural fibre composites.

Hybridization on which many researchers are working, incorporating two or more types of fibre. In order to achieve better utilization of one type of fibre that could balance with what is missing in other. Hybridization strategies also help in making composite materials economical. Mercedes S class presently manufacture 27 of its components (with a total weight of 43 kg) from natural fibre-based composites. In Europe, the end-of-life vehicle (ELV) directive conditions that the vehicles must be manufactured with 95% recyclable materials in which 85% recoverable through reuse or mechanical recycling (Jawaid and Khalid, 2011; Prasad et al., 2018a). Many car companies such as BMW, Audi, and GM are using non-wood fibre composites for designing various exterior and interiors parts of automobiles.

Natural fibre-based composite will increase in near future in various applications (Li et al., 2020). But still, a number of problems is there which have to be solved to make them strong competitor to synthetic fibre composites. If 100% biodegradable composite is manufactured, then it is difficult to control biodegradation. Natural fibre-based composite shows non-linear mechanical behaviour, poor long-term application, and low strength application. To overcome these problems numbers of natural fibre surface modification techniques has been developed to increase water resistance, flame resistance, mechanical properties and thermal stability (Faruk et al. 2012).

4.9 SUMMARY

Rising prices of fossil fuel-based products, consumer acceptance, and strong government support for eco-friendly products drive the use of natural fibre in composites to new levels. Natural fibre is an abundantly available resource, which otherwise goes as waste and creates pollution. The fibre industries are rurally oriented and increase employment in the rural sector. Natural fibre-polymer composites outperform in most parameters such as energy consumption, weight, recyclability, and price except strength. Strength is lower compared to synthetic glass fibres. Significant performance at lower prices makes them more attractive for numerous applications. Moisture absorption, mechanical performance, durability, fire resistance, and fibre quality are the major challenges associated with natural fibre composites. Mechanical properties and moisture absorption can be reduced by fibre's surface modification techniques. There are many researchers who are working worldwide to address and overcome the mentioned challenges.

Despite of the aforementioned challenges, various manufacturers have launched several commercial products in various sectors. Automotive industry is the most active and leading user of natural fibre-based products.

REFERENCES

Akil, H., Omar, M.F., Mazuki, A.M., Safiee, S.Z.A.M., Ishak, Z.M. and Bakar, A.A., 2011. Kenaf fiber reinforced composites: A review. *Materials & Design*, *32*(8–9), pp.4107–4121.

Alvarez, V.A., Ruscekaite, R.A. and Vazquez, A., 2003. Mechanical properties and water absorption behavior of composites made from a biodegradable matrix and alkaline-treated sisal fibers. *Journal of composite materials*, *37*(17), pp.1575–1588.

Alvarez, V.A. and Vázquez, A., 2004. Thermal degradation of cellulose derivatives/starch blends and sisal fibre biocomposites. *Polymer Degradation and Stability*, *84*(1), pp.13–21.

Azwa, Z.N., Yousif, B.F., Manalo, A.C. and Karunasena, W., 2013. A review on the degradability of polymeric composites based on natural fibres. *Materials & Design*, *47*, pp.424–442.

Bavan, D.S. and Kumar, G.M., 2010. Potential use of natural fiber composite materials in India. *Journal of Reinforced Plastics and Composites*, *29*(24), pp.3600–3613.

Bismarck, A., Baltazar-Y-Jimenez, A. and Sarikakis, K., 2006. Green composites as panacea? Socio-economic aspects of green materials. *Environment, Development and Sustainability*, *8*(3), pp.445–463.

Bledzki, A.K. and Gassan, J., 1999. Composites reinforced with cellulose based fibres. *Progress in Polymer Science*, *24*(2), pp.221–274.

Bledzki, A.K., Mamun, A.A., Lucka-Gabor, M. and Gutowski, V.S., 2008. The effects of acetylation on properties of flax fibre and its polypropylene composites. *Express Polymer Letters*, *2*(6), pp.413–422.

Bogoeva-Gaceva, G., Avella, M., Malinconico, M., Buzarovska, A., Grozdanov, A., Gentile, G. and Errico, M.E., 2007. Natural fiber eco-composites. *Polymer Composites*, *28*(1), pp.98–107.

Chawla, K.K., 1987. Excerpts from "Composite Materials". *Science and Engineering*, pp.89–92.

Dhakal, H.N., Zhang, Z.Y. and Richardson, M.O.W., 2007. Effect of water absorption on the mechanical properties of hemp fibre reinforced unsaturated polyester composites. *Composites Science and Technology*, *67*(7–8), pp.1674–1683.

Dittenber, D.B. and GangaRao, H.V., 2012. Critical review of recent publications on use of natural composites in infrastructure. *Composites Part A: Applied Science and Manufacturing*, *43*(8), pp.1419–1429.

Elanchezhian, C., Ramnath, B.V., Ramakrishnan, G., Rajendrakumar, M., Naveenkumar, V. and Saravanakumar, M.K., 2018. Review on mechanical properties of natural fiber composites. *Materials Today: Proceedings*, *5*(1), pp.1785–1790.

Eichhorn, S.J., Baillie, C.A., Zafeiropoulos, N., Mwaikambo, L.Y., Ansell, M.P., Dufresne, A., Entwistle, K.M., Herrera-Franco, P.J., Escamilla, G.C., Groom, L. and Hughes, M., 2001. Current international research into cellulosic fibres and composites. *Journal of materials Science*, *36*(9), pp.2107–2131.

Faruk, O., Bledzki, A.K., Fink, H.P. and Sain, M., 2012. Biocomposites reinforced with natural fibers: 2000–2010. *Progress in Polymer Science*, *37*(11), pp.1552–1596.

Faruk, O., Bledzki, A.K., Fink, H.P. and Sain, M., 2014. Progress report on natural fiber reinforced composites. *Macromolecular Materials and Engineering*, 299(1), pp.9–26.

Food and Agricultural Organization (FAO) of the United Nations, 2006. www.fao.org/newsroom/en/news/2006/1000472/index.html.

Gholampour, A. and Ozbakkaloglu, T., 2020. A review of natural fiber composites: Properties, modification and processing techniques, characterization, applications. *Journal of Materials Science*, 55(3), pp.829–892.

Godara, S.S., 2019. Effect of chemical modification of fiber surface on natural fiber composites: A review. *Materials Today: Proceedings*, 18, pp.3428–3434.

Gowda, T.Y., Sanjay, M.R., Bhat, K.S., Madhu, P., Senthamaraikannan, P. and Yogesha, B., 2018. Polymer matrix-natural fiber composites: An overview. *Cogent Engineering*, 5(1), p.1446667.

Ho, M.P., Wang, H., Lee, J.H., Ho, C.K., Lau, K.T., Leng, J. and Hui, D., 2012. Critical factors on manufacturing processes of natural fibre composites. *Composites Part B: Engineering*, 43(8), pp.3549–3562.

Holbery, J. and Houston, D., 2006. Natural-fiber-reinforced polymer composites in automotive applications. *Jom*, 58(11), pp.80–86. www.lucintel.com/LucintelBrief/PotentialofNaturalfibercomposites-Final.pdf

Jaafar, J., Siregar, J.P., Salleh, S.M., Hamdan, M.H.M., Cionita, T. and Rihayat, T., 2019. Important considerations in manufacturing of natural fiber composites: a review. *International Journal of Precision Engineering and Manufacturing-Green Technology*, 6(3), pp.647–664.

Jawaid, M.H.P.S. and Khalil, H.A., 2011. Cellulosic/synthetic fibre reinforced polymer hybrid composites: A review. *Carbohydrate Polymers*, 86(1), pp.1–18.

John, M.J. and Anandjiwala, R.D., 2008. Recent developments in chemical modification and characterization of natural fiber-reinforced composites. *Polymer Composites*, 29(2), pp.187–207.

John, M.J. and Thomas, S., 2008. Biofibres and biocomposites. *Carbohydrate Polymers*, 71(3), pp.343–364.

John, M.J., Varughese, K.T. and Thomas, S., 2008. Green composites from natural fibers and natural rubber: effect of fiber ratio on mechanical and swelling characteristics. *Journal of Natural Fibers*, 5(1), pp.47–60.

Joseph, P.V., Rabello, M.S., Mattoso, L.H.C., Joseph, K. and Thomas, S., 2002. Environmental effects on the degradation behaviour of sisal fibre reinforced polypropylene composites. *Composites Science and Technology*, 62(10–11), pp.1357–1372.

Kabir, M.M., Wang, H., Lau, K.T. and Cardona, F., 2012. Chemical treatments on plant-based natural fibre reinforced polymer composites: An overview. *Composites Part B: Engineering*, 43(7), pp.2883–2892.

Karthi, N., Kumaresan, K., Sathish, S., Gokulkumar, S., Prabhu, L. and Vigneshkumar, N., 2020. An overview: Natural fiber reinforced hybrid composites, chemical treatments and application areas. *Materials Today: Proceedings*, 27, pp.2828–2834.

Ku, H., Wang, H., Pattarachaiyakoop, N. and Trada, M., 2011. A review on the tensile properties of natural fiber reinforced polymer composites. *Composites Part B: Engineering*, 42(4), pp.856–873.

Kushwaha, P.K. and Kumar, R., 2011. Influence of chemical treatments on the mechanical and water absorption properties of bamboo fiber composites. *Journal of Reinforced Plastics and Composites*, 30(1), pp.73–85.

Li, M., Pu, Y., Thomas, V.M., Yoo, C.G., Ozcan, S., Deng, Y., Nelson, K. and Ragauskas, A.J., 2020. Recent advancements of plant-based natural fiber–reinforced composites and their applications. *Composites Part B: Engineering*, p.108254.

Li, X., Tabil, L.G. and Panigrahi, S., 2007. Chemical treatments of natural fiber for use in natural fiber-reinforced composites: a review. *Journal of Polymers and the Environment*, *15*(1), pp.25–33.

Lucintel, 2011, Opportunities in Natural Fiber Composites.

Marais, S., Gouanvé, F., Bonnesoeur, A., Grenet, J., Poncin-Epaillard, F., Morvan, C. and Métayer, M., 2005. Unsaturated polyester composites reinforced with flax fibers: effect of cold plasma and autoclave treatments on mechanical and permeation properties. *Composites Part A: Applied Science and Manufacturing*, *36*(7), pp.975–986.

Markets and Markets, research report on Natural Fiber Composites Market: Global Forecasts to 202 (Published date: April 2016, Report code: CH 2984). www.marketsandmarkets.com/Market-Reports/natural-fiber-composites-market-90779629.html (Accessed on 04.03.2021)

Martin, A. R., Manolache, S., Mattoso, L.H.C., Rowell, R.M., Dense, F., 2000. Plasma modification of sisal and high-density polyethylene composites: effect on mechanical properties. In: Natural polymers and composites proceedings. pp. 431–436.

Milosevic, M., Valášek, P. and Ruggiero, A., 2020. Tribology of natural fibers composite materials: An overview. *Lubricants*, *8*(4), p.42.

Mohanty, A.K., Drzal, L.T. and Misra, M., 2002. Engineered natural fiber reinforced polypropylene composites: influence of surface modifications and novel powder impregnation processing. *Journal of Adhesion Science and Technology*, *16*(8), pp.999–1015.

Mohanty, S., Verma, S.K. and Nayak, S.K., 2006. Dynamic mechanical and thermal properties of MAPE treated jute/HDPE composites. *Composites Science and Technology*, *66*(3–4), pp.538–547.

Nair, K.M., Thomas, S. and Groeninckx, G., 2001. Thermal and dynamic mechanical analysis of polystyrene composites reinforced with short sisal fibres. *Composites Science and Technology*, *61*(16), pp.2519–2529.

Pereira, P.H.F., Rosa, M.D.F., Cioffi, M.O.H., Benini, K.C.C.D.C., Milanese, A.C., Voorwald, H.J.C. and Mulinari, D.R., 2015. Vegetal fibers in polymeric composites: a review. *Polímeros*, *25*(1), pp.9–22.

Petroudy, S.D., 2017. Physical and mechanical properties of natural fibers. In *Advanced high strength natural fibre composites in construction* (pp. 59–83). Woodhead Publishing.

Pizzi, A., Kueny, R., Lecoanet, F., Massetau, B., Carpentier, D., Krebs, A., Loiseau, F., Molina, S. and Ragoubi, M., 2009. High resin content natural matrix–natural fibre biocomposites. *Industrial Crops and Products*, *30*(2), pp.235–240.

Potluri, R. and Krishna, N.C., 2020. Potential and applications of green composites in industrial space. *Materials Today: Proceedings*, *22*, pp.2041–2048.

Prasad, N., Agarwal, V.K. and Sinha, S., 2015. Physico-mechanical properties of coir fiber/LDPE composites: Effect of chemical treatment and compatibilizer. *Korean Journal of Chemical Engineering*, *32*(12), pp.2534–2541.

Prasad, N., Agarwal, V.K. and Sinha, S., 2016. Banana fiber reinforced low-density polyethylene composites: effect of chemical treatment and compatibilizer addition. *Iranian Polymer Journal*, *25*(3), pp.229–241.

Prasad, N., Agarwal, V.K. and Sinha, S., 2018a. Hybridization effect of coir fiber on physico-mechanical properties of polyethylene-banana/coir fiber hybrid composites. *Science and Engineering of Composite Materials*, *25*(1), pp.133–141.

Prasad, N., Agarwal, V.K. and Sinha, S., 2018b. Thermal degradation of coir fiber reinforced low-density polyethylene composites. *Science and Engineering of Composite Materials*, *25*(2), pp.363–372.

Ragoubi, M., Bienaimé, D., Molina, S., George, B. and Merlin, A., 2010. Impact of corona treated hemp fibres onto mechanical properties of polypropylene composites made thereof. *Industrial Crops and Products*, *31*(2), pp.344–349.

Saheb, D.N. and Jog, J.P., 1999. Natural fiber polymer composites: a review. *Advances in Polymer Technology: Journal of the Polymer Processing Institute*, *18*(4), pp.351–363.

Bavan, D.S. and Kumar, G.M., 2010. Potential use of natural fiber composite materials in India. *Journal of Reinforced Plastics and Composites*, *29*(24), pp.3600–3613.

Seki, Y., Sarikanat, M., Sever, K., Erden, S. and Gulec, H.A., 2010. Effect of the low and radio frequency oxygen plasma treatment of jute fiber on mechanical properties of jute fiber/polyester composite. *Fibers and Polymers*, *11*(8), pp.1159–1164.

Swolfs, Y., Gorbatikh, L. and Verpoest, I., 2014. Fibre hybridisation in polymer composites: a review. *Composites Part A: Applied Science and Manufacturing*, *67*, pp.181–200.

Tserki, V., Matzinos, P., Zafeiropoulos, N.E. and Panayiotou, C., 2006. Development of biodegradable composites with treated and compatibilized lignocellulosic fibers. *Journal of Applied Polymer Science*, *100*(6), pp.4703–4710.

Van de Velde, K. and Kiekens, P., 2001. Thermoplastic pultrusion of natural fibre reinforced composites. *Composite Structures*, *54*(2–3), pp.355–360.

Venkateshwaran, N., Perumal, A.E. and Arunsundaranayagam, D., 2013. Fiber surface treatment and its effect on mechanical and visco-elastic behaviour of banana/epoxy composite. *Materials & Design*, *47*, pp.151–159.

Vigneshwaran, S., Sundarakannan, R., John, K.M., Johnson, R.D.J., Prasath, K.A., Ajith, S., Arumugaprabu, V. and Uthayakumar, M., 2020. Recent advancement in the natural fiber polymer composites: a comprehensive review. *Journal of Cleaner Production*, p.124109.

Xie, Y., Hill, C.A., Xiao, Z., Militz, H. and Mai, C., 2010. Silane coupling agents used for natural fiber/polymer composites: A review. *Composites Part A: Applied Science and Manufacturing*, *41*(7), pp.806–819.

Yan, L., Chouw, N. and Jayaraman, K., 2014. Flax fibre and its composites–A review. *Composites Part B: Engineering*, *56*, pp.296–317.

Yousif, B.F., Shalwan, A., Chin, C.W. and Ming, K.C., 2012. Flexural properties of treated and untreated kenaf/epoxy composites. *Materials & Design*, *40*, pp.378–385.

Zahran, M.K. and Rehan, M.F., 2006. Grafting of acrylic acid onto flax fibers using Mn (IV)-citric acid redox system. *Journal of Applied Polymer Science*, *102*(3), pp.3028–3036.

5 Biomass Accretion and Control Strategies in Gas Biofiltration

Rahul[1], Vivek Kumar[2],
Amarendra Kumar Dash[3] and
Jay Mant Jha[4]

[1]Chemical Engineering Department, Jaipur National
University, Jaipur, Rajasthan – 302017, India

[2]Chemical Engineering Department, Rajiv Gandhi
University of Knowledge Technologies, R.K. Valley,
Andhra Pradesh – 516330, India

[3]English Department, Rajiv Gandhi University
of Knowledge Technologies, Nuzvid,
Andhra Pradesh – 521202, India

[4]Chemical Engineering Department, Maulana Azad
National Institute of Technology, Bhopal, Madhya
Pradesh – 462003, India

CONTENTS

DOI: 10.1201/9781003196358-5

5.1   INTRODUCTION

Microbiological activity transforms pollutants into harmless products in biofilters. The population densities and the type of species present, the metabolic pathways, and their interdependence among each other and with their environment are treated as fundamental parameters in a biofilter operation. Studying a diverse collection of organisms interacting with each other and their environment is ecology; the microorganisms in a biofilter can only be understood by considering them as part of an ecosystem. Our knowledge of ecosystems' working is at best fragmentary, and our understanding of microbial ecosystems is weakest of all. Even so, an ecosystem approach is the best guide for efforts to develop better microbial communities for pollutant treatment.

5.2   MICROBIAL SPECIES IN BIOFILTERS

5.2.1   SELECTION AND PROLIFERATION

Biofilters are inevitably biologically open systems. The enormous amounts of air that pass through them carry aerosols and carry the cells, spores, and cysts of a tremendous variety of microorganisms. When compost is used as a biofilter medium, it brings an initial inoculum, including thousands of species. As biofiltration proceeds, these species will thrive or fail according to their abilities to find a place in the biofilter ecosystem. Even in a biofilter treating a single contaminant, there will be many logical niches to occupy (Webster et al., 1996). Complex compounds, in some cases, may take multiple metabolic steps in transforming finally into carbon dioxide and water, and diverse species may specialize in different steps of the process (Idaq et al., 2021).

Whether it is the original contaminant or some metabolite, species consuming the same substrate will compete fiercely. The less capable species may die out. Nevertheless, there may also be specialization based on the biofilter microenvironment: some microorganisms might do well at the surface of the biofilm, while others succeed deep within it, and yet others swim in the water film outside of it. One species could do well in deep pores where water is more abundant, while another succeeds on the support medium's convex surfaces, where the thinner water film excludes predators. If the biofilter is able to remove multiple pollutants from the air, the number of such species multiply accordingly.

Predators are common in biofilters. In the microbial ecosystem, the contaminants serve as the food source for degrading microorganisms. However, their population growth attracts protozoa which feed on them, bacteria that parasitize them, and viruses that kill them and release more viruses. Thus, each degrader species is prone to be a food source for several predators. Predators too fall in the lower place in the food chain.

Biofilters have abundance of protozoa: de Castro et al. (1997) saw numerous protozoa in bench-scale biofilters, ranging in size from 5 to 50 μm. More giant cells were seen early in the acclimation process, while smaller species were seen as steady-state systems.

Ecologists have proposed that most ecosystems developing in a new habitat pass through a succession of structures before reaching a climax community in which species populations roughly stabilize. There are indications that the microbial ecosystem in a biofilter may not reach a steady-state until long after it has reached high biodegradation rates. Webster et al. (1996) characterized a biofilter microbial ecosystem using phospholipid fatty acid analysis and found that its characteristics became approximately constant only after hundreds of days.

5.2.2 Inoculation of Biofilters

The open nature of biofilters limits the control designers and operators might wish to have. Many investigators have suggested using a single ideal species, known to vigorously degrade the compound of interest, as an inoculum for a biofilter. Such an approach may be successful, but only if the species is also "ecologically viable." The species must succeed under the biofilter's environmental conditions, growing at pH and temperature prevailing in the reactor and withstanding the inevitable variations. It must survive competitive organisms, avoid predators, and grow fast enough to make up for its losses to both. Swanson and Loehr (1997) have noted reports of inoculation which did not affect compost biofilters and suggested that the inoculum species could not compete. The ability to degrade the pollutant is a necessary trait, and rapid utilization of the pollutant will give a species a competitive advantage, but it does not guarantee success. Indeed, some investigators have suggested that microorganisms grown in the lab and released for bioremediation of soils may fail because they have become "lab adjusted." In as few as 25 generations, the species can develop ideal traits for the petri dish environment and lose those that provide fitness in the wild. Bohn (1996) was particularly blunt in saying, "Expecting an introduced laboratory or bioengineered microbial culture to flourish in field conditions is naive. The cultures are originally derived from soils and fed special food sources in the laboratory. Food sources and predator relations are much different in the field. The new microbes are just another food source for the native population."

While there are many reasons why inoculation with a single chosen species may fail, it is unlikely to do any harm. The choice and preparation of a proper inoculum remain an essential matter for future research.

An advantage of using compost for the biofilter medium is that it brings a well-developed community of microorganisms. The organisms which have been doing the composting are abundant and well adapted to their environment. The microbial consortia present can degrade the multitude of compounds found in leaves and grass. When the compost is placed in a biofilter and air is passed through, the contaminant will become a dominant substrate. Those microorganisms which can degrade will proliferate and become abundant. However, even compost may benefit from inoculation in some cases. Wright et al. (1997) found that acclimation in compost biofilters treating gasoline vapors was much more rapid when they were inoculated with a culture that had been grown on gasoline. Leson and Smith(1997) came to the same conclusion, suggesting that inoculation will speed acclimation but will not affect ultimate removal efficiencies. The delayed acclimation of the species found on the compost may result from gasoline compounds that are uncommon in the environment and possibly toxic at high concentrations. The investigators also suggested that acclimation may be more successful if it has begun with a more dilute waste stream that is less challenging to the microorganisms. Van Langenhove and Smet (1996) compared organic sulfide treatment in an uninoculated compost biofilter with treatment in one that had been inoculated with a culture of microorganisms enriched by growth with dimethyl sulfide. The inoculated biofilter was far more effective, with an elimination capacity of 28 g m^{-3} h^{-1} rather than 0.42 g m^{-3} h^{-1}.

If a non-compost medium is used or suspected that the compost might not contain species that degrade the contaminant of interest, general inocula are often added. Activated sludge from a wastewater treatment plant is commonly employed because it contains an immense variety of rugged organisms exposed to the typical wastes of civilization (however, concern over disease organisms that may also be present may complicate ultimate disposal of the medium). While the bulk of the inoculum species will die out, there are usually a few species present that will degrade the contaminant. Other general inocula are chosen with a more specific rationale. A biofilter intended to treat chlorinated hydrocarbons might be inoculated with an extract from the soil at a site contaminated with chlorinated hydrocarbons for many years. The presumption is that species that can degrade compounds are likely to have become established at the site during the many years of exposure.

Van Groenestijnet al. (1995) developed a biofilter to treat ethylene and 1,3-butadiene, using soil found at the side of a road as an inoculum. The sample, exposed to various hydrocarbons for many years, contained microorganisms that degraded the contaminants sufficiently. When they started a biofilter for treating gases at high temperatures, they used an inoculum taken from compost during the composting cycle's high-temperature portion. In duo cases, nothing was known about the species present, but it was presumed that an effective inoculum would be found if a mixed culture were taken from a natural environment similar to that expected in the biofilter.

The relationship between the chosen inoculum and the ultimate microbial ecosystem characteristics in the steady-state biofilter is likely complex and indeed very poorly understood. General ecological observations suggest that different groups of species often combine to make ecosystems with similar structures and capabilities, and there is some indication that this occurs in biofilters. De Castro et al. (1997) compared the performance of biofilters that received three different inocula. Chipped wood mixed with mushroom compost was used to treat α-pinene after extracts from spent mushroom compost, activated sludge—Orpine forest soil. Initially, the biofilter with activated sludge extract was most effective. After seven weeks, all three were performing equally well; however, fatty acid methyl ester data analyzed by principal components analysis indicated that the biofilter inoculated with activated sludge supported different species than the other two.

In a second experiment, de Castro et al. (1997) used wood chips, compost, and perlite and inoculated with pine soil extract, pulp mill activated sludge, and municipal activated sludge to make media for treating α-pinene. Even though the cultures were grown in α-pinene-enriched liquid media for three weeks and were added to the biofilters at equal cell densities, differences in biofilter acclimation were seen. The pine soil extract biofilter took four days, the pulp mill sludge biofilter took 14 days, and the municipal activated sludge biofilter took 37 days. It was consistent with suggesting that the degree of previous exposure the microbial consortium has toα-pinene indicates how rapidly the inoculated biofilter will acclimate. In this case, too, the initial differences in biofilter performance disappeared as time passed. Fatty acid methyl ester analysis was not successful because the microbial consortium's nature was obscured by high concentrations of fatty acids in the medium.

Webster et al. (1997) used phospholipid fatty acid analysis on biofilters' microbial ecosystems used on wastewater plant off-gases. While the biofilters reached a steady-state in terms of treatment effectiveness within 100 days, the fatty acid profile was still changing in some after 300 days. Biofilters with different profiles achieved similar results. Thus, in both of these studies, ecosystems with different development histories and different species compositions functioned with equal effectiveness. The differences between them could not be seen without a sophisticated analysis.

5.3 SUBSTRATE UTILIZATION

Biofilters succeed (except cometabolism) because many microorganisms present that use the contaminant as food or substrate. The compound may serve as an energy source or building material, or both. When used solely for energy, a simple organic compound is converted to carbon dioxide and water discharged from the biofilter (Deshusses et al., 1995). However, microorganisms strive to grow and reproduce, and some of the carbon from the compound will often end up as part of the microorganisms. Indeed, a vigorously growing microorganism in an environment where food is abundant may convert half of its substrate carbon to biomass. In the long run, this constitutes a problem for biofilters, and the biomass may accumulate and clog the reactor.

The fraction of contaminant converted to biomass is controlled by several factors. Starving microorganisms are forced to use a more significant fraction of their energy food and may have none to spare for growth. Stress may reduce the amount of growth, and some biotrickling filter operators have controlled biomass by adding salts to increase the ionic strength to stressful levels (Diks et al., 1994).

Even when the contaminant-degrading microorganisms are proliferating, it is conceivable that biomass growth in the biofilter as a whole could be slow because predators are rapidly consuming the degraders.

Some compounds treated in biofilters do not contribute their elements to growth. The microorganisms in biofilters that remove hydrogen sulfide convert it almost entirely to sulfate, remaining in the water. The conversion provides energy, but very little of the sulfur is incorporated into the cells. The organism uses the energy to fix atmospheric carbon dioxide for growth (Thakur et al., 2011).

5.3.1 INDUCTION

The cell can only utilize a compound when the appropriate enzymes are present. The genes which code for the enzyme must be active. Some enzymes are "constitutive"; they are always present in the cell and ready for use whenever the substrate is present. Others are "induced" and are synthesized by the organism only when the substrate appears at concentrations above an "induction threshold." This second type may limit biofilters' performance and, indeed, in all biological treatment systems. The induction will not occur if the substrate is present in concentrations below the threshold. It may also not occur if a second substrate is a better energy source because of its simpler molecular structure or higher concentration. For the biofilter as a whole, this means a low concentration pollutant may not degrade or degrade in the presence of another pollutant.

5.3.2 SUBSTRATE INTERACTION

The performance will depend upon the characteristics of contaminants and operation conditions of the biofilter. These contaminant characteristics may include absorptivity, solubility, bond structure, and potential biodegradability. When treating pure waste streams, performance on a mixed stream depends on operating conditions such as loading rates, temperature, nutrient availability, moisture content, and pH. Contaminants also compete in securing active adsorption sites in the biofilter bed, changing their microbes' availability. The contaminants will be more or less adsorbed depending on filter medium moisture content, temperature, and pH.

Competitive inhibition may occur because the first substrate is toxic to the organisms which consume the second. Indirect ecological mechanisms may contribute. First substrate, in case, having high concentration, being readily degradable, providing abundant energy actually supports the organisms consuming it. It enables them to compete vigorously for other resources such as nutrients, space besides preventing the growth of organisms using the second substrate (Laconi et al, 2006).

Veir et al. (1996) studied the interaction of dichloromethane and toluene. A compost biofilter was first used to treat dichloromethane and, after 53 days of acclimation, it was essentially removing all of it. The addition of toluene caused an immediate decline in dichloromethane removal efficiency, but considerable recovery was seen in the following three weeks. The authors suggested that the species which degraded the more energy-rich toluene at first competed strongly with the dichloromethane-degrading species for another controlling resource (Sorial et al, 1995). Over time, however, the species each dominated in spatially separate volumes within the biofilter.

Leson and Smith (1997) noted that greater time might be necessary to complete acclimation in biofilters treating complex mixtures. It suggests that the success of some species may be detrimental to others. Deshusses (1994) found that methyl ethyl ketone (MEK) and methyl isobutyl ketone (MIBK) were simultaneously removed in a compost biofilter, but when the concentration of MEK was increased, the removal of MIBK decreased, and vice versa.

Generally, multiple contaminants will be adequately removed when they have similar properties. The same microbial enzymes can often degrade various aromatic compounds. Several examples of biofilters treating gasoline or other mixes of benzene, toluene, ethylbenzene, and xylene (BTEX) vapors have been successfully demonstrated at both the laboratory scale and pilot scale (Rahul et al., 2013; Chang et al., 1995, 1996; Wright et al., 1997). In some cases, combining similar compounds can have a positive synergistic effect on removing the contaminants. The presence of xylene has been shown to significantly improve the removal rate for toluene (Paca et al., 1997; Turala et al., 2020); however, the reverse effect did not occur. Oxygenated compounds have also been effectively removed in biofilters when they were treated in mixed waste streams.

An extreme case of negative interaction between pollutants was reported by van Langenhove et al. (1989), who observed that removing an undefined mixture of aldehydes was reduced from 85 to 40% by the addition of 40 ppm SO_2 to the treated air stream. Ultimately, aldehyde biodegradation was irreversibly inhibited when the biofilter was exposed to 90 ppm SO_2 for longer than three days.

Contaminants with very different chemical properties may still be simultaneously biodegraded in a biofilter. When overall loading rates in a biofilter are low, a diverse microbial population may thrive and utilize many contaminants. It is often the case for biofilters used at wastewater treatment facilities. In addition to the hydrogen sulfide and mercaptans found in wastewater off-gases, hundreds of organic compounds at low concentrations will also be present. Effective simultaneous removal of these contaminants from such a waste stream has been demonstrated (Ergas et al., 1995; Webster et al., 1997).

5.3.3 ACCLIMATION

It is observed that an acclimation period of a few minutes is of no consequence to biofilter operation, but if it is necessary to wait for a year before treatment begins, the system will be impractical. The organism presented with a new substrate must go through a series of biochemical changes to begin using the compound for food. First, a chain of reactions must occur, giving the signal to "turn on" the genes that code for the needed enzymes. The genes must create the transfer RNAs, which then produce the proteins which make up the enzymes. A sufficient supply of the enzyme must accumulate. During this period, called the "lag phase," the cell often grows substantially without dividing. When it is fully ready, it begins to utilize the substrate and divides to produce more cells.

These cellular processes occur fairly rapidly for common substrates, and the delay for this type of acclimation will likely be hours or days. It is commonly seen in the biofiltration of petroleum fuel vapors (Leson and Smith, 1997; Wright et al., 1997). For unusual substrates, however, much longer delays have been seen. A biofilter operated to degrade methyl tertiary-butyl ether (MTBE), for example, showed no activity for a year, then suddenly became very effective (Eweis et al., 1997, Smet et al., 1996). Degradation may not have been possible until a cell underwent a random mutation or until a specific gene transfer occurred from one species to another. It is these processes that may take the most extended time.

Even where cellular acclimation is fast, processes that might be called ecosystem acclimation may also take time. Cell division must occur for a while before the degrading organisms become abundant and the compound is transformed at a high rate. Competition or predation may slow the process. If the competent cells were not well distributed in the medium, it might take some time to colonize the whole volume. This form of acclimation is sometimes followed by less effective treatment. This could be because the microorganisms consume substrate rapidly as they grow but are active only at maintenance levels when they reach densities at which nutrients are limiting.

Success in degradation may be delayed until the correct organism is deposited in the biofilter on a dust particle. There is a notable difference between biofilters for air and biological systems used for water treatment. High concentrations of a wide variety of microorganisms are always present in a wastewater. If the correct species is not present, it will indeed arrive soon. Inadvertent inoculation of air biofilters, however, is much less effective. The density of microorganisms in the air is much lower, and the conditions are much less benign. The best biofilter species do well in a wet, dark environment where food is abundant. Traveling through the air exposes the cell to an arid environment with no food and ultraviolet light intensity. Many kinds of cells can produce cysts or spores to propagate through the air, but the ideal organism for the biofiltration of a given contaminant may not be among them. The species expected to arrive frequently in a wastewater treatment system might make the trip to an air biofilter only very rarely. In the MTBE work (Eweis et al., 1997), it may have been the arrival of the organism's correct species, which caused degradation to start suddenly after a year of no treatment.

The presence of another may delay acclimation to one substrate. Seed and Corsi (1996) found that a biofilter could acclimate and degrade toluene alone in one day but require nine days to start efficient degradation when the toluene was presented as part of the mixture benzene and xylene. It may have occurred because microorganisms were first attracted to the benzene, which is presumably more easily degraded. Only when benzene concentrations were depleted in portions of the biomass did microorganisms begin to degrade the toluene.

Loy et al. (1997) showed that acclimation and growth of a biofilm were more rapid on polyurethane foam coated with carbon than on uncoated foam, suggesting that medium characteristics and local microenvironmental conditions may be necessary for acclimation rates.

In a biofilter treating toluene, Kinney et al. (1996) found that organisms starved for three days recovered in seven hours, and organisms starved for 27 days recovered in 30 hours.

The re-acclimation time can be minimized by taking a good care of the microorganisms for the period of shutdown. Indeed, sufficient air flow avoids the onset of anaerobic conditions which is a completely different and undesired microbial ecosystem. (Some soil biofilters can be shut down each night. They are lightly loaded and less adsorptive so that biodegradation does not exhaust the oxygen in the medium.) Water should be provided to prevent drying, which will kill many organisms. In cold climates, recovery may be more rapid if the biofilter is kept warm. Kinney et al. (1996) and Wright et al. (1997) have even recommended providing a small amount of the contaminant to maintain those species that utilize it will be maintained in a healthy condition. Of course, the need for re-acclimation will be reduced if variations in contaminant load can be minimized.

5.3.4 Uptake of Dissolved Compounds

The cell membrane is a barrier to many compounds, which may only pass when they encounter specific sites on the cell surface and are allowed through the biochemical machinery. This control allows the cell to choose those valuable compounds and exclude many that are harmful.

Some compounds are beneficial to the cell and are present in low concentrations in the environment and drawn into the cell with higher concentrations. It requires that the cell have a specific catalytic apparatus in the membrane, and energy must operate it. As compounds reach low concentrations in the environment around the cells, it may become energetically uneconomical to use the substances, and treatment will stop. For the most recalcitrant compounds, degradation may fail at concentrations that are unacceptable for release from the biofilter, and treatment will not be successful.

5.3.5 Phagocytosis

Microorganisms will often encounter substrate, which is in the form of small particles. Some are capable of phagocytosis, a process in which a particle is engulfed by a cell and physically taken inside. Some protozoa use this approach to prey on bacteria and consume bits of organic matter. It is reasonable to presume that this process helps clear away organic particulate matter deposited in biofilters, but there have been no investigations of the phenomena. Phagocytosis is undoubtedly essential in biofilters' microbial ecology because predators in the food chain use it. Cox and Deshusses (1997) used protozoa to reduce the accumulation of biomass in a biotrickling filter were no doubt consuming bacteria in this way.

5.3.6 EXOENZYMES

Some microorganisms obtain their food by releasing exoenzymes, which operate outside the cell. The fungus Phanerocaetechrysosporium, for example, releases a peroxidase that helps it to consume complex biological materials such as the polymers that make up wood (Braun-Lullemann et al., 1995). Lignin, hemicellulose, and cellulose are insoluble polymers and so cannot be drawn into the fungal cell as single molecules. The exoenzyme attacks the surface of the wood, breaking off short polymers and monomers, which then can be absorbed by the cells. Like phagocytosis, this is a process that has been widely observed in natural environments but never explicitly confirmed in biofilters. However, biofilters successfully treat some strongly adsorbed compounds on the support medium, and fungi are common. Exoenzymes may be degrading adsorbed compounds.

5.3.7 AEROBIC AND ANAEROBIC METABOLISM

The chemical reactions that cells use as energy sources and any reaction in which they process large amounts of material must be thermodynamically favored. Gasification (Idaq et al., 2021) helps convert the combustible portion of biomass into fuel gases (H_2, CO, and CH_4) at high-temperature. Photosynthetic organisms obtain energy from the sun. However, the microorganisms which occupy the dark recesses of a biofilter must have chemical sources of energy. They support some synthetic reactions that consume energy, such as those which make DNA, proteins, and other cell components, but they must carry out at least one reaction in which relatively large amounts of "food" are converted to products in a way that generates and captures energy. Any reaction that consumes energy must be used in relatively minor amounts to maintain the energy balance of the cell.

The energy that a compound "contains" depends on its environment. For energy-yielding reactions, microorganisms act as biological catalysts, converting compounds that are not at equilibrium with their environment to products that are. As a reaction moves toward equilibrium, it releases energy, and the cell links the reaction to processes that allow it to make use of this energy, but different parts of the environment have different background conditions (Molino et al., 2018).

We can expect the microorganisms' chemical conversions to depend on the species present and the environmental conditions. Within a single biofilter, organic compounds may be converted to carbon dioxide and water at the biofilm surface, while anaerobic metabolism occurs deeper where the oxygen is depleted. If the airflow short-circuits, passing around some isolated volumes within the biofilter, they may become anaerobic. Organisms in these areas may create sulfide, sulfur-containing organics, or other foul-smelling compounds, causing biofilter operation problems.

5.3.8 TOXICITY

High concentrations of a contaminant or a sudden increase in concentration can adversely affect the microbial population. In general, toxic shocks are more likely to occur with pollutants that have low Henry's coefficients (Deshusses, 1997). It is also possible that high concentrations may not themselves prove toxic to the microbial population, but the degradation intermediates will (i.e., because of seduced pH or high salt concentration). For hydrogen sulfide treatment in a bench biofilter, the maximum treatment capacity (above which the system's overloading was indicated by a sudden decrease in H_2S removal efficiencies) was found to be 80 g m^{-3} h^{-1} (Allen and Yang, 1991). At higher concentrations, the hydrogen sulfide itself was toxic to the microbes.

Additionally, as the hydrogen sulfide was degraded, sulfuric acid formed, reducing the medium pH. This declining pH eventually inhibited the system. Pollutant loading of more than 2 g m^{-3} of ethanol resulted in the degradation intermediates acetaldehyde, acetic acid, and ethyl acetate. The acid reduced the biofilter medium pH, inhibiting pollutant removal (Leson et al., 1993; Devinny and Hodge, 1995). Hence, care must be taken to size the biofilter correctly and dilute the contaminants when toxic concentrations.

5.4 THE MICROBIAL COMMUNITY

5.4.1 LONGITUDINAL STRATIFICATION

Biofilters operate essentially as plug flow devices. The contaminants are much higher at the bottom end (inlet), and it becomes lower at the top (outlet) of the biofiltration unit. The influent end supports dense biomass, possibly not substrate limited and more likely to grow until clogging occurs or nutrients become limiting. This region will require more nutrients, generate more heat, and be susceptible to high acid accumulation and upset. At the effluent side, microorganisms are prone to starving and hence produce negligible or no biomass. Hugler et al. (1996), in studies of a biotrickling filter, found biofilms 5 mm thick near the inlet and only 2 mm thick two-thirds of the way through.

It suggests several strategies for modified biofilter operation. If it is necessary to replace the medium because of clogging, it may be possible to replace only the medium near the influent end. Some biofilters are constructed as a series of movable trays so that trays near the influent can be treated differently. The order of the biomass trays may be changed from the influent end to the effluent end so that the healthy biomass will provide suitable treatment. At the same time, the biomass will be starving, reducing clogging and heat loss. Similarly, biofilters can be operated in series, with the delivery piping arranged so that it is easy to change their order in the flow stream. Farmer et al. (1995) arranged bench-scale biofilters this way and found modest reductions in the head loss.

Kinney et al. (1996) operated a toluene biofilter in two phases. Firstly, the biofilter was run in the usual manner until clogging had become a substantial problem in portions of the bed near the inlet. In the second phase, the feed direction was switched every three days. The clogging and short-circuiting present at the end of phase one gradually disappeared during phase two. By the 71st day of alternating treatment, concentration profiles during the up-flow period were essentially identical (but reversed) to those seen during the down-flow period. During the first phase, 29 to 38% of the carbon entering the biofilter as toluene left the biofilter as carbon dioxide. The remainder was presumably accumulating as biomass. In the second phase, the carbon dioxide emissions' carbon dioxide levels rose to contain 100% and more of the incoming toluene carbon, indicating steady-state or even declining biomass. While work was done only in the laboratory, it suggests a promising strategy for biomass control. However, the authors suggest that two or three factors should increase the reactor size to accommodate the starvation periods. Thus may not be economically feasible.

5.4.2 Biofilms in Biofilters

In microbial ecosystems everywhere, cells commonly arrange themselves at the solid surfaces in the form of films. There are many ecological benefits to this tactic. While they may seem very active when observed in a microscope, microorganisms swim at low rates at the macroscopic scale. The fastest typically move only a few centimeters per minute. They are thus essentially planktonic, unable to swim against the current in even a gentle flow. A microorganism that finds itself in a desirable environment can avoid being carried away by attaching itself to some surface. Some microorganisms can "visit" the surface and leave at any time.

The microorganisms become a film because many of them exude a polysaccharide gel. This gel helps them in coming in such form when they grow to higher population at the solid surface. It protects from predators, which often cannot penetrate the film. It also provides some protection against toxic substances, possibly because they are adsorbed on polysaccharides. In medical practice, this is a severe problem: implanted medical devices often develop a film of infectious organisms protected from antibiotics.

As the biofilm thickens, it also becomes a barrier to chemical transport. The water within the polysaccharide gel is stationary so that advection is suppressed, and molecular diffusion is the only mode of transport. If the microorganisms are active, they may consume the contaminant more rapidly than diffuse inwards, leaving the cells more profound in the biofilm without substrate (Gerrard A.M., 1997). Under these conditions, the rate of treatment is controlled by the biofilm diffusion rate rather than by the amount of biomass present. If the substrate is abundant and readily degraded, oxygen may not penetrate at a sufficient rate to supply the microorganisms, and the deeper portions of the biofilm may become anaerobic.

These chemical gradients have been the subject of extensive modeling efforts and are presumably associated with gradients in microbial species densities. Certainly, aerobic species must dominate near the biofilm's surface, while others occur where the biofilm has become anaerobic. The concentration of substrate varies with depth, and the surface species will be those capable of degrading high concentrations of contaminant. In contrast, species deeper in the biofilm can survive on lower concentrations of the contaminant, transformation by-products, or waste materials and dead cells. Mirpuri et al. (1997) proposed a model including three categories of cells: those capable of degrading toluene at high concentrations, those that can degrade toluene at lower concentrations under favorable conditions, and those that cannot degrade toluene at all. Experiments indicated that the toluene degraders are abundant near the gas-liquid interface, while the other types are abundant near the bottom of the biofilm.

Kinney et al. (1996)have noted that increasing biomass thickness may have other effects on phase transfer. A biofilm will clog any pore with an opening of maximum diameter less than twice the biofilm's thickness. As the biofilm thickens, it will fill or occlude more pores, reducing the total surface area available for gas transfer. In the extreme case, the flow will be restricted to a few channels, the transfer will be prolonged, and the biofilter will fail.

The growth of the biofilm also reduces the total surface area available for gas transfer. Alonso et al. (1997) presented a model relating growth of biofilm which fills the available openings around the point of contact between two support particles and the surface area.

In water treatment systems, biofilms are subject to "sloughing." It refers to the tendency of sections of the biofilm to break off and be carried away with the water flow. It may occur because the biofilm base becomes anaerobic or is starved for food and because the shear stress from the flowing water is significant. The pieces of biofilm are collected as sludge. Efficient sloughing is vital to overall performance because it prevents clogging of the treatment system and continually exposes new surfaces for more biofilm growth and the attendant contaminant consumption. Sloughing may occur in biotrickling filters but is not significant in biofilters, so it does not help mitigate clogging there.

Biofilm models have been aggressively developed for many years for reactors treating wastewater. Typically, they assume a planar biofilm of uniform thickness and model the inward diffusion of substrate and oxygen and the outward diffusion of metabolic products. In sophisticated models, the biofilm growth rate may also be simulated, affecting contaminant degradation and diffusion rates. Such models have a significantly advanced understanding of fixed-film treatment systems.

Many investigators have applied similar models to biofilms in biofilters used for air treatment (Mathur et al., 2012). While this is certainly appropriate, some cautions should be observed. The biofilm in an air-phase reactor differs from that in a water-phase reactor. In the water phase reactor, water movement through the porous medium exerts a considerable hydrodynamic shear. It smoothes the surface of the biofilm and carries away any microorganisms that are not attached. In an air-phase biofilter, water clings to the medium's surface by surface tension and does not flow rapidly (Gallardo-Rodríguez et al., 2019). Thus, it is possible for elaborate microbial structures, such as elevated fungal hyphae, to grow undisturbed. Many workers have observed fuzzy growth on media in real biofilters. Such growth cannot be adequately modeled as a uniform planar layer because the microorganisms' physical conformation is so different. The biodegradation of pollutants in the biofilm of a biofiltration system is a combination of physico-chemical and biological phenomena. Basically following three mechanisms (Figure 5.1) are responsible for the transfer and subsequent biodegradation within the bed (Swanson et al., 1997, Adler 2001, Mathur et al., 2006).

The number of cells in the biofilter might grow by a factor of 10^4 during the startup period. If it is presumed that the seed culture was well distributed throughout the biofilter volume, the result would be many colonies of 10,000 cells each. The colonies may produce local nutrient depletion or waste product accumulation. Some colonies may develop earlier and others later, smoothing the variation associated with acclimation and nutrient limitation.

A simple calculation can be made to demonstrate this effect. A theoretical medium of spherical particles 0.2 cm in diameter arranged in neat rows would have a specific surface area of 7.8 $cm^2 cm^{-3}$. If inoculum is added to provide 10^4 cells cm^{-3} and one μm in diameter, they occupy a specific surface area of 7.7×10^{-7} cm^2 cm^{-3}. If each cell grows and produces a colony that occupies 10^4 as much as the original cell area, the total area covered is 0.77 cm^2 cm^{-3} or about 10%. Thus, for most of

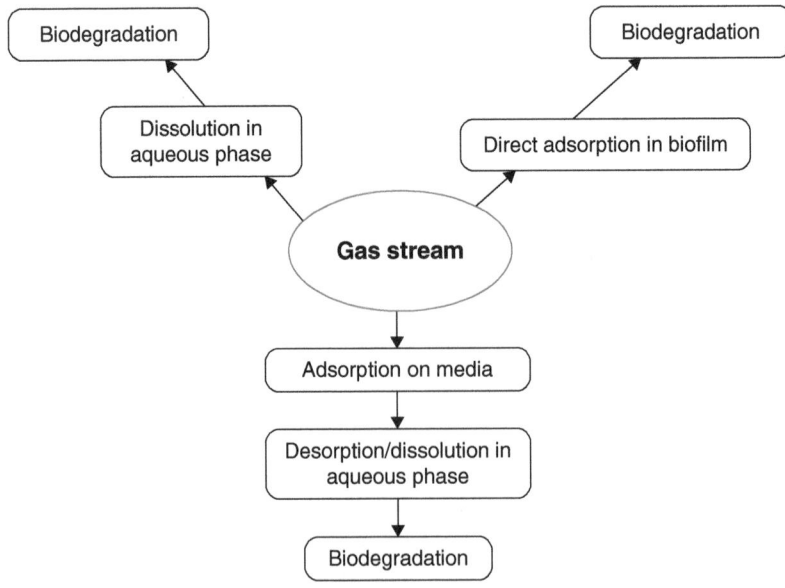

FIGURE 5.1   Mechanisms for biodegradation of VOCs.

the acclimation period, the cells exist as individual colonies, and completion of a biofilm only comes when cell counts are very high.

The water phase in the air biofilter is relatively stationary; it can support a population of swimming microorganisms. These will be free to move near the surface of the water, see higher concentrations of substrate because of the lesser diffusion limitation, and contribute significantly to the biofilter activity.

Hugler et al. (1996) measured bacterial concentration in the biofilm and water phases of a biotrickling reactor used for treating sulfides. They found the bacterial density in the water to be about one-third that in the biofilm. They suggested that this was not an essential factor for treatment in biotrickling filters; however, it remains notable that a substantial fraction of the cells were suspended in the water. The standing water in a biofilter may hold more. The same investigators developed a remarkably complete and exciting picture of the biofilm in their biotrickling filter. The biofilm may be quite different in biofilters or systems treating organic substances, but this is a well-developed example illustrating biofilm complexity. They characterized the biofilm as a "vast mixture" containing fungi, bacteria, yeasts, ciliated protozoa, amoebae, nematodes, and algae. Their biofilm sections showed three distinct layers, as the characteristics of the microbial community varied with depth in the film. At the interface between the biofilm and the water, phase was a thin whitish zone of dense bacteria and the fungal filaments' tips or hyphae. There were pockets observed which may have been channeled which allowed the flow of water. The intermediate layer was denser, with more hyphae and fewer bacteria, and a vigorous nematode population. In the basal layer, next to the solid surface, hyphae were again abundant, but many were collapsed as if they were dead or inactive. Other dormant fungi were seen, with a few bacteria. The investigators suggested that this deeper layer may have been deprived of substrate or oxygen, so the organisms were not healthy. They noted with understatement that "… rigorous mathematical modeling of this biological process is not trivial …"

The complexity of biofilms' ecology is illustrated by discussing the abundance of fungi in their biotrickling filter. Fungi might not be expected in a biotrickling filter treating inorganic sulfides, which are rapidly degraded by bacteria. Hugler et al. (1996) suggested four possible explanations: (1) the fungi could be autotrophs living on organic matter created by the sulfide oxidizing bacteria, (2) they could themselves be sulfide oxidizers (this is observed in a few fungi), (3) organic substrate for the fungi may be present in small quantities in the air, and (4) the relative resistance of the fungal hyphae to degradation may mean that they accumulate in significant amounts even if they grow very slowly. While it is clear that the fungi are essential organisms in this biotrickling filter, there is still a fundamental lack of knowledge of their microbial ecosystem's function.

The erect filamentous organisms may grow into the air space. Their hydrophobic surface remains dry, and they absorb substrate directly from the air stream. The high surface area of the filament can greatly facilitate phase transfer. Braun-Lullemann et al. (1995) suggested the importance of direct adsorption of contaminants on the filamentous white-rot fungus used in a biofilter.

The convoluted surface of the natural biofilm may also provide an extra surface for mass transfer. Indeed, some investigators (Potera, 1996, Thakur et al., 2013) have seen pores and channels within biofilms and suggest that the microbial community organizes itself in this manner in order to promote the transfer of substrate and oxygen from the water phase by advection rather than diffusion.

Modeling the effects of such intricate biofilm microstructure currently exceeds our ability. Direct observation is difficult because of the delicacy of the biofilm. Drying the material destroys its shape, and it cannot be easily sectioned for microscopic study. Models which assume simple planar biofilms still provide the best insight possible; however, their limitations should be kept in mind, and efforts to make them more realistic should continue.

5.4.3 HIGHER ORGANISMS IN BIOFILTERS

Bacteria and fungi are the major pollutants degraders in biofilter (van Groenestijn et al., 1995. de Castro et al., 1997). As biofilters are open to colonization by new bacteria and fungi, they are also open to colonization by higher organisms. De Castro et al. (1996) made microscopic observations of three biofilters that had received inocula from various sources. In the system inoculated with activated sludge, ciliated protozoa were observed on day 26 of operation, rotifers on day 54, and nematodes on day 70. All of these are common in sewage sludge. In a column inoculated with soil extract, nematodes were observed sooner. Thus, both the biofilters developed longer food chains, but the nature and timing of the development depended on the inoculum.

Insect larvae may be seen. The compost used in many biofilters will bring a host of worms and insects, and many of these may thrive in the system if the pollutant is not too toxic. In some cases, the insect larvae have metamorphosed to flying insects, a minor nuisance. In one closed biofilter, the upper headspace was colonized by thousands of spiders preying on the flying insects.

While a little careful study has been done, all of this suggests that biofilters can support complex ecosystems with long food chains. Flying insects may occasionally be an irritation, but the establishment of these food chains is desirable. As the predators consume the bacteria and fungi, they use much of what they eat for energy and use only a small part of their growth biomass. More than 90% of the biomass that is eaten will typically be converted to carbon dioxide and water. At least 90% of the predator biomass will be degraded when their predators eat them, so the organic material which enters the food chain is rapidly eliminated from the biofilter. It likely has a substantial influence on reducing clogging in biofilters where a low, relatively non-toxic load allows a complete ecosystem to develop.

Cox and Deshusses (1997) utilized this phenomenon in the control of biomass in experimental biotrickling filters. Two biotrickling filters used for the degradation of toluene were identical except that an inoculum of various protozoa species was added to one. The biotrickling filter which received the protozoa had less biomass, a lower pressure drop, and better treatment efficiencies after 80 days of operation.

5.5   BIOMASS CLOGGING

When the incoming organic carbon produces more biomass exceeding endogenous respiration, it is collected in high amounts in the biofilter. When loads on the system are extensive, and mineral nutrients are abundant, the biomass may block the filter bed. Media clogging, in turn, produces large pressure drops and channeling starts (Bihan et al., 2000). Back pressures on the blower equipment increase wear on the system and increase electrical demand. Air channeling will limit the amount of contaminant being treated and negatively affect a biofilter's performance. Excessive growth may also lead to other, more chronic problems. Examples include enhanced deterioration of the filter bed packing material by the large consortium of microbes and filter bed compaction as biomass weight builds. As clogging increases, anaerobic zones of activity may develop, and odorous end products may be generated. Limiting excessive biomass growth is essential for biofilter success.

Numerous studies have focused on predicting excessive microbial growth and devising means to control clogging (Angelidaki et al., 2018, Covarrubias-Garcia et al., 2017). Biomass growth is commonly concentrated near the inlet of the biofilter. As the biofilm thickens at the inlet because of the higher concentrations of the organic substrate, clogging conditions begin to prevail. This clogging causes treatment effectiveness to decline near the biofilter's inlet, and further growth occurs more profound in the reactor.

Research on one biofilter found 10^{10} viable heterotrophic bacteria per gram of medium at the filter bed's inlet and 10^8 viable bacteria per gram of medium at the outlet (Medina et al., 1995). The growth of microorganisms along the filter bed's length will generally be proportional to the removal rates. Zero-order kinetics may be seen at the system's inlet, while first-order pollutant removal may occur more profound in the medium.

Clogging control strategies have ranged from rudimentary methods, such as mechanical tilling or water injection, to more advanced nutrient limitation or biomass starving methods through flow reversal. In the earlier days of biofilter reactor operation, the predominant biofilter growth control method was to remove the material, rotate or till it, and place it back into the reactor (Souza et al., 2021). This approach works well for small biofilters but becomes costly and labor-intensive for larger units. The addition of water or backwashing at high flow rates may effectively shear off biomass from the packing material but should only be used on inorganic media (Sorial et al., 1994, 1997). Application of flowing water to organic filter beds is not advisable because of the probability of compaction, leaching of nutrients, and enhancement of air channeling.

The nutrient limitation may control biomass effectively on inorganic filter beds. Because the stoichiometry of biomass is relatively constant, reducing the total nutrient concentration limits the biomass amount that can form. Various studies have looked at controlling nitrogen or phosphorus concentrations for this purpose, with good results (Morgenroth et al., 1995; Smith et al., 1996; Weber and Hartrnans, 1996; Govind et al., 1997). However, maintaining nutrient concentrations at a level that will limit biomass growth with minimum damage to treatment efficiency is a tricky balancing act.

Switching the direction of airflow in a biofilter has proven to be one effective means to control excessive growth without long lag times in performance (Kinney et al., 1996). The initial flow direction (either top- or bottom-loaded) provides a carbon-rich area near the inlet and a carbon-deprived area at the outlet. After some time, the flow is reversed, reversing the carbon-rich and deprived areas as well. However, starving a large portion of the biofilter by overdesigning the reactor might not always be economical.

For all clogging control methods, some limitations exist. Biofilter media may be clogged in some portions of the bed, but not in others. It may be necessary to perform "smoke" or tracer test experiments on the system to determine if clogging conditions contribute to channeling problems. If clogging is evident from such tests, removal of the medium may be warranted.

5.6   CONCLUSIONS

A biofilter is only successful when the microbial ecosystem it contains is healthy and vigorous. To a considerable extent, biofilter's proper operation consists of maintaining this biological activity; it is essential to remember the microorganisms are part of an ecosystem, subject to environmental conditions and interactions among the species.

Bohn (1996) has emphasized this point in comparing biofilters using artificial media and complex engineered control systems with simpler reactors using compost or soil. He notes, "Minimizing size by maximizing control can create instability. Microbial populations fluctuate with time for reasons which are difficult to determine. Microbes release growth inhibitors and bactericides, which slow and stop the microbial activity. The smaller size accumulates these problems. A system with more internal self-regulation is likely to function better longer."

This philosophy deserves close consideration, especially when it is noted that elaborate engineering control is expensive and sometimes unreliable, and natural controls are accessible. Control of the biofilter environment may allow process optimization, a small footprint, and economical operation, but only the operator knows the optimal environmental conditions.

Ecosystems consist of large numbers of organisms working diligently to make efficient use of resources. When that resource is the contaminant being fed to a biofilter, allowing an ecosystem similar to those found in nature to flourish in its way may be a convenient approach.

Whether biofilters reproduce ecosystems found in nature or create highly artificial ones, it is essential to remember that the organisms present will continue to follow ecology laws. It is a milieu in which simple rules combine to govern very complex systems, and a fundamental understanding of the system will contribute to good biofilter management.

REFERENCES

Adler, S.F., (2001), 'Biofiltration- a Primer', Chemical Engineering Progress, 97(4), pp. 33–41.

Allen, E.R. and Yang, Y, (1991) 'Biofiltration control of hydrogen sulfide emissions', In: Proceedings of the 84th Annual Meeting of the Air and Waste Management Association, Air and Waste Management Association, pp. 16–21.

Alonso, C., Suidan, M.T., Sorial, G.A., Smith, F.L., Biswas, P., Smith, P.J. and Brenner, R.C., (1997) 'Gas treatment in trickle-bed biofilters: biomass, how much is enough?', Biotechnology and Bioengineering, 54(6), pp. 583–594.

Angelidaki, I. Treu L., Tsapekos P., Luo, G., Campanaro, S., Wenzel H., Kougias, P.G. (2018) 'Biogas upgrading and utilization: Current status and perspectives', Biotechnology Advances 36(2), pp. 452–466.

Bohn, H.L., (1996) 'Biofilter media' In: Proceedings of the 89th Annual Meeting & Exhibition of the Air and Waste Management Association, Air and Waste Management Association, Pittsburgh, PA.

Braun-Lullemann, A., Johannes, C., Majcherczyk. A., and Hutterman, A., (1995) 'The use of white-rot fungi as active biofilters', In: Proceedings of the Third International In Situ and On-Site Bioreclamation Symposium, Hinchee, R.E., Skeen. R.S., and Sayles, G.D., (Eds.)Battelle Press, Columbus. O.H., p. 235.

Bihan, Y.L. and Lessard P., (2000) 'Monitoring biofilter clogging: biochemical characteristics of the biomass', Water Research, 34(17), pp. 4284–4294.

Chang, A. and Yoon, H., (1995) 'Biofiltration of gasoline vapors', In: Proceedings of the 1995 Conference on Bioftltration (an Air Pollution Control Technology). Hodge. D.S. and Reynolds. F.E., Eds., The Reynolds Group, Tustin, CA, pp. 123–130.

Chang, A.N. and Devinny, J.S. (1996) 'Biofiltration of JP4 jet fuel vapors', In: Proceedings of the 1996 Conference on Biofiltration (an Air Pollution Control Technology). Reynolds, F.E. , Ed., The Reynolds Group, Tustin, CA, pp. 142–148.

Covarrubias-García, I., Aizpuru A., Arriaga S., (2017) 'Effect of the continuous addition of ozone on biomass clogging control in a biofilter treating ethyl acetate vapors', Science of The Total Environment, (584–585), pp. 469–475.

Cox, H.H., Deshusses, M.A., (1997) 'Biological waste air treatment in biotrickling filters', Current Opinion in Biotechnology, 9(3), pp. 256–262.

Devinny, J.S., Hodge, D.S., (1995) 'Formation of acidic and toxic intermediates in overloaded ethanol biofilters'. 45, pp. 125–233.

Deshusses, M.A., (1997) 'Biological waste air treatment in biofilters', Current Opinion in Biotechnology, 8(3), pp. 335–339.

Deshusses, M.A., Hamer, G., Dunn, I.J., (1995) 'Behavior of biofilters for waste air biotreatment. 2. Experimental evaluation of a dynamic model', Environmental Science and Technology, 29(4), pp. 1059–1068.

de Castro, A. , Allen, D.G.G. , and Fulthorpe, R.R. (1996) 'Characterization of the microbial population during biofiltration and the influence of the inoculum source', In: Proceedings of the 1996 Conference on Biofiltration (an Air Pollution Control Technology), Reynolds, F.E. , Ed., The Reynolds Group, Tustin, CA, pp. 164.

de Castro, A., Allen, D.G., Fulthorpe, R.R., (1997) 'Characterization of the microbial population during biofiltration and the influence of the inoculum source', In: Proceeding of the Air and waste Management Association's 90th Annual Meeting and Exhibition, Paper 97-WA71A.04. Air and Waste Management Association, Pittsburgh, PA.

Diks, https://onlinelibrary.wiley.com/action/doSearch?ContribAuthorRaw=Diks%2C+ R+M+M R. M. M., Ottengraf, https://onlinelibrary.wiley.com/action/doSearch?C ontribAuthorRaw=Ottengraf%2C+S+P+P S. P. P., Vrijlnad, https://onlinelibrary. wiley.com/action/doSearch?ContribAuthorRaw=Vrijlnad%2C+S S. (1994), 'The existence of a biological equilibrium in a trickling filter for waste gas purification', Biotechnology Bioengineering, 44(11), pp. 1279–1287.

Ergas, S.J., Schroeder, E.D., Chang, D.P.Y., and Morton, R.L., (1995) 'Control of volatile organic compound emissions using a compost biofilter', Water Environment Research, 67(5), pp. 816–821.

Eweis, J.B., Chang, D.P.Y., Schroeder, E.D., Scow, K.M., Morton, R.L., and Caballero, R.C., (1997) 'Meeting the challenge of MTBE biodegradation', In: Proceedings of the 90th Annual Meeting and Exhibition of the Air and Waste Management Association, 8–13 June, Toronto, Canada.

Farmer, R.W., Chen, J.S., Kopchynski, D.M., and Maler, W.J., (1995) 'Reactor switching: proposed biomass control strategy for the biofiltration process', In: Biological Unit Processes for Hazardous Waste Treatment, Hincheet R.E., Skeen. R.S., and Sayles, G.D. (Eds.), Battelle Press, Columbus, OH, pp. 243–248.

Gallardo-Rodríguez, J.J., Rios-Rivera, A.C. Bennevitz, M.R. (2019), 'Living biomass supported on a natural-fiber biofilter for lead removal', Journal of Environmental Management, (231), pp. 825–832.

Gerrard, A.M., (1997) 'Economic design of biofilter systems', Journal of Chemical Technology and Biotechnology, 68(4), pp. 377–380.

Govind, R., Desai, S., and Bishop, D.F., (1997) 'Control of biomass growth in biofilters', In: Proceedings of the Fourth International In Situ and On-Site Bioremediation Symposium, Vol. 5, Battelle Press, Columbus, OH, pp. 195.

Hugler, W.C., Cantu-De la Garza, J.G., and Villa-Garcia, M., (1996) 'Biofilm analysis for an odor-removing trickling filter', In: Proceedings of the 89th Annual Meeting and Exhibition of the Air & Waste Management Association, Pittsburgh, PA.

Idaq, H. and Dincer, I., (2021) 'A novel biomass gasification based cascaded hydrogen and ammonia synthesis system using Stoichiometric and Gibbs reactors', Biomass and Bioenergy (145), pre proof.

Kinney, K., du Plessis, C., Schroeder, E.D., Chang, D.P.Y. and Scow, K.M., (1996) 'Optimizing microbial activity in a directionally-switching biofilter', In: Proceedings of the 1996 Conference on Biofltration (an Air Pollution Control Technology), Reynolds, FE, Ed., The Reynolds Group, Tustin, California, 150.

Laconi C.D., Ramadori, R., Lopez, A., Passino R. (2006) 'Influence of hydrodynamic shear forces on properties of granular biomass in a sequencing batch biofilter reactor', Biochemical Engineering Journal, 30(2), pp. 152–157.

Leson G, and Smith, B.J., (1997) 'Petroleum Environmental Research Forum field study on biofilters for control of volatile hydrocarbons', Journal of Environmental Engineering, 123(6), pp. 556–562.

Loy, J., Heinrich, K., Egerer, B., (1997) 'Influence of filter material on the elimination rate in a biotrickling filter bed', In: Proceedings of the 90th Annual Meeting and Exhibition of the Air and Waste Management Association, Pittsburgh, PA.

Molino, A., Larocca V., Chianese S., and Musmarra D., (2018) 'Biofuels Production by Biomass Gasification: A Review', Energies, 11(181), pp. 1–31.

Mathur, A.K., Rahul, Majumder, C.B., Gautam, S.B., McNaught, I.J., (2012) 'Modelling and kinetic aspects of a BTEX contaminated air-treating biofilter', Journal of Environmental Studies, 69(3), pp. 475–489.

Mathur, A.K., Sundaramurthy, J., Balomajumder, C., (2006) 'Kinetics of the removal of mono-chlorobenzene vapour from waste gases using a trickle bed air biofilter', Journal of Hazardrous Material, 137(3), pp. 1560–1568.

Medina. V.F., Devinny, J.S., and Ramaratnam, M., (1995) 'Biofiltration of toluene vapors in a carbon-medium biofilter', In: Biological unit Processes for Hazardous Waste Treatment, Proceedings of the Third International In Situ and On-Site Bioreclamation Symposium, Hinchee, RE, Skeen, RS., and Sayles, G.D., Eds., Battelle Press, Columbus, pp. 257.

Mirpuri, R., Sharp, W., Villaverde, S., Jones, W., Lewandowski, Z., and Cunningham, Al., (1997) 'Predictive model for toluene degradation and microbial phenotypic profiles in flat plate vapor phase bioreactor', Journal of Environmental Engineering, 123(6), pp. 586–592.

Morgenroth, E., Schroeder, E.D., Chang, D.P.Y., Scow, K.M., (1996) 'Nutrient limitation in a compost biofilter degrading hexane', Journal of the Air & Waste Management Association, 46(4), pp. 300–308.

Paca, J., Marek, J., Weigner, P., and Koutsky, B., (1997) 'Biofilter characteristics at high xylene and toluene loadings' in 'Biological Waste Gas Cleaning', Prins, W.L. and van Ham, J. (Eds.)VDI Verlag GmbH, Dusseldorf, Germany, pp. 123–129.

Potera, C., (1996) 'Biofilms invade microbiology', Science, 273(5283), pp.1795–1797.

Rahul, Mathur A.K., Balomajumder C. (2013) 'Biological treatment and modeling aspect of BTEX abatement process in a biofilter', Bioresource Technology, (142), 9–17.

Seed, L.P. and Corsi. R.L., (1996) 'Biofilteration of benzene, toluene and o-xylene: substrate effects and carbon balancing', In: Proceedings of the 89th Annual Meeting and Exibition of the Air and Waste Management Association, Pittsburgh, PA.

Smet, E., van Langehove, H., and Verstraete, W., (1996) 'Long-term stability of a biofilter treating dimethyl sulphide', Applied Microbiology and Biotechnology, 46(2), pp. 191–196.

Smith, F.L. , Sorial, G.A. , Suidan, M.T. , Breen, A.W. , and Biswas, P. (1996) 'Development of two biomass control strategies for extended stable operation of highly efficient biofilters with high toluene loadings', Environ. Sci. Technol., 30(5), 1744.

Sorial, G.A., Smith, F.L, Suiden, M.T., Biswas, P., Brenner, R.C., (1995) 'Evaluation of Trickle Bed Biofilter media for toluene Removal', Journal of Air & Waste Management Association, (45), pp. 801–810.

Sorial, G.A., Smith, F.L., Suidan, M.T., Pandit, A., Biswas, P., Brenner, R.C., (1997) 'Evaluation of trickle bed biofilter performance for BTEX removal', Journal of Environmental Engineering, (123), pp. 530–538.

Souza F.H.de, RoeckerP.B., Silveira D.D., Sens, M.L., Campos, L.C., (2021) 'Influence of slow sand filter cleaning process type on filter media biomass: backwashing versus scraping', Water Research, (189), pre proof.

Swanson, W.J. and Loehr, R.C., (1997) 'Biofitration: Fundamentals, Design and Operations Principle and Applications', Journal of Environmental Engineering, 123(6), pp. 536–546.

Thakur, P.K., Mathur, A.K., Gautam, S.B., Balomajumder, C., Rahul (2013) 'Determination of mass transfer coefficients for mixture of compost, sugarcane bagasse and granulated activated carbon as packing media in biofilter', Bioremediation Journal, 17(1), pp. 61–70.

Thakur P.K., Rahul, Mathur A.K., Balomajumder C., (2011) 'Biofiltration of Volatile Organic Compounds (VOCs) – An Overview', Research Journal of Chemical Sciences, 1(8), pp.83–92.

Turala, A. and Wieczorek, A., (2020), 'Biomass Growth and Its Control in the Process of Biofiltration of Air Contaminated with Xylene on a Biotrickling Column Filled with Expanded Clay', Sustainability, 12(5412), 1–14.

van Groenestijn, J., Harkes, M., Cox, H., and Doddema, H., (1995) 'Ceramic materials in biofiltration', In: Proceedings of the 1995 Conference on Biofiltration (an Air Pollution Control Technology), Hodge, D.S. and Reynolds, F.E., Jr, Eds., The Reynolds Group, Tustin, CA, pp. 317–324.

van Langenhove, H., Lootens, A., and Schamp, N., (1989) 'Inhibitory effects of SO_2 on biofiltration of aldehydes', Water Air Soil Pollution, 47(1), 81–86.

van Langenhove, H. and Smet, E., (1996) 'Biofiltration of organic sulfur compounds', In: Proceedings of the 1996 Conference on Biofiltration (an Air Pollution Control Technology), Reynolds, F.E., Ed., The Reynolds Group, Tustin, CA, p. 206.

Vier, J.K., Schroeder, E.D., Chang, D.P.Y., and Scow, K.M., (1996) 'Interaction between toluene and dichloromethane degrading populations in a compost biofilter', In: Proceeding of the 89th Annual Meeting ad Exhibition of the Air and Waste Management Association, Pittsburgh, PA.

Weber, F.J. and Hartmans, S., (1996) 'Prevention of clogging in a biological trickle-bed reactor removing toluene from contaminated air', Biotechnology and Bioengineering, 50(1), pp. 91–97.

Webster, T.S., (1996) 'Control of Air Emissions from Publicly Owned Treatment Works Using, Biological Filtration', Ph.D. thesis, The University of Southern California, Los Angeles.

Webster, T.S., Devinny, J.S., Torres, E.M., and Basrai, S.S., (1996) 'Biofiltration of odors, toxics and volatile organic compounds from publicly owned treatment works', Environmental Progress, 15(3), pp. 141–147.

Webster, T.S., Devinny, J.S., Torres, E.M., and Basrai, S.S., (1997) 'Microbial ecosystems in compost and granular activated carbon biofilters', Biotechnology and Bioengineering, 53(3), pp. 296–303.

Wright, W.F., Schroeder, E.D., Chang, D.P.Y., and Romstad, K., (1997) 'Performance of a pilot-scale compost biofilter treating gasoline vapor', Journal of Environmental Engineering, 123(6), pp. 547–555.

6 Enzymatic Biodiesel Production from Biomass

Nisha Gupta[1], Dilip Kumar[2] and Bhawna Verma[3]
[1,3] Department of Chemical Engineering, Indian Institute of Technology, Banaras Hindu University, Varanasi, India
[2]Department of Biochemical Engineering, Harcourt Butler Technical University, Kanpur, India

CONTENTS

DOI: 10.1201/9781003196358-6

6.1 INTRODUCTION

Regular exhaustion of energy is unavoidable for the existence of human beings. Several reasons are there for hunt of a possible substitute source of energy as a fuel which is economically feasible, technically optimized, environmentally friendly, easily available and competitive from the economic view point. The prime purpose is the continuous rise in the demand of fossil fuels in every other field of daily life, such as power and electricity generation, transportation, consumption on residential level, and industrial operations (Alebian-Kiakalaieh et al., 2013). This increasing demand is not only causing the depletion of energy sources which are non-renewable but in addition to this, fossil fuels burning is leading towards the emission of harmful green house gases such as carbon mono-oxide, carbon dioxide and methane, which are the prime factors for global warming and are severely affecting the Earth (Norjannah et al., 2016). The total energy consumption of world was found to be increased twice from the year 1971 to 2001 and the total energy demand of whole world by year 2030 will grow to 53%. For example, the consumption of petroleum in U.S.A. till the year 2030 will increase from 84.4 to 116 million barrels per day (Alebian-Kiakalaieh, et al., 2013). Several alternative sources of energy naming hydro, wind, geothermal, solar, nuclear, and biomass are there which successfully fulfil the criteria of sustainability but are not completely feasible from the economic point of view. The best way to attain economic sustainability as well as environmental protection is biofuel that are particularly made from biomass feedstock which is readily available. All the plant or animal matter that can be used to produce power such as electricity or heat is biomass. Biomass can be converted into different types of gaseous, liquid as well as solid fuels through several chemical, thermochemical, and biological conversion methods. The energy derived through biomass is considered by far the largest renewable source of energy, accounting 10.4% of the total primary supply of energy of the world or 77.4% of the global renewable supply of energy (Yusuf et al., 2011).The idea of using biofuels was invented by Rudolf Karl Diesel during the nineteenth century. He illustrated the utilisation of oils obtained from vegetable feedstock as an alternative for diesel fuel during the world exhibition in 1900 held in Paris. In his opinion, the biomass fuel consumption will certainly become reality as he developed and designed the future versions of his engines accordingly (Alebian-Kiakalaieh et al., 2013).

A biofuel is a kind of gaseous or liquid fuel which can be obtained predominantly from substrates of biomass. Familiar examples include ethanol, methanol and biodiesel. Ethanol can be produced by bacteria or yeast-mediated sugar crops fermentation, for example sweet sorghum, sugar-beet and sugar cane, or of starchy crops, for example cassava and corn. It can also be produced at lower yields through enzymatic or acid hydrolysis followed by the process of fermentation from other feedstock such as cellulose, polymer of sugar available through woody crops. Methanol is generally produced by the gasification process of wood followed by compaction and synthesis of methanol through woody crops or wood. Lastly, almost 90% biodiesel fuels are commonly produced through oil crops, namely soybean, palms, sunflowers, and rapeseed by the oil extraction with mechanical pressing or using appropriate solvents and then performing transesterification process for finally obtaining diesel fuel from the oil (Giampietro et al., 1997).

The biofuels generation and consumption is very likely to increase steadily in upcoming decades. Speaking of the present scenario, biofuels including biodiesel and bioethanol are already being used by various countries namely the United States, Brazil, Italy, Germany, Austria, and Australia. It is being predicted that this ongoing trend is more likely to rise in the upcoming future and more and more countries will contribute towards the power generation using biofuel and its consumption.

Considering all biofuels, the most prominent one is biodiesel because it has similar chemical characteristics and structure and same content of energy as that of the conventional diesel. In addition to this, any kind of modification in the diesel engine is not required because biodiesel is already adaptable with ongoing recent models of engine and can easily be blended with diesel on commercial level as type of fuel used for transportation (Yusuf et al., 2011).

6.2   BIODIESEL

Biodiesel is referred as the liquid fuel consisting of mono alkyl esters of long chain fatty acids. It is a renewable source of liquid fuels used for transportation because it is obtained from several types of renewable sources and lipids of nature, for example, animal fat, edible and non-edible vegetable oils and vegetable oils derived from waste (WVOs) (Watanabe et al., 2000). We can obtain biodiesel from different kinds of sources including (A) edible oils (for example, palm oil, soyabean, canola and sunflower), (B) non-edible oil (for example, mahua oil, linseed oil, Daturametel Linn, Jatropha curcas, Calophyllum inophyllum, and Hevea brasiliensis oils), (C) animal fats as well as waste oils (for example, poultry fat, chicken fat, and beef tallow) (Tan et al., 2019).

Various feedstocks which are used in the biodiesel production are mention under Table 6.1 (Alebian-Kiakalaieh et al., 2013).There are several advantages of biodiesel in comparison to that of common fuels which includes its biodegradability, non- toxicity and environment friendliness. It can be easily bendable with common diesel fuel and represents its physical and chemical characteristics very close to that of standard fuel.

Biodiesel has an upper hand when it comes to technical features as compared to standard fuel or diesel because biodiesel exhibits an intrinsic lubricant ability, generates less emission, comparatively greater flash points than diesel or other

TABLE 6.1
Feedstocks Used for Biodiesel Production

Conventional feedstock		Non-conventional feedstock
Mahua	Brassica carinata	Lard
Soyabean	Rubber plant	Tallow
Nile tilapia	Brassica napus	Sea mango
Rapeseed	Rice bran	Fish oil
Palm	Copra	Bacteria
Poultry	Sesame	Palanga
Canola	Groundnut	Fungi
Babassu	Cynaracardunculus	Micro-algae
Tobacco seed	Barley	Tarpenes
Sunflower	Olive	Okra
Corn	Jojoba oil	Pongaminapinnata
Coconut	Pumpkin	Algae
Used cooking oil	Peanut	Jatropha curcas
Mustard	Camelina	Poultry fat
Linseed	Cottonseed	Latexes

petroleum-based products and minor sulphur content. Biodiesel is also quite safer to handle as well as store and might be generated in domestic aggregate very easily (Tacias-Pascacio et al., 2019).

It is predicted that biodiesel can significantly replace about 10% of Europe's total consumption of diesel approximately as well as approximately 5% of that of Southeast Asia. However, the major obstruction which lies in front for the production as well as commercialization of biodiesel is its higher expenditure and cost in comparison to that of petroleum-based fuels. It has been estimated that biodiesel production costs about 1.5 times greater than those of petroleum diesel or gasoline which also depends upon various feedstock oils sources. For the production of biodiesel, it has been found that around 70–95% of the total cost appears from raw materials' cost. Hence, the feedstocks which have greater extent to obtain low cost and economical biodiesel are animal fat, cooking oil, etc. (Hoque et al., 2011). Biodiesel can be used in different kinds of concentrations on blending. The most general forms are B5 (up to 5% biodiesel) and B20 (6–20% biodiesel). The most typical form is B100 which is purely biodiesel and is useful as blendstock to obtain lower blends. B100 is rarely useful as a fuel for transportation.

6.2.1 Physical Properties

Biodiesel is a transparent amber-yellow fuel which is liquid in nature and has viscosity as same as that of the petroleum-based diesel fuel. Contrary to petroleum-based fuels, biodiesel is non- explosive and non-flammable and having a flash

TABLE 6.2
Physical Properties of Biodiesel

General name	Biodiesel
General chemical name	Fatty acid methyl ester (FAME)
Range of chemical formula	C14–C24 methyl esters
Range of kinematic viscosity (mm²/s, at 313 K)	3.3–5.2
Range of density (kg/m³, at 288 K)	860–894
Range of flash point (K)	420–450
Range of boiling point (K)	>475
Range of distillation (K)	470–600
Reactivity	Avoid strong oxidizing agents
Solubility in water	Insoluble in water
Biodegradability	Higher biodegradable as compared to that of petroleum diesel
Odour stable	Soapy odour /light musty
Physical appearance	Clear liquid, light to dark yellow in colour
Vapour pressure (mm hg, at 295 K)	<5

Source: Saxena et al. (2013).

point at 423K while the flash point of standard diesel is 337K. It does not produce harmful emissions while burning as fuel and reduces toxicity (Yusuf et al., 2011). The physical properties of biodiesel has been presented in table 6.2.

6.2.2 CHEMICAL PROPERTIES

Fats and natural oils are esters of fatty acids and glycerol known as triglycerides or glycerides. Fatty acids are divided into two types: the fatty acids which are polarized and consists of single or one carbon bond are known as saturated fatty acids, while the fatty acids which contain single or more than one carbon-to-carbon double bonds and which are also polarized are known as unsaturated fatty acids. Some common examples are oleic, stearic, palmitic and linolenic fatty acids.

6.2.3 BIODIESEL STANDARDS

Biodiesel Specifications are known by IS in India, in USA by ASTM, in European Union by EN, etc. These specifications must be fulfilled by all the biodiesels. These standards play vital role in determining the relevant physical as well as chemical characteristics of oil and the examination of biodiesel for its use into the engine system as a fuel. These specifications balances all the important properties of fuel including oxidation stability, sulphur content, cetane number, kinematic viscosity (mm^2 /s), glycerine (% m/m), acid number (mg KOH/g), heating value (MJ/kg), pour point, cloud point (°C), boiling point (°C), density (kg/m^3), flash point, etc. (Singh et al., 2019).

6.3   BIODIESEL PRODUCTION PROCESSES

Various processes have been reported for the transformation of fats and oils into reliable biodiesel fuels. More habitually, methyl alcohol is used for producing biodiesel but sometimes ethyl alcohol can also be used for the same. Generally, any alcohol whether it is primary, secondary or tertiary is useful for biodiesel production but the only restriction is that cost becomes higher as higher the alcohol becomes. However methyl alcohol is more reactive and cheaper as compare to that of ethyl alcohol but ethyl alcohol represents good properties in field of toxicity and renewability (Kumar et al., 2016).

6.3.1   DIRECT USE BLENDING

This method is the easiest one. Diesel fuel and biodiesel are taken in a certain proportion and are then mixed simultaneously for using it in the CI engine directly. Various reports have been found on performing different experiments successfully related to the blending or mixing of conventional/ standard diesel fuel with the vegetable oils. 80% of diesel oil was completely mixed with 10% of vegetable oil by a Brazilian company, Caterpillar and this mixture was subjected to testing for chamber engines for pre-combustion. It was then revealed that without any changes into the engine, this mixed fuel was capable of retaining its total power. Later in 1982, 5% of diesel and 95% of cooking oil were blended together and tested. After these experiments, it was concluded that only with slight modifications or changes into the engine, 100% vegetable oil is also useful as a fuel (Kumar et al., 2016).

For direct diesel engines as well as indirect ones, the straight application of these blends of oil or vegetable oil have found to be dissatisfying and unsuitable due to very obvious causes which was its higher viscosity as compared to that of standard diesel fuel, thickening of lubricating oil, carbon deposits, composition of acid in oil, content of free fatty acid, oxidation leading to the formation of gum and while combustion and storage polymerization takes place (Yusuf et al., 2011).

6.3.2   MICROEMULSION PROCESS

To avoid higher viscosity problem of vegetable oils, there are some experiments conducted taking microemulsions into account using methanol, butanol-1 or ethanol as solvents. Microemulsions are referred to colloidal dispersion in equilibrium of microstructures of fluid which are optically isotropic with dimensions ranging between 1 and 150 nm in diameter generally. It is formed instinctively of two liquids which are immiscible with one or more than one non-ionic or ionic amphiphiles (Ma and Hanna, 1999).Microemulsions posses low volumetric heating value as comparison to the standard diesel fuel because of the presence of alcohol content in it. The alcohol present in it also have higher latent heat of vaporization which leads to the cooling of chamber of combustion and lowers down nozzle coking. A microemulsion formed between vegetable oil and methanol can execute almost as same as the diesel fuel (Yusuf et al., 2011).

Ziejewski et al.,1984 conducted an experiment where the emulsion containing 33.4% (by vol) of 1-butanol, 53% (vol) of winterized and alkali refined sunflower oil and 13.3% (vol) of 190-proof ethanol was prepared. This prepared emulsion was non-ionic and at 40 degree Celsius it showed the viscosity 6.31 cSt. The cetane number of this emulsion was 25 and ash content was less than 0.01%. On increasing the content of butanol-1, more even spray patterns with lower viscosity was observed.

6.3.3 THERMAL CRACKING (PYROLYSIS)

The process of converting the complicated structures of hydrocarbons into the simplest form either in the company of catalyst or without using catalyst is known as thermal cracking. The viscosity and the density of oil get decreased on performing this process because these two are major characteristics of a vegetable oil which directly affect the engine atomization.

Hence, fuel obtained by this method is directly useful in any diesel engine without undergoing any sort of modifications into the engine system. For the production of biodiesel, the most common used catalyst for this process are red mud, alumina and zeolite. This process occurs at a temperature range of 250–350C. The plant for performing thermal cracking to produce biodiesel consists of a reactor

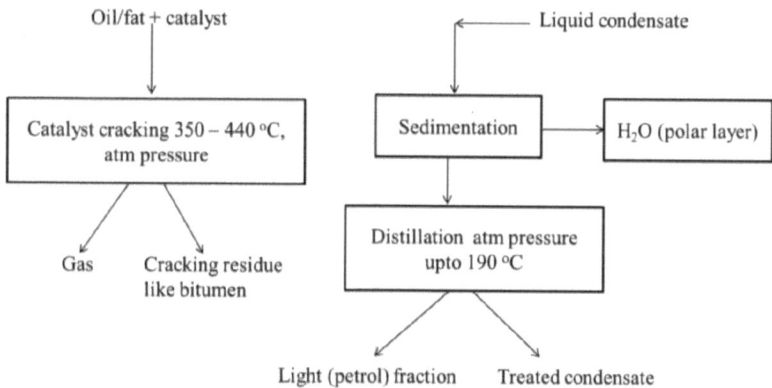

FIGURE 6.1 Schematic representation of thermal cracking process.

having safety valve, indicator for temperature, drain pipe, etc. The animal fat or oil which is essential to get transformed into biodiesel is deposited inside this reactor and then the suitable amount of heat is given to it due to which vaporization of animal fat or oil takes place and through the pipe it reaches the condenser unit. The condenser unit is used to cool down the vapour inside the liquid and the final liquid product is accumulated in a beaker which is known as biodiesel. The flow process of production of biodiesel using this process is represented in Figure 6.1 (Rajalingam et al., 2016).

6.3.4 REACTIVE DISTILLATION PROCESS

Reactive distillation (RD) is an important process which plays vital role in successfully manufacturing biodiesel because the end products produced at the end of reaction are commanded by chemical equilibrium. The process in which distillation separation and chemical reactions proceeds simultaneously in one single equipment is known as reactive distillation. For some particular reactions, this method has special benefits: for example, when one or more than one reactant is needed in a reaction, when some products are needed to be removed in a reaction for its completion and the recovery process of a product or recycling of co-product is complicated or has some chances of forming azeotrope. In reactive distillation catalyzed (homogeneous/heterogeneous) and non-catalyzed reactions can occur, the temperature and pressure of conversion being similar to distillation conditions. (Alebian-Kiakalaieh et al., 2013).

This distillation column compromises of a reactive zone (core) added with separation sections of rectification and stripping whose degree rely upon the behaviour of separation of reaction mixture. It is necessary to provide optimized operating pressure which is useful in determining the temperature of reaction and must be adaptable with the activity of catalyst. In mixed oxides, when we employ methanol as alcohol, the optimum temperature range is 130–150C, which leads to 6–10 bar pressure. Also, the temperature of reboiler must not be above than the limit decided by product thermostability, which is generally 200C. This restriction can be overcome by keeping a particular quantity of methanol into the bottoms which can be simply recovered by simple flash followed by further recycling. This distillation is also useful in the stage of esterification. But there is basic dissimilarity in esterification process naming, both the products i.e., glycerol and fatty ester, are heavier and their separation through distillation cannot be performed.

For such case, by applying alcohol amount in larger excess by increasing two times (or more) the internal reflux according to the required stoichiometry, shifting equilibrium happen to be ensured. In addition to this, by using this process, better environment can be established for integration of heat, driving of the reboiler can be done with the energy derived through process waste whereas a condenser is used to recover it as hot or raised stream which can be useful for driving several other operations (Kiss et al., 2008a).

6.3.5 DUAL REACTIVE DISTILLATION

This method permits more chances of producing biodiesel and special chemicals. The major difficulties to overcome are removal of water for solid catalyst protection against deactivation as well as escaping high cost of recovery of excess alcohol (Dimian et al., 2009) reported in a paper, related to fatty acid dual esterification containing heavy and light alcohols. Methanol as well as 2-ethyl hexanol (long-chain) represents same reactive function. This reaction undergoes at a temperature range of 130–200C and a moderate pressure is applied. Esterification behaves the same way as the reactive azeotrope distillation merged with reactive absorption with separation agent as water and the co-reactant is heavy alcohol. Later on, the researchers proposed an original mode of process for controlled strategy application in order to obtain operating conditions in optimized way. The primary benefits include lower costs of equipment by high integrated modelling, high flexibility while performing an operation and a reactive distillation unit which is multi-functional.

Furthermore, some researchers worked in a way of finding more integrated processes which can be proved helpful in reducing consumption of energy for generation of biodiesel. Reactive distillation, dual as well as catalytic RD show major effects in the cost reduction and consumption of energy in the production of fatty acid (Kiss et al., 2008b). It was investigated and found out that reactive distillation supported with heat integration consumes less energy which for heating is about 43% and for cooling is about 47% in comparison to those of standard reactive distillation. The prime benefits for performing this process include:

1. without providing alcohol in excess amount, reactants can be fed into stoichiometric ratios;
2. as compared to conventional units, productivity of process will get raised about five to ten times higher;
3. neutralization of catalyst, treatments of salt water, and production of soap are removed;
4. a high range of fatty acids as well as alcohols are suitable;
5. sulphur is being removed from the end product;
6. conversion of reactants and yield of products is very high;
7. cost of investment decreases; and
8. consumption of energy majorly reduces (Alebian-Kiakalaieh et al.,2013).

6.3.6 MEMBRANE TECHNOLOGY

Using a membrane reactor happens to be the most suitable method for producing biodiesel because it is capable to overcome all the limitations and restrictions faced by other conventional methods of production. A device or unit which compromises reaction as well as a separation in a single unit is known as membrane reactor. Membrane reactors are classified on the basis of four parameters (Thomas, 2009): the design of the reactor; the membrane used in the process (porous, organic, dense, or inorganic); the reactor of membrane is (whether catalytic or inert); and the reaction process occurring inside membrane reactor (for example, esterification, dehydrogenation, water dissociation, etc.) A membrane reactor not only provides separation but also increases the yield and selectivity of a reaction. This method is very helpful in biodiesel production because it removes glycerol from end product stream and retains the triglycerides which untreated within membrane (Shuit et al., 2012).

A new technology involving porous membrane which is hydrophobic in nature has also proved helpful in the production of biodiesel. Various benefits of this process are:

1. several raw materials, namely vegetable oil (pure), cooking oil (recycled), solid oils; and animal fat with any FFA level can be used;
2. for various types of raw materials, several kinds of membranes have already been developed;
3. a membrane system involving purification of glycerine fulfils standards of USP;
4. no extra additives or chemicals are needed; and
5. the cost is reduced (Alebian-Kiakalaieh et al., 2013).

Sdrula (2010) performed an experiment involving separation of biodiesel and crude glycerides through membrane technology. Millions of tons of crude glycerides in the form of waste is being produced by biodiesel industries and with significant rise in the generation of biodiesel, this quantity of glycerine is predicted to grow more rapidly in near future. Keeping this problem in mind, the researcher found an economical solution combining the electro-dialysis of high efficiency and nanofiltration for purification of glycerine and its recovery. The High Efficiency Electro Pressure Membrane (HEEPM) is useful and operational in continuous, batch and semi-batch processes. The general HEEPM technology is represented in Figure 6.2 and a prototype of biodiesel plant production through membrane separation is shown by Figure 6.3.

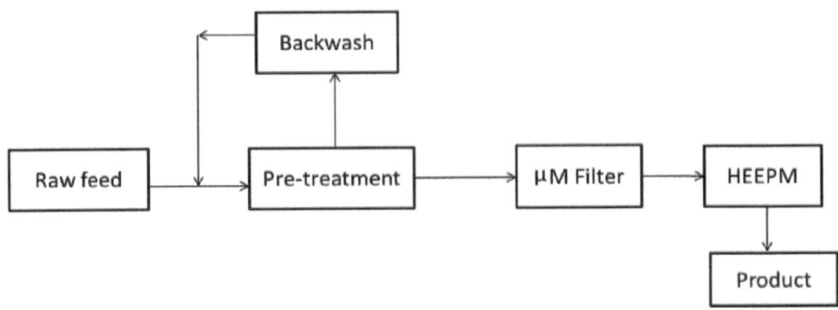

FIGURE 6.2 Schematic representation of HEEPM technology.

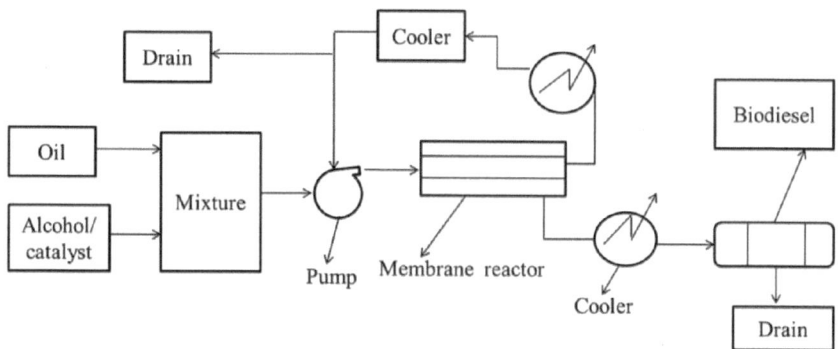

FIGURE 6.3   Representation of prototype of biodiesel plant through membrane separation.

6.3.7 The Transesterification Process

The process in which vegetable oil gets reacted with alcohol while utilizing of a catalyst is known as transesterification reaction. The catalyst is used in this process is helpful in improving yield of product and increasing rate of reaction. This process is commonly referred as alcoholysis. Generally, ethanol and methanol are being used in this process because of their low cost and good chemical and physical properties. In general oil and alcohol are taken in mole ratio of 3:1 for this process (Kumar et al., 2016).

The quantity and kind of catalyst used in this process is taken on the basis of presence of free fatty acid (FFA) in feedstock oil. The higher the quantity of FFA the higher the conditions become unfavourable for production of biodiesel effecting directly the yield of biodiesel and leading towards the soap formation.

6.3.7.1 The Transesterification Process Using Alkali Catalyst

In this method various bases such as sodium hydroxide (NaOH), potassium hydroxide (KOH) or sodium methoxide work in the place of catalyst. The most efficient one in all the three is sodium methoxide but not economically feasible. In this process, reaction between oil and alcohol takes place using base as a catalyst. The temperature for the reaction must be about 50–60C in order to obtain the biodiesel of highest yield and it should be less than the boiling point of methanol so that no vaporization of methanol should take place to prevent its wastage. Continuous stirring at 1,300 rpm is done throughout the whole procedure for carrying out dynamic mixing which eventually increases the reaction rate. The total reaction will get completed in about an hour. The end products resulting from this process are glycerol and biodiesel which are further sent to the separating chamber for 12–24 hours. After separation, biodiesel is collected as the upper layer and glycerol will be the lower layer (Rajalingam et al., 2016).

6.3.7.2   The Transesterification Process Using Acid Catalyst

When the content of free fatty acid is greater than 1% in oil feedstock, then biodiesel can not be produced from this fatty oil. Hence biodiesel yield through this transesterification method is very low. In this process acid is used as a catalyst and conversion of fatty acid into ester takes place. This method is also useful in the conversion of triglycerides in biodiesel but this will require more time, hence it's usually not preferable. In this method, catalyst used are either sulphuric acid or phosphoric acid involving reaction between oil and alcohol resulting into production of biodiesel and water is also produced but the water produced must be immediately get rid of as the presence of water will result into the soap formation (Rajalingam et al., 2016).

6.3.7.3   Two-step Transesterification Process

When feedstocks compromising higher amount of free fatty acids are dealt with, then this method comes to use. According to various researches it was found that catalysis through alkaline media alone directly can't catalyse the transesterification of feedstock oils of those having FFAs in higher amount in the range of 0.5% to nearly lower than 3% of mass fraction of oil. Thus, transesterification regarding higher FFAs can only be done by applying two step transesterification methods. In this method, the primary step is esterification of FFAs through acid catalysis and converting them into FAMEs as mentioned in section 2.7.2, and the second step is transesterification through alkaline catalysis as described in section 2.7.1. Using this method, various reports have been found of obtaining biodiesel with higher yields (Aransiola et al., 2014).

The melting and boiling points of several methyl esters, mono, di, triglycerides, and fatty acids grow with respect to carbon atoms present in carbon chain of a compound but lowers down as the quantity of the double bonds increases. The melting point is also get affected due the presence of hydrogen bonding as well as polarity of molecules (Ma and Hanna, 1999).

6.3.8   TRANSESTERIFICATION THROUGH ENZYMATIC TECHNOLOGY

Biodiesel production through enzymatic technology using lipase has already been proved useful, efficient in terms of consumption of energy and nature friendly in the FAAE production from potential feedstock oil. Although this process has several benefits yet due to certain drawbacks i.e., conversion of low-quality feedstock, lower consumption of energy and glycerol production of food-grade, it cannot be applied for large scale application. The major restrictions while applying this enzymatic process for production of biodiesel on an industrial scale are efficiency of conversion and cost of enzyme (Pourzolfaghar et al., 2016).

Key factors which play crucial role in the production of biodiesel by directly affecting the conversion efficiency, yield and economic feasibility are enzymes, bioreactor design, acyl acceptors and substrate. The kind of enzyme which is used in enzymatic production of biodiesel is lipase which is used for converting oil into

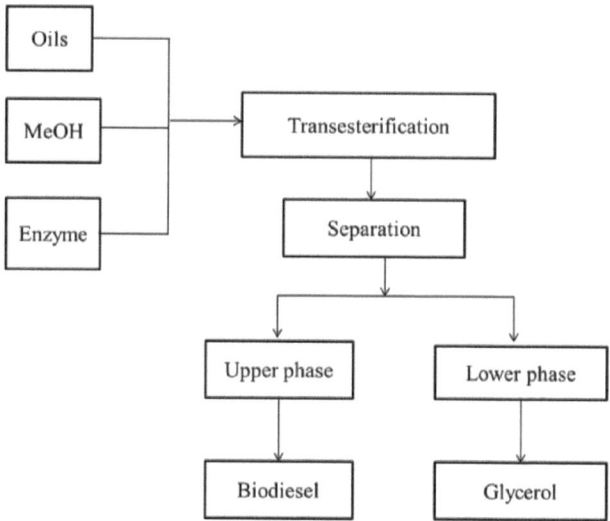

FIGURE 6.4 Biodiesel production through enzymatic process.

biodiesel as fatty acid alkyl esters producing glycerol as by-product. The extraction of lipase can be done through various sources namely yeast, bacteria and fungi (Norjannah et al., 2016).

Production of biodiesel through enzymatic technology is possible by using both intracellular as well as extracellular lipases. While considering situation of extracellular lipase, the recovery of enzyme is already been done previously from the broth of live producing microbe and further purified, whereas intracellular lipases reside in cell producing walls or inside cells. In both these situations immobilization of enzyme takes place and then it is used for eventually reducing downstream operations such as recycling and separation. The flow process of production of biodiesel using this process is represented in Figure 6.4.

6.3.8.1 Extracellular Lipase

The main microorganisms which are used to produce extracellular lipase are *Rhizopus oryzae, Mucormiehei, Pseudomonas cepacia* (Kumar et al., 2019), and *Candida antarctica.*In several studies, it has been found that the main objective of production of biodiesel using free lipases was focused around lipase screening and the keys factors that play vital role in influencing the rate of reaction. Soluble lipases are useful when it comes to easy preparation process and lower costs of preparation but in several situations, they are used only when they are inactivated. Immobilization technologies once improvised has offered lipases with modified and high level of operational stability, temperature optimum and reusability, leading to higher rates of conversion and shorter span of time of reaction respectively. After several researches and investigations, the most commercial immobilized lipases named as Lipozyme TL IM, Novozym 435, Lipase PS-C and Lipozyme RM IM. For production of biodiesel, various methods of immobilization of lipases are applied: for example, cross-linkage, covalent bonding, adsorption, encapsulation, and entrapment (Gog et al., 2012).

Mittlebach (Mittelbach, 1990) performed experiment involving transesterification procedure of oil derived from sunflower and primary alcohols for example butanol, ethanol and methanol by using *C. antarctica* and *M. miehei* (Novozym 435) in both the absence as well as presence of solvent i.e., petroleum ether. The final yield was high when on using butanol and ethanol without even using solvent but when methanol was used without the presence of solvent then only the traces of methyl esters were obtained. Lipase-catalysed production of biodiesel fuel from some Nigerian lauric oils. Biochem. Soc. Trans. performed an experiment for converting palm kernel oil by using ethanol and methanol as solvents and the final conversion yield was 72% and 15% respectively. The transesterification procedure containing soyabean oil using ethanol and methanol in company of immobilized *Pseudomonasflourescens* resulted in the final conversion yield to 65% and 67% for ethanol and methanol respectively (Noureddini, et al. 2005). But by further investigation done by Linko and group (Linko et al., 1998) the conversion of as high as 97% is also possible for performing transesterification process of rapeseed oil by using 2-ethyl-1-hexanol.

6.3.8.2 Intracellular Lipase

The major drawback of obtaining biodiesel by the process using extracellular enzymes was the comparatively higher cost of lipase because of the purification processes as well as complex separation. On the contrary to this, utilizing microbial cells which produce intracellular lipase in the form of whole-cell biocatalysts, ester yields of acceptable characteristics are achieved at relatively lower cost (Gog et al., 2012).

Technology involving use of biomass support particles which were porous in nature was introduced by (Atkinson et al., 1979) for using immobilized whole-cell lipases. (Ban et al., 2002) used whole-cell *R. oryzae* (immobilized) for undergoing transesterification procedure on vegetable oils and several suitable conditions for culture media, water content effect on the process of production and effects of cell pre-treatment were examined. For enhancing the activity of methanolysis of immobilized cells, different kinds of substrate were mixed into the media and from them oleic acid and olive oil were determined to be the effective ones. As comparison to the extracellular method, in this report higher conversion of 90% was obtained when water content was 15% and methanol was added stepwise.

6.3.8.3 Substrate

Feedstock for biodiesel production is selected appropriately on the basis of the physical and chemical characteristics of oil, process economy, process of the chemistry, agriculture potential, geography and climatic atmosphere of the area of production. For biodiesel production, the input components employed are divided into particular three categories: waste oils (waste cooking oils, animal fats, lard, and yellow grease), oleaginous microorganisms (fungi, microalgae, yeast, and bacteria) and oils derived from plants (edible as well as non-edible oils). Production of biodiesel is generally divided into three types of generations: first, second, and third generations, depending upon several factors, such as the development level, processing type technology, and feedstock type. In biodiesel of the first generation, feedstock normally compromises of vegetable oils (edible) for example soyabean oil in South America, the United States, and China, rapeseed oil (Europe), palm oil in tropical countries, and sunflower oil (Ukraine). Because of several drawbacks in edible oils use namely, higher cost, competition with consumption of food and limited sources, biodiesel feedstock from second generations are helpful in increasing the security of food and reducing biodiesel production cost. The sole drawback of using second generation feedstocks is the requirement of land with larger areas having moist soil. This issue can be resolved completely by using biodiesel feedstock of third generation obtained from oleaginous microorganisms. On the basis of area, this type of generation can be useful in obtaining oil 15–300 times higher than those conventional crops for production of biodiesel (Pourzolfaghar et al., 2016).

6.3.8.4 Acyl Acceptor

Alcohol is the most general kind of acyl acceptor which is used for synthesis of biodiesel. Ethanol as well as methanol are most commonly used alcohols for this kind of reaction as they are relatively cheaper than other alcohols. FAME (fatty-acid methyl ester) and FAEE (fatty-acid ethyl esters) are the main products of biodiesel when methanol as well as ethanol are used respectively. The short length of chain and high polarity of methanol are the traits which make methanol the best efficient kind of alcohol for the transesterification reaction. The major drawback of the use of methanol is the deactivation of lipase, its inhibition or denaturation. The reason behind this drawback is conformational change, entry of triglycerides getting blocked, enzyme unfolding, immiscibility in between alcohol and triglycerides or alcohol adsorption into the surface of immobilized polar material. In order to overcome this problem, various researches have been performed and showed that the addition of methanol in stepwise manner or continuous manner into the system can be very effective.

Ethanol is commonly obtained from the renewable resources hence it is greener alternative as comparison to methanol which is derived from fossil fuels. It even has lower inhibitory effects on the lipase as that of methanol. In addition to this, ethanol also does not need for the stepwise addition in order to get substrate high conversion but this stepwise addition may be found helpful in increasing the rate of reaction.

Instead of using ethanol and methanol individually, they can be in several proportions mixed up together and can be used in one single reaction (Norjannah et al., 2016). Dai et al. (2014) performed experiments involving various combinations (proportion-wise) of these two alcohols i.e., ethanol plus methanol, using

Novozym 435 for soyabean oil conversion. It was reported that 0% (ethanol is 100 mol%), 20%, 40%, as well as 60% methanol in mixed alcohols lead to ample yield of about 95% (one step addition and the molar ratio of the oil to the alcohol is 1: 3). Further it was also found that methanol was getting consumed at a faster rate than that of ethanol, hence ethanol was left out in the system in the end which results towards improving the solubility in between alcohol and oil as well as the deactivation of lipase through methanol was also getting reduced.

6.3.8.5 Bio-reactor Design

For industrialization and commercialisation, the development of bio-reactors is essential in order to determine the enzymatic processes potential. The probable types of bio-reactors design are the packed-bed reactor (PBR), membrane reactor (MBR), fluidized-bed reactor (FBR), and stirred-tank reactor (STR). STR operates in batch-wise method and contains a vessel where reaction mixture is kept and was manually stirred. It is operated quite easily. Instead of these benefits, the major problem in STR is comparatively less volumetric productivity. For catalytic reactions at large scale, packed bed reactor has been used conventionally as it possesses high efficiency of reaction, lower cost as well as easy to operate and maintain (Kumar et al., 2020a; Hama and Kondo, 2013).Various studies reported successfully using packed bed reactors for the production of biodiesel through enzymatic technique involving several setups, some of them were operating nine PBRs serially having a hydro cyclone unit attached right next to PBRs for the splitting of glycerol, single recirculating packed bed reactor, single packed bed reactor involving stepwise addition of methanol and three packed bed reactors connected serially and removing glycerol intermediately. Experiment was performed by operating PBR for production of biodiesel through enzymatic process involving Shirashima oil alcoholysis using *Aspergillus oryzae* as a biocatalyst (Yoshida et al., 2012). Some 96.1% of conversion of oil was obtained including 140 minutes per pass time of residence and adding methanol stepwise having 4.25 M equivalents to oil for the 6 passes in optimized manner (Chen et al., 2011) operated PBR for continuous production of biodiesel using soyabean oil alcoholysis having Novozym 435 as a catalyst.

The optimized conditions included 0.1 ml/min of flow rate at 52C and a molar ratio of oil to methanol of 1:4 and finally 83% of total molar conversion was obtained with having any loss in activity of the lipase for operating continuously for 30 days. Table 6.3 represents various researches related to the production of biodiesel through enzymatic process.

6.4   EFFECT OF SOLVENTS ON THE PRODUCTION OF BIODIESEL

Generation of biodiesel from alcohol and oil with the help of enzymes has been performed in both presence as well as absence of the solvent. Tallow oil methanolysis was done with the help of lipase i.e., *Mucormehei* in hexane which

TABLE 6.3
Comparison of Various Researches on Enzymatic Biodiesel Production

Enzyme/oil	Acyl acceptor	Production cost	Technique applied	Conversion (%)	Authors/year
Soyabean oil, (R. oryzae)	CH_3OH	Low	Stepwise methanol addition in a packed bed reactor	90	(Hama et al., 2007)
Soyabean oil, (Novozyme 435)	CH_3OH	High	Stepwise addition of methanol and enzyme preincubation into soyabean oil as well as methyl oleate	97	(Samukawa et al., 2000)
Triolein, (P. flourescens)	Butanol	Moderate	In place of acyl acceptor, butanol was utilized and none solvent was required	90	(Iso et al., 2001)
Vegetable oil, (Novozyme 435)	CH_3OH	Moderate	Stepwise addition of methanol	90–93	(Watanabe et al., 2000)
Sunflower oil, (Novozyme 435)	CH_3OH	High	Stepwise methanol addition and glycerol removal through dialysis	97	(Belafi-Bako et al., 2002)
Rapeseed oil, (Novozyme 435 &Lipozyme TL IM)	CH_3OH	High	Combined utilization of Lipozyme TL IM as well as Novozyme 435 using tert-butanol as solvent	95	(Li et al., 2006)
Soyabean oil, (Novozyme 435)	Methyl acetate	High	Methyl acetate is used as a novel acyl acceptor, which did not have any inhibitory effects.	92	(Du et al., 2004)
Cotton seed oil, (Novozyme 435)	CH_3OH	High	The solvent used here was tert-Butanol	97	(Royon et al., 2007)
Soyabean oil, (Novozyme 435)	CH_3OH	High	Stepwise methanol addition plus glycerol removal using iso-propanol as a solvent	98	(Xu et al., 2004)
Vegetable oil, (R. oryzae)	CH_3OH	Low	Stepwise methanol addition and glutaraldehyde application for enzyme stability	90	(Ban et al., 2002)
Jatropha oil, (Novozyme 435)	Ethyl acetate	High	Ethyl acetate was used which was not having any kind of inhibitory effects	91.3	(Modi et al., 2007)
Waste cooking oil, (Novozyme 435)	CH_3OH	Low	Stepwise methanol addition	90	(Watanabe et al., 2000)

TABLE 6.4
Effect of Solvent (Absence/Presence) on Production of Biodiesel through Enzymatic Technique

Alcohol	Source of oil	Solvent	Lipase	Yield	References
Methanol	Tallow	Hexane	*Mucormiehei*	94.8	(Mittelbach, 1990)
	Sardine	Water	*Aspergillus niger* (immobilized)	94.55	(Arumugam and Ponnusami, 2017)
	Waste cooking palm oil	Tert-butanol	*Candida antartica* B	19.4	(Halim et al., 2009)
	Cotton Seed	None	*C. antarctica*	92	(Jordanov et al., 2007)
	Nonedible Castor oil	None	*Bacillus aerius* (immobilized)	78.13	(Narwal et al., 2015)
Ethanol	Soyabean oil	Hexane/solvent free	*Thermomyceslanuginosa*	70–100	(Rodrigues and Ayub, 2011)
	Spent Coffee Grounds(SCG)	Hexane	*LipozymesRM IM,TL100L* and *CALB L*	97.2	(Caetano et al., 2017)
	Sludge palm oil (SPO)	Ethanol, tert-butanol	*Candida cylindracea*	20.94	(Nasaruddin et ai., 2014)
Propanol	Tallow	Hexane	*Mucormiehei*	98.6	(Nelson et al., 1996)
	Cotton seed oil	None	*Candida antarctica*	91.5	(Kose et al., 2002)
Isopropanol	Tallow	Hexane	*C.antartica*	51.7	(Mittelbach, 1990)
	Restaurant grease	None	*Thermomyces lanuginose* (immobilized)	61	(De et al., 1999)
	Karanja, Castor	None	*P.cepecia* (immobilized)	75	(Kumar *et al.*, 2018)(Kumar et al., 2020b)
				76	
Butanol	Restaurant grease	None	*Thermomyces lanuginose* (immobilized)	90	(De ei al., 1999)
	Butyric acid	None	*Candida rugosa*	187	(Gomes et al., 2004)

has led towards the ester yield of 77.8%. The transesterification reaction with the absence of solvent is supported by several workers as it is economical. In addition to this, flammability as well as toxicity of the organic solvents plus recovery of the product without anymore evaporation of organic solvent supports the solvent free reaction. The yields of ester attained from alcoholysis by the use of several lipases in both absence as well as the presence of the solvent is shown in Table 6.4 (Shah et al., 2003).

6.5 MERITS OF BIODIESEL

The prime importance of biodiesel is that it is renewable and environment friendly instead of those petroleum-based products. Biodiesel can be used similar to that of diesel fuel because of its readily availability, higher cetane number, portability, renewability, high efficiency of combustion, high biodegradability and less aromatic as well as sulphur content (Demirbas, 2009). It can be used very easily: No such modification of vehicle is required or no other equipment for fuelling is needed.

- Economy, performance and power: High efficiency in terms of cost as well as performance makes biodiesel a suitable fuel with proved high power generation.
- Nature friendly: Use of biodiesel prevents the emissions of greenhouse gases into the atmosphere which are responsible for increased global warming, thus reducing environment pollution and improving health of the natural beings.
- Use of biodiesel helps in reducing the usage of oils imported from foreign countries.
- Because of its low toxicity, biodiesel handling is quite safer as well as easy storage is possible as comparison to those of petroleum-based fuels.

6.6 DE-MERITS OF BIODIESEL

The main drawbacks of using biodiesel is its high cloud as well as pour point, higher viscosity, higher emissions of nitrogen oxide, low content of energy, higher cost, low engine power and speed, engine compatibility and injector coking (Demirbas, 2009).

- Considering the recent scenario, the cost of biodiesel fuel is approximately 1.5 times to the cost of petroleum-based diesel fuel.
- Ample amount of energy is required to obtain biodiesel from crops as well as for fertilization, sowing and harvesting.
- In certain engines, use of biodiesel as a fuel can harmfully effect their rubber houses.

- The infrastructure for proper distribution of biodiesel needs to be improved and well planned.

6.7 CONCLUSION

In the upcoming years, biodiesel shows great potential in the supply of energy globally around the world. By raising the share of energy mix of biodiesel under relevant conditions, the great contribution can be done towards the reduction of emissions of greenhouse gases, enhancement of global energy security, most specifically in the developing countries, marching towards the sustainable development in rural areas (Lin et al., 2011).The transesterification reaction was also carried out using oil. The long chain fatty acid ester formed by this reaction is expected to be utilized as a diesel fuel that does not give out sulphur oxide and diminish the soot particulate.

Biodiesel is obtained from several kinds of edible as well as non-edible vegetable oils, waste-cooking oil, animal fats, etc. At present, the edible oils used to derive biodiesel are palm, soyabean, rapeseed and sunflower. And the inedible oils used to produce biodiesel include *F. elastica, J. curcas, A. indica, M. indica, P. pinnata, C. inophyllum jatropha*, rubber seed, neem, silk, mahua, waste cooking, cotton tree, microalgae, etc. (Demirbas, 2009).

Enzymatic transesterification reaction for the production of biodiesel is more beneficial as compared to other chemical reactions as it possesses various suitable parameters such as mild conditions of reactions, no generation of wastewater as well as no saponification, high product quality and easy recovery of products. Despite of all these positive factors, further improvement is required for carrying out enzymatic mechanism because of the high enzyme cost, slow rate of reaction and inhibition of enzyme. On improving particular factors, it is quite certain that the use of enzymatic reaction for producing biodiesel can be used in industries on large scale and also leads towards the renewable and nature friendly production of energy.

REFERENCES

Alebian-Kiakalaieh, A., Amin, N. A. S. and Mazaheri, H., 2013. A review on novel processes of biodiesel production from waste cooking oil. *Applied Energy, 104*, pp. 683–710.

Aransiola, E.F., Ojumu, T.V., Oyekola, O.O., Madzimbamuto, T.F. and Ikhu-Omoregbe, D.I.O., 2014. A review of current technology for biodiesel production: State of the art. *Biomass and Bioenergy, 61*, pp.276–297.

Arumugam, A. and Ponnusami, V., 2017. Production of biodiesel by enzymatic transesterification of waste sardine oil and evaluation of its engine performance. *Heliyon, 3*(12), p. e00486.

Atkinson, B., Black, G.M., Lewis, P.J.S. and Pinches, A., 1979. Biological particles of given size, shape, and density for use in biological reactors. *Biotechnology and Bioengineering, 21*(2), pp.193–200.

Ban, K., Hama, S., Nishizuka, K., Kaieda, M., Matsumoto, T., Kondo, A., Noda, H. and Fukuda, H., 2002. Repeated use of whole-cell biocatalysts immobilized within biomass support particles for biodiesel fuel production. *Journal of Molecular Catalysis B: Enzymatic*, *17*(3–5), pp.157–165.

Belafi-Bako, K., Kovacs, F., Gubicza, L. and Hancsok, J., 2002. Enzymatic biodiesel production from sunflower oil by *Candida antarctica* lipase in a solvent-free system. *Biocatalysis and Biotransformation*, *20*(6), pp. 437–439.

Caetano, N.S., Caldeira, D., Martins, A.A. and Mata, T.M., 2017. Valorisation of spent coffee grounds: production of biodiesel via enzymatic catalysis with ethanol and a co-solvent. *Waste and Biomass Valorization*, *8*(6), pp.1981–1994.

Chen, H.C., Ju, H.Y., Wu, T.T., Liu, Y.C., Lee, C.C., Chang, C., Chung, Y.L. and Shieh, C.J., 2011. Continuous production of lipase-catalyzed biodiesel in a packed-bed reactor: optimization and enzyme reuse study. *Journal of Biomedicine and Biotechnology*, *2011*.

Dai, J.Y., Li, D.Y., Zhao, Y.C. and Xiu, Z.L., 2014. Statistical optimization for biodiesel production from soyabean oil in a microchannel reactor. *Industrial & Engineering Chemistry Research*, *53*(22), pp.9325–9330.

De, B.K., Bhattacharyya, D.K. and Bandhu, C., 1999. Enzymatic synthesis of fatty alcohol esters by alcoholysis. *Journal of the American Oil Chemists' Society*, *76*(4), pp.451–453.

Demirbas, A., 2009. Progress and recent trends in biodiesel fuels. *Energy Conversion and Management*, *50*(1), pp.14–34.

Dimian, A.C., Bildea, C.S., Omota, F. and Kiss, A.A., 2009. Innovative process for fatty acid esters by dual reactive distillation. *Computers & Chemical Engineering*, *33*(3), pp.743–750.

Du, W., Xu, Y., Liu, D. and Zeng, J., 2004.Comparative study on lipase-catalyzed transformation of soyabean oil for biodiesel production with different acyl acceptors. *Journal of Molecular Catalysis B: Enzymatic*, *30*(3–4), pp.125–129.

Giampietro, M., Pimentel, D. and Ulgiati, S., 1997 Feasibility of large-scale biofuel production. Does an enlargement of scale change the picture. *Bio Science*, *47*(9), pp. 587–600.

Gog, A., Roman, M., Toşa, M., Paizs, C. and Irimie, F.D., 2012. Biodiesel production using enzymatic transesterification–current state and perspectives. *Renewable Energy*, *39*(1), pp.10–16.

Gomes, F.M., Pereira, E.B. and de Castro, H.F., 2004. Immobilization of lipase on chitin and its use in nonconventional biocatalysis. *Biomacromolecules*, *5*(1), pp.17–23.

Halim, S.F.A., Kamaruddin, A.H. and Fernando, W.J.N., 2009. Continuous biosynthesis of biodiesel from waste cooking palm oil in a packed bed reactor: Optimization using response surface methodology (RSM) and mass transfer studies. *Bioresource Technology*, *100*(2), pp.710–716.

Hama, S. and Kondo, A., 2013. Enzymatic biodiesel production: An overview of potential feedstocks and process development. *Bioresource Technology*, *135*, pp.386–395.

Hama, S., Yamaji, H., Fukumizu, T., Numata, T., Tamalampudi, S., Kondo, A., Noda, H. and Fukuda, H., 2007. Biodiesel-fuel production in a packed-bed reactor using lipase-producing *Rhizopus oryzae* cells immobilized within biomass support particles. *Biochemical Engineering Journal*, *34*(3), pp.273–278.

Hoque, M.E., Singh, A. and Chuan, Y.L., 2011. Biodiesel from low cost feedstocks: The effects of process parameters on the biodiesel yield. *Biomass and Bioenergy*, *35*(4), pp.1582–1587.

Iso, M., Chen, B., Eguchi, M., Kudo, T. and Shrestha, S., 2001. Production of biodiesel fuel from triglycerides and alcohol using immobilized lipase. *Journal of Molecular Catalysis B: Enzymatic*, *16*(1), pp.53–58.

Jordanov, D.I., Dimitrov, Y.K., Petkov, P.S. and Ivanov, S.K., 2007. Biodiesel production by sunflower oil transesterification. *Oxidation Communications*, *30*(2), pp.300–305.

Kiss, A. A., Dimian, A. C. and Rothenberg, G., 2008a. Biodiesel by catalytic reactive distillation powered by metal oxides. *Energy and Fuels*, *22*(1), pp. 598–604.

Kiss, A.A., Dimian, A.C. and Rothenberg, G., 2008b. Biodiesel production by heat-integrated reactive distillation. In *Computer Aided Chemical Engineering* (Vol. 25, pp. 775–780). Elsevier.

Kose, O., Tuter, M. and Aksoy, H.A., 2002. Immobilized *Candida antarctica* lipase-catalyzed alcoholysis of cotton seed oil in a solvent-free medium. *Bioresource Technology*, *83*(2), pp.125–129.

Kumar, A., Shukla, S.K. and Tierkey, J.V., 2016. A review of research and policy on using different biodiesel oils as fuel for CI engine. *Energy Procedia*, *90*, pp.292–304.

Kumar, D., Das, T., Giri, B.S., Rene, E.R. and Verma, B., 2019. Biodiesel production from hybrid non-edible oil using bio-support beads immobilized with lipase from *Pseudomonas cepacia*. *Fuel*, *255*, p.115801.

Kumar, D., Das, T., Giri, B.S. and Verma, B., 2018. Characterization and compositional analysis of highly acidic karanja oil and its potential feedstock for enzymatic synthesis of biodiesel. *New Journal of Chemistry*, *42*(19), pp.15593–15602.

Kumar, D., Das, T., Giri, B.S. and Verma, B., 2020a. Optimization of biodiesel synthesis from nonedible oil using immobilized bio-support catalysts in jacketed packed bed bioreactor by response surface methodology. *Journal of Cleaner Production*, *244*, p.118700.

Kumar, D., Das, T., Giri, B.S. and Verma, B., 2020b. Preparation and characterization of novel hybrid bio-support material immobilized from *Pseudomonas cepacia* lipase and its application to enhance biodiesel production. *Renewable Energy*, *147*, pp.11–24.

Lin, L., Cunshan, Z., Vittayapadung, S., Xiangqian, S. and Mingdong, D., 2011. Opportunities and challenges for biodiesel fuel. *Applied Energy*, *88*(4), pp.1020–1031.

Li, L., Du, W., Liu, D., Wang, L. and Li, Z., 2006. Lipase-catalyzed transesterification of rapeseed oils for biodiesel production with a novel organic solvent as the reaction medium. *Journal of Molecular Catalysis B: Enzymatic*, *43*(1–4), pp.58–62.

Linko, Y.Y., Lamsa, M., Wu, X., Uosukainen, E., Seppala, J. and Linko, P., 1998. Biodegradable products by lipase biocatalysis. *Journal of Biotechnology*, *66*(1), pp.41–50.

Ma, F. and Hanna, M.A., 1999. Biodiesel production: a review. *Bioresource Technology*, *70*(1), pp.1–15.

Mittelbach, M., 1990. Lipase catalyzed alcoholysis of sunflower oil. *Journal of the American Oil Chemists' Society*, *67*(3), pp.168–170.

Modi, M.K., Reddy, J.R.C., Rao, B.V.S.K. and Prasad, R.B.N., 2007. Lipase-mediated conversion of vegetable oils into biodiesel using ethyl acetate as acyl acceptor. *Bioresource Technology*, *98*(6), pp.1260–1264.

Narwal, S.K., Saun, N.K., Dogra, P., Chauhan, G. and Gupta, R., 2015. Production and characterization of biodiesel using nonedible castor oil by immobilized lipase from *Bacillus aerius*. *BioMed Research International*, *2015*.

Nasaruddin, R.R., Alam, M.Z. and Jami, M.S., 2014. Evaluation of solvent system for the enzymatic synthesis of ethanol-based biodiesel from sludge palm oil (SPO). *Bioresource Technology*, *154*, pp.155–161.

Nelson, L.A., Foglia, T.A. and Marmer, W.N., 1996. Lipase-catalyzed production of biodiesel. *Journal of the American Oil Chemists' Society*, *73*(9), pp.1191–1195.

Norjannah, B., Ong, H.C., Masjuki, H.H., Juan, J.C. and Chong, W.T., 2016. Enzymatic transesterification for biodiesel production: a comprehensive review. *RSC Advances*, *6*(65), pp.60034–60055.

Noureddini, H., Gao, X. and Philkana, R.S., 2005. Immobilized Pseudomonas cepacia lipase for biodiesel fuel production from soyabean oil. *Bioresource Technology*, *96*(7), pp.769–777.

Pourzolfaghar, H., Abnisa, F., Daud, W.M.A.W. and Aroua, M.K., 2016. A review of the enzymatic hydroesterification process for biodiesel production. *Renewable and Sustainable Energy Reviews*, *61*, pp.245–257.

Rajalingam, A., Jani, S.P., Kumar, A.S. and Khan, M.A., 2016. Production methods of biodiesel. *Journal of Chemical and Pharmaceutical Research*, *8*(3), pp.170–173.

Rodrigues, R.C. and Ayub, M.A.Z., 2011. Effects of the combined use of *Thermomyceslanuginosus*and *Rhizomucormiehei* lipases for the transesterification and hydrolysis of soyabean oil. *Process Biochemistry*, *46*(3), pp.682–688.

Royon, D., Daz, M., Ellenrieder, G. and Locatelli, S., 2007. Enzymatic production of biodiesel from cotton seed oil using t-butanol as a solvent. *Bioresource Technology*, *98*(3), pp.648–653.

Samukawa, T., Kaieda, M., Matsumoto, T., Ban, K., Kondo, A., Shimada, Y., Noda, H. and Fukuda, H., 2000. Pretreatment of immobilized *Candida antarctica* lipase for biodiesel fuel production from plant oil. *Journal of Bioscience and Bioengineering*, *90*(2), pp.180–183.

Saxena, P., Jawale, S. and Joshipura, M.H., 2013. A review on prediction of properties of biodiesel and blends of biodiesel. *Procedia Engineering*, *51*, pp.395–402.

Sdrula, N., 2010. A study using classical or membrane separation in the biodiesel process. *Desalination*, *250*(3), pp.1070–1072.

Shah, S., Sharma, S. and Gupta, M. N., 2003. Enzymatic transesterification for biodiesel production. *Indian Journal of Biochemistry and Biophysics*, *40*(6), pp. 392–399.

Shuit, S.H., Ong, Y.T., Lee, K.T., Subhash, B. and Tan, S.H., 2012. Membrane technology as a promising alternative in biodiesel production: A review. *Biotechnology Advances*, *30*(6), pp.1364–1380.

Singh, D., Sharma, D., Soni, S.L., Sharma, S. and Kumari, D., 2019. Chemical compositions, properties, and standards for different generation biodiesels: A review. *Fuel*, *253*, pp.60–71.

Tacias-Pascacio, V.G., Torrestiana-Sánchez, B., Dal Magro, L., Virgen-Ortíz, J.J., Suárez-Ruíz, F.J., Rodrigues, R.C. and Fernandez-Lafuente, R., 2019. Comparison of acid, basic and enzymatic catalysis on the production of biodiesel after RSM optimization. *Renewable Energy*, *135*, pp.1–9.

Tan, S.X., Lim, S., Ong, H.C. and Pang, Y.L., 2019. State of the art review on development of ultrasound-assisted catalytic transesterification process for biodiesel production. *Fuel*, *235*, pp.886–907.

Thomas, J.M., 2009. Handbook Of Heterogeneous Catalysis. 2., completely revised and enlarged Edition. Vol. 1–8. Edited by G. Ertl, H. Knozinger, F. Schüth, and J. Weitkamp. *Angewandte Chemie International Edition*, *48*(19), pp. 3390–3391.

Watanabe, Y., Shimada, Y., Sugihara, A., Noda, H., Fukuda, H. and Tominaga, Y., 2000. Continuous production of biodiesel fuel from vegetable oil using immobilized *Candida antarctica* lipase. *Journal of the American Oil Chemists' Society*, *77*(4), pp.355–360.

Xu, Wei Du, Jing Zeng and Dehua Liu, Y., 2004. Conversion of soyabean oil to biodiesel fuel using lipozyme TL IM in a solvent-free medium. *Biocatalysis and Biotransformation*, *22*(1), pp.45–48.

Yoshida, A., Hama, S., Tamadani, N., Fukuda, H. and Kondo, A., 2012. Improved performance of a packed-bed reactor for biodiesel production through whole-cell biocatalysis employing a high-lipase-expression system. *Biochemical Engineering Journal*, *63*, pp.76–80.

Yusuf, N.N.A.N., Kamarudin, S.K. and Yaakub, Z., 2011. Overview on the current trends in biodiesel production. *Energy Conversion and Management*, *52*(7), pp.2741–2751.

Ziejewski, M., Kaufman, K.R., Schwab, A.W. and Pryde, E.H., 1984. Diesel engine evaluation of a nonionic sunflower oil-aqueous ethanol microemulsion. *Journal of the American Oil Chemists' Society*, *61*(10), pp.1620–1626.

7 Catalytic Cracking of Jatropha curcas Non-Edible Oil to Hydrocarbons of Gasoline Fraction

Optimization Studies through the Box-Behenken Method

Saurabh Yadav[1], Neeru Anand[2], Dinesh Kumar[2] and Suantak Kamsonlian[1]

[1]Department of Chemical Engineering, Motilal Nehru National Institute of Technology Allahabad, Prayagraj – 211004, India

[2]University School of Chemical Technology, GGS Indraprastha University Delhi, Delhi – 110078, India

CONTENTS

DOI: 10.1201/9781003196358-7

7.1   INTRODUCTION

With concerns regarding release of GHG gases in the atmosphere due to burning of fossil fuels use of new and renewable energy sources for producing power and energy has gained lots of attention both politically and economically [1]. An increase in the population and rise in living standards in developing and developed countries has forced government to form the rules to reduce the emission of GHG in the atmosphere [2]. Release of these gases as well as other pollutants have increased both the human health risk as well as atmosphere temperature which have brought undesirable change in the world weather leading the unwanted rain trends, floods, draughts, and increase of sea level. Considering the above problems research on production of alternative fuels is going on in almost every part of the world. India imports around twice the number of oils produced by itself, hence it is dependent on many countries such as Saudi Arabia, Iran, Iraq, Kuwait, UAE, Yemen, and Malaysia. Hence for being self-dependent, research on production of high calorific value liquid fuel from non-edible oils (palm oil, canola oil, jatropha oil, buriti oil) is being carried out widely. Progress, 2015 has proposed three different models Energy Technology Perspective (ETP) 2015 via 2, 4, and 6 degree scenario (DS) to reduce the CO_2 emission by reducing the use of fossil fuel and promoting the use of sustainable energy sources [3]. An increase in the various renewable sources e.g., solar, tidal, wind, and biomass. Solar, tidal, wind etc. have been proved beneficial in terms of generating electricity which can be used both industries as well as transportation sector particularly in developed countries due to availability of technology [2]. In developed countries as well as in one-third world countries transportation sector still depends on the fuels obtained from fossil resources. The use of bio-ethanol and biodiesel has been widely accepted in the transportation sector but to meet the huge demand more ways have to be researched to fulfil the transportation sector demand [1]. Production of alternative fuels has become a necessity because of rise in prices of petroleum and diesel and deple-tion of their sources. The main objective of this work is to produce hydrocarbons of liquid fuel range from non-edible oil. Fuels are any materials that are capable of producing heat energy that can be used for various purposes. The major source of energy includes diesel, petroleum, natural gas and coal. With declining fossil fuel resources and increasing demand of conventional fuels, production of alternative fuels that are economical, environment friendly, energy efficient is much required.

7.2 LITERATURE REVIEW

Oil whether edible or non-edible can't be directly used as fuel in the engine and has to be converted to biodiesel using transesterification. These oils may be converted to gasoline range hydrocarbons using catalysts [4]. Thermochemical conversion of various oils has been reported by many authors inside a fixed bed reactor [5]. The impact of various parameters e.g. Temperature, catalyst weigh, time, catalysts type and load have been reported [6] but the optimization of these parameters for the cracked liquid products yields as well as quality of liquid products in terms of gasoline hydrocarbon fraction have not yet reported. The effect of hydrogen donor solvent to increase the calorific value of the fuel as well as reduction of the viscosity has been reported by few authors [7, 8]. Catalytic cracking of these oils can provide fuel with better standards in terms of reduced oxygen content and acidic fraction and have been reported earlier. Extensive research is being carried out globally on catalytic cracking of vegetable oils to produce biofuels. Several types of catalysts such as zeolites, SAPO, siliceous material, and oxides of Ti have been studied by various researchers [9]. Zeolites have been known for their high selectivity due to size, shape, and acidity have shown excellent performance. Several kinds of zeolite catalysts utilized in the catalytic cracking for production of liquid fuel have been reported in the literature. For the optimization of process parameters two methods namely composite centre design (CCD) [10] and Box-Behenken method [11–14] have been reported in various works. Box-Behenken method (BBM) is a rotatable second-order. It can be observed as a globular, rotating hexahedron with each variable settled at one of three equally spaced values. The method yields the vital prominence on central points of the variable's range and is appraised as "safer" way; after all the reaction doesn't requires to carry out at the utmost points. BBD doesn't accommodate an embedded factorial outline and needs little treatment blends than CCD [12].

India is being rich in terms of land, rain etc. and non-edible oil seeds e.g. Jatropha, Pongamia, Tung etc can be grown in the vast waste land available with it [15]. The oil obtained from Jatropha oil contains curcine and deterpine which are toxic in nature hence can't be used as edible oil [16]. Jatropha curcas oil contains mainly unsaturated acids e.g. C18:1 (40%), linoleic C18:2 (37%) and palmitic C16:1 (1%) [16]. These acids though falls in the range of C_{14}-C_{18} hydrocarbons suitable to be used as fuel but corrosive in nature due to acids and also have high oxygen content leading to lower calorific value. The removal of these acids in from of H_2O and CO_2 can be achieved by thermochemical conversion in the presence of suitable catalyst. In a country like India, where the demand of fuel is increasing day by day, obtaining alternative fuels by catalytic cracking of non-edible oils can be one of the option. The fuels obtained by this method are eco-friendly as they don't emit nitrogen oxides, sulphur oxides, sulphuric, carbonic, and nitric acid. Production of alternative fuels by such methods in India will reduce our dependency on the other countries. Apart from this setting up plants and using these methods on large scale production of alternative fuels will provide employment also. Many countries are working on alternative fuels. Lots of researches are being carried out in countries such as Germany, China, the United States, UK, etc. Experiments on vegetable oils such as jatropha curcas, pongamia pinata, madhuca indica, jajoba, palm, buriti, and canola using different catalyst are being carried out. Usually non-edible oils are used for production of alternative fuels, because they are comparatively cheaper than edible oils (used for cooking purposes). The present study reports the cracking of non-edible Jatropha oil using ZSM-5 as a catalyst to produce liquid hydrocarbons in the gasoline range. Optimization of operating conditions was carried out using Box-Behenken method to obtain the mathematical relation and response surface diagrams to understand the effect within the range of parameters used in the work.

7.2.1 JATROPHA CURCAS

Jatropha curcas is shrub or small tree of euphorbiaceae family [17]. The seed and plants are inedible to human and even animals also. Due to presence of deterpine and curcine the seeds are toxic. Well-drained soils with good aeration are required for its growth and can also be adapted to low nutrient content marginal soils. Research over the past decade has shown that oils extracted from jatropha plant can be used for biodiesel production by transesterification. But in this work, we have carried out catalytic cracking of Jatropha oil. Cetane number and calorific value of Jatropha oil matches that of diesel but properties such as viscosity and density are much higher. It is usually found growing in subtropical and tropical region. Even in wasteland Jatropha plants can grow and also in terrain, saline, sandy and gravelly soils. It can even bloom in stony and poor soils. Within 9 days complete germination is accomplished. Addition of manure during these 9 days before germination is not favourable for its growth but after that it is favourable. The flower development is at the end of the stem hence a good ramification leads to production of greatest amount of fruits. Jatropha plans are self-adaptive. It flourishes with even 2.5 cm rain in a year, planting and ploughing are no required regularly as Jatropha has life expectancy of around 40 years. Treatment with pesticides is not required at all as plant itself has fungicidal and pesticidal properties. The plants start yielding after nine months to an year time but best yield is often achieved after two to three years' time. If planted in flip flop way the productivity of jatropha can be between 0.8 kg to 1 kg of seed per metre of live fence. The average production of seed is around 3.5 ton per hectare which ranges from 0.4 tons per hectare in first year to over 5 tons per hectare after 3 years. Plantation in one hectare gives between 400 and 600 litres of oil if grown in average soil.

Some physical properties of jatropha oil are as follows: -

1. Density is 0.93g/cm^3.
2. Kinematic viscosity is 52.76cSt.

3. Cetane number is 38.
4. Flash point is 210C.
5. Calorific value is 38 MJ/kg.

7.2.2 CATALYSTS

A catalyst is any substance that alters rate of reaction without itself being consumed. In case of heterogeneous catalysis, the phase of catalyst varies from the products or reactants. The procedure varies with homogeneous catalytic activity where the catalysts, products and reactants are in identical form. Form not only differentiates between gas, liquid and solid components. It even differentiates immiscible combination (examples are water and oil) or anywhere an associate is present [18]. Catalysts are beneficial as they usually escalate the reaction rate without their being consumed and hence, they are recyclable. Heterogeneous catalytic activity typically involves solid form catalysts and gaseous form catalysts. Usually in catalytic cracking there is a loop of reaction, molecular adsorption, and desorption occurring at the catalyst surface. Heat transfer, mass transfer, and thermodynamics affect the kinetics (rate) of reaction. They are inactive regarding reactants covering across the whole surface, only few spots have catalytic active sites called active sites. The overall surface area of a solid phase catalysts has a powerful impact on the number of accessible active sites. During industrial application solid phase catalysts are usually permeable to increase surface area, usually getting near to 50–400 m²/g. Few mesoporous silicates like mobil composition of matter no. 41 (MCM-41) have surface area more than thousand m²/g. Porous material are cost effective due to their more surface area to mass ratio and increased activity of catalysis.

In most of the cases a solid phase catalyst is dissipated on aiding component to enhance surface area and maintain stability [19]. Normally catalysts supports are dormant with high melting point materials. Remarkably catalysts support is permeable (frequently alumina-based carbon, zeolite, or silica) and elected for their low mass to surface area ratio. In case of any given reaction permeable supports must be chosen like that reactants and product can exit or enter. Generally, materials are deliberately added to supplement to the feed of the reaction or the catalytic substance to influence catalytic stability, selectivity, and activity. Such substances are termed as promoters. e.g., Al_2O_3 is supplemented during synthesis of ammonia to provide balance by lowering actions on the iron catalysts. Deactivation of catalyst can be explained as loss of activity of catalysts or selectivity also. Substances which reduce rate of reaction are called poisons. They chemisorb to surface of catalysts and decrease the active number for molecules of reactants. Some commonly known poisons are elements of GROUP 5, 6 and 7 like arsenic and lead and consuming breed with numerous bonds e.g. unsaturated hydrocarbons and carbon monoxide. E.g., sulphur rattles the methanol production by poisoning the carbon zinc oxide catalyst. Substances which boost the rate of reaction are termed as promoters. For example, the presence of alkali metals in synthesis of ammonia increases nitrogen rate breaking. The existence of poisons and promoters can change the activation energy of the step of rate limiting and alter selectivity of catalysts for the creation of few compounds. Depending on the quantity a material can be unfavourable or favourable for a synthetic technique. Catalysts can also be deactivated by sintering, fouling, coking, vapour solid reactions, solid state transformation and erosion.

7.2.3 Catalytic Cracking

Catalytic cracking is an alteration action which can be enforced to a variation of reactants which can be heavy crude oil to gas oil [20]. The approach of thermal catalytic cracking is mostly similar to normal thermal cracking however it contradicts by utilization of a catalyst consumed in the transformation and it is one of the few efficient operation utilized in refining process which utilizes a catalyst to boost efficiency of process. Catalytic cracking in economic process associates connecting a reactant which can be a gas oil fraction with a catalyst under suited position of heat (temperature), tension (pressure) and residence time. Through this process a considerable part (more than 50%) of the reactant is transformed into gasoline or lower-boiling-point products normally during a single-pass application. While carrying out the cracking process carbonaceous substance is usually accumulated on the catalyst, considerably weakening its action and replacement such deposition is necessary. The carbonaceous content accumulates arises from the thermal decaying of polar species of high molecular weight in the feed supplied. The elimination of the retentate from the catalysts is normally done by heating in the company of air as far as activity of catalyst is restored. Weight hourly space velocity (WHSV) is defined as the weight of feed flowing per unit weight of the catalyst per hour.

7.2.4 Biofuels

Biofuels are sustainable power source (energy source) derived from wastes or biological substances, which can play beneficial act cutting down carbon dioxide discharge [21]. They are one of the biggest sources of sustainable power in today world. In the transport sector, biofuels are blended with diesel and gasoline. Biodiesels are alkyl esters long chain fatty acids. Such esters are received by vegetable oil through transesterification with ethanol or methanol. Biofuels which are obtained by inedible oil have good capacity of being alternate fuel. Biodiesels can be enticing alternate fuel as they are eco-friendly and can be obtained by treatment of non-edible or edible oils. A wide collection of plants which produce inedible oil are treated by chemical process for production of biodiesel. Few inedible oils are silk cotton tree, jatropha, mahua oil or madhuca indica, microalgae, rubber seed, karanja, neem etc. which are easily accessible in advance nations and are cheaper comparatively than edible oils. A wide collection of plants which yield inedible oil can be treated chemically for production of biofuel are found everywhere in India. Since edible oils are used for cooling purposes hence their price is high but non-edible oils are not used frequently in common man life hence their cost is low. The price of biofuel and need of edible oil can be decreased by inedible oil instead of cooking oil. Inedible oils are genuinely crucial for flourishing and crude shortage countries [22]. The essential asset sources for production of biofuel by inedible oils can be achieved by plant breed such Ratanjyote, jatropha, Pongamia, Karanja, Nagchampa, neem, rubber seed tree, mahua, silk cotton tree, babassu tree, Euphorbia tirucalli, microalgae, etc. These are conventional in most of the part of India and are more affordable than cooking oils used for domestic purpose. The high prices of fuel lead to demand of alternate fuels which ultimately lead to research on new techniques and sources of alternate fuels. Considering the above problems research on production of alternative fuels is going on in almost every part of the world. India imports around twice the number of oils produced by itself, hence it is dependent on many countries such as Saudi Arabia, Iran, Iraq, Kuwait, UAE, Yemen, and Malaysia. Hence for being self-dependent, research on production of high calorific value liquid fuel from non-edible oils (palm oil, canola oil, jatropha oil, buriti oil) is being carried out widely. In a country like India, where the demand of fuel is increasing day by day, obtaining alternative fuels by catalytic cracking of non-edible oils can be one of the options. The fuels obtained by this method are eco-friendly as they don't emit nitrogen oxides, sulphur oxides, sulphuric, carbonic, and nitric acid. Production of alternative fuels by such methods in India will reduce the country's dependency on the other countries. Apart from this setting up plants and using these methods on large scale production of alternative fuels will provide employment also. Production of alternative fuels from non-edible oils will lead to consumption of useless non-edible oils. Usually in this field research is focussed of synthesis and characterization of catalysts, thermal cracking of oil, thermal catalytic cracking of oil and after that the organic liquid product (OLP) is characterized or analysed using GCMS, FTIR, CHNO analysis, etc. If the properties of organic liquid products match the properties of any fuel characteristics, then they can be termed as biofuel. Many countries are working on alternative fuels. Widespread research is being carried out in countries such as Germany, China, USA, UK. Experiments on vegetable oils such as jatropha curcas, pongamia pinata, madhuca indica, jajoba, palm, buriti, canola using different catalyst are being carried out. Usually, non-edible oils are used for production of alternative fuels, because they are comparatively cheaper than edible oils (used for cooking purposes).

Investigation on cracking kinetics of n-hexane and n-decane was performed, it was observed that the gas phase composition contacted with zeolite catalysts HY and HZSM5 in a properly mixed batch reactor [23]. Variation in operating parameters was obtained as 200°C to 500°C in case of catalyst temperature and 0.5 to 20 mbar for reactant paraffin pressure, respectively. Further, the effect of crystal size and the ratio of Si/Al on the kinetics were studied. The cracking rate of HZSM5 at above 400°C, it was observed that pseudo-first order in partial pressure of paraffin. Around 300C, maximum rate of reaction of pure n-decane on HZSM5 was achieved, however minimum temperature was attained around 360C.

Initiation duration along with autocatalysis was also obtained with a reactant (n-decane) in this ranges of temperature based on Si/Al ratio and deactivation degree of the zeolite.

Researchers also carried out the catalytic cracking of n-heptane onto HZSM5 catalysts with different Si/Al ratios at 450–650C for forming selectivity of light olefins [24]. The results revealed that HZSM5 zeolites exhibits similar selectivity of the same conversion with different acid site densities. The increased in selectivity of ethylene plus propylene takes place, whereas decreased in ratio of ethylene/propylene was observed with the rise of reaction temperatures. The experimental data showed that an elevated temperature is necessary to achieve a high yield of ethylene plus propylene. Finally, they had concluded from the activation energies and selectivity, the mono-molecular cracking is leading at an increase in temperature (>650C).

The effect of mixtures of silica-alumina catalysts and HZSM-5 for the catalytic conversion of biofuel to hydrocarbons was investigated [25]. The results of preliminary study revealed that production of different variety of liquid hydrocarbons is possible and by mixing with catalysts like silica-alumina and HZSM5 in various proportions gives extremely variation in hydrocarbon content (from aromatic to aliphatic hydrocarbons). Further, used biofuel was produced by a fast thermal processing of wood in a micro-reactor at atmospheric pressure of 1.8–7.2 WHSV and temperature of 330–410°C. Interestingly, the data showed that the maximum yields of total hydrocarbons and organic liquid product (OLP) were achieved at 370C for all catalyst mixtures. It was also found that an organic liquid product fraction was produced using silica-alumina which mainly consists of aliphatic hydrocarbons. Characterization of catalysts was conducted by using various techniques such as FT-IR analysis, NMR spectroscopy analysis, X-ray powder diffraction analysis, BET surface analysis for pore sizes and desorption test with ammonia at programmed temperature.

7.3   MATERIALS AND METHODS

7.3.1   MATERIALS AND CHARACTERIZATION

Jatropha curcas oil was purchased from the jatropha Vikas Sansthan. Zeolite with SAR 27 was purchased from Sud Chemie. Zeolite was characterized using BET for the estimation of specific surface area, pore volume, pore diameter and total surface area utilizing Micromeritrics ASAP-2010 (NY, USA) and XRD analysis was performed on Philips X'pert diffractometer.

Parameters on which catalytic cracking process depends are: -

1. temperature,
2. catalyst, and
3. weight hourly space velocity (WHSV).

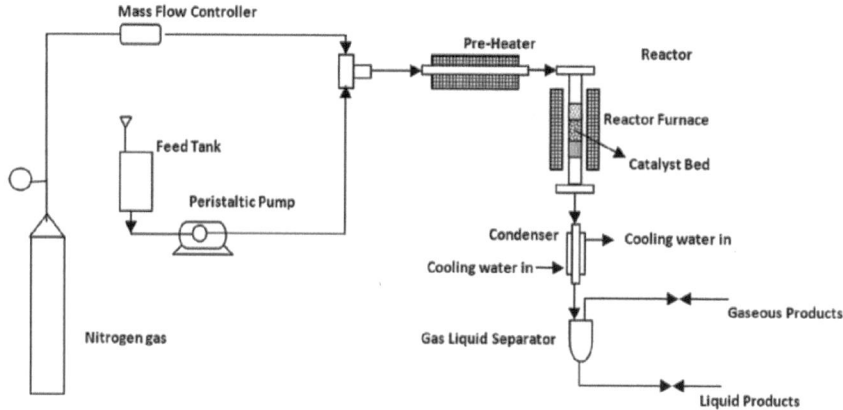

FIGURE 7.1 Schematic diagram of experimental setup.

7.3.2 EXPERIMENTAL SET-UP AND METHODOLOGY

Catalytic cracking of Jatropha curcas oil was carried out on SS-316 (Stainless Steel) fixed bed reactor. The schematic diagram is shown in Figure 7.1. The experimental rig consists of peristaltic pump, nitrogen cylinder, pre-heater with furnace, reactor with furnace, condenser and a separation funnel.

Cracking of jatropha oil was carried out over a reaction temperature of 400–450 C, with catalyst amount between 1–2 g and tetralin fraction between 0–0.2. The reactor was electrically heated with digital temperature controller. Feed flow rate (0.5 millilitre/minute) was managed using peristaltic pump. Rate of flow of nitrogen was kept constant at 30 ml/minute. The volume of feed is 30 ml in each experiment. Products obtained after cracking were flowed through a condenser. This condensed liquid products and gas were separated using a separation funnel. Liquid product was analysed using gas chromatography. Initially experiments were done without catalyst i.e. thermal cracking. In thermal cracking, OLP yield was 70% (21 ml). Hydrocarbons of C6-C8 range were 18%, C9-C13 range hydrocarbons were 17%, hydrocarbons of C14-C18 range were 25%. After this, catalytic cracking was carried out at temperature 400°C with 1g catalyst. OLP yield was 66% (20 ml). Yield of hydrocarbons of C6-C8 range were 40%, hydrocarbons of C9-C13 range were 28% and hydrocarbons of C14-C18 range were 10%. After this, tetralin was added to enhance the yield of C9-C13 range hydrocarbons.

7.3.3 MATHEMATICAL MODEL AND DESIGN OF EXPERIMENTS

Design of experiments is widely used to optimize the process parameters for maximizing or minimizing the desired response. The advantage of design of experiments is to reduced number of total numbers of runs which save both time and resources. In this work Box-Behenken design (BBD) is used to optimize the parameters as it

TABLE 7.1
Level of Variables Chosen for Box-Behenken Design

Variable	Symbol	Variable Level		
		Low (-1)	Centre (0)	High (+1)
Temperature	X_1	400	425	450
Catalyst Weight	X_2	1	1.5	2
Tetralin-oil ratio	X_3	0	0.1	0.2

TABLE 7.2
Design Matrix for Oil Catalytic Cracking Process Parameter Optimization Using the Box-Behenken Method

S. No.	Actual/Coded values of variables		
	Temp (°C)	Catalyst Wt (gm)	T/O ratio
1	400	1	0.1
2	450	1	0.1
3	400	2	0.1
4	450	2	0.1
5	400	1.5	0
6	450	1.5	0
7	400	1.5	0.2
8	450	1.5	0.2
9	425	1	0
10	425	2	0
11	425	1	0.2
12	425	2	0.2
13	425	1.5	0.1
14	425	1.5	0.1
15	425	1.5	0.1
16	425	1.5	0.1
17	425	1.5	0.1

has lesser number of experiments to conduct as compared centre composite design. Three variables namely temperature (X_1), catalyst weight (X_2) and tetralin to oil ratio (X_3) with the feed were studied. The lower levels as well as upper levels chosen for the study are presented in Table 7.1.

The number of runs in BBD is given as $2k*(k-1) + m$, where k is the number of parameters and m is the replicate quantity of central points. The design matrix as suggested by the Box-Behenken method with 5 runs as central point run and total 17 runs is shown in Table 7.2.

Design Expert Software, version 8.0.7.1 (Stat-Ease Inc., Silicon Valley, CA) was used to build the regression model. Three responses namely gas phase yield, liquid (OLP) yield and gasoline hydrocarbon fraction were selected. The effect of these variables was investigated on the response Y given as:

$$Y = \beta_0 + \beta_1 X_1 + \beta_2 X_2 + \beta_3 X_3 + \beta_{11} X_1^2 + \beta_{22} X_2^2$$
$$+ \beta_{33} X_3^2 + \beta_{12} X_1 X_2 + \beta_{13} X_1 X_3 + \beta_{23} X_2 X_3 \quad (1)$$

Where Y is the predicted response, β_0 model constant, β_1 to β_3 are linear coefficients and β_{12}, β_{13}, β_{23} are cross product coefficients and β_{11}, β_{22}, β_{33} are the quadratic coefficients.

7.3.4 PRODUCT CHARACTERIZATION

Analysis of liquid products was performed on Newchrom 6700 gas chromatogram fitted with Rtx-5 column (specs) staring from 70 C to 250C at a ramp rate of 2.5 C/min using N_2 as carrier gas. Before the analysis the liquid phase was mixed with known amount of CS_2, filtered using micro filters and injected with a ratio of 50:1. For FT-IR analysis was performed on Thermo Nicolet Model: 6700 with the help of KBr pellets. For oxygen content analysis Elementar CHNS analyser was used. The estimation of calorific value, calorimeter was used as per ASTM D3286-91a.

7.4 RESULTS AND DISCUSSIONS

7.4.1 CATALYST CHARACTERIZATION

The pore volume and surface area of the catalyst ZSM-5 (SAR=27) was obtained as $0.65 m^3/g$ and 381 m^2/g respectively. The XRD of the catalyst was done and 2 theta peaks were observed at 7.92, 9.8, 13.4, 22.8 and 24.1 which resembles with the standard XRD of ZSM-5.

7.4.2 STATISTICAL ANALYSIS FOR GAS, LIQUID AND GASOLINE FRACTION HYDROCARBONS

Statistical significance of parameters was analysed to estimate the analysis of variance (ANOVA) and adequacy of the empirical model. The correlation linking response (Y) and variables for all levels were obtained for gas, Liquid and GHF.

$$Y_{gas} (\%) = 16.2 + 0.03 * X_1 + 5.37 * X_2 - 2.75 * X_3 \quad (2)$$

$$Y_{OLP} (\%) = 83.51 - 0.03 * X_1 - 5.4 * X_2 + 1.25 * X_3 \quad (3)$$

$$Y_{GHF} (\%) = -2717.65 + 13.21 * X_1 - 45.1 * X_2 - 494.75 * X_3$$
$$+ 0.50 * X_1 * X_2 + 0.60 * X_1 * X_3 + 15.0 * X_2 * X_3$$
$$- 0.016 * X_1^2 - 51.8 * X_2^2 + 780 * X_3^2 \quad (4)$$

The response for the defined levels of variables in our experimental plan could be calculated from equation 2–4. The experimental and forecasted values of Y_{Gas} and Y_{Liquid} are presented in Figures 7.2 and 7.3. The results of ANOVA for response surface linear model to predict the gas and liquid yield and response surface quadratic model to predict the gasoline range are summarized in Table 7.3.

It is observed from the plot that experimental values match with the predicted values (R^2 values of 0.96 for liquid yield, 0.95 for gas yield). The significance of the coefficients can be checked from F value and p value. The larger F value (80.28 for gas, 95.34 for liquid and 12.25 for gasoline range) indicates that the model is notable. p-value (probability of error) below 0.05 show prototype are important. Response surface linear model to predict gas yield indicate that temperature and catalyst weight are significant model terms with p value lower than 0.05. Also, catalyst weight has a more significant effect on bio-oil yield than temperature. Hence Catalyst weight >Temperature > Tetralin fraction is the order of significance. Also, coefficient of variation value (CV) of 1.31% shows that the outcomes of the fitted model are genuine. The larger F value of 80.28 shows that the prototype is noteworthy. p-value (probability of error) less than 0.0001 shows model terms are noteworthy. The significance of all coefficients was established

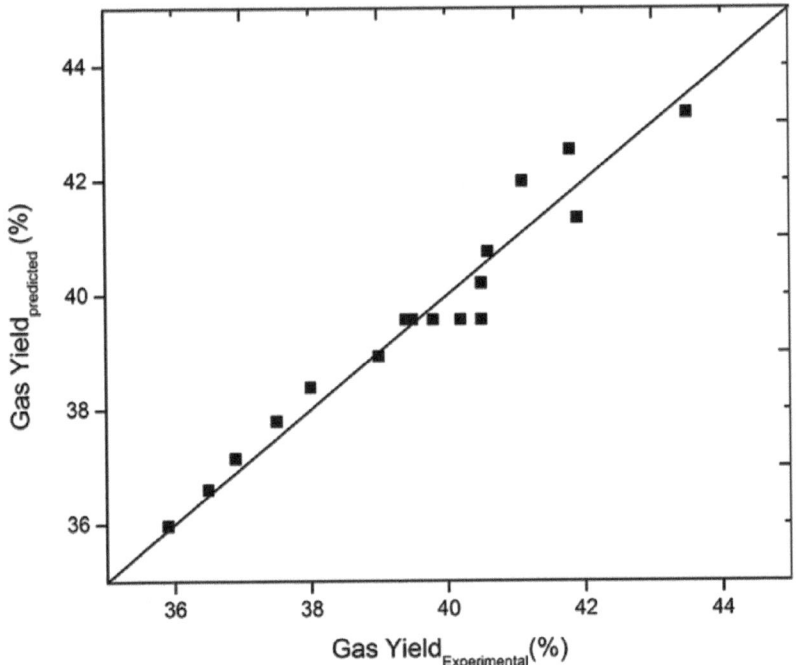

FIGURE 7.2 Plot for the experimentally obtained gas yield and theoretically predicted gas yield.

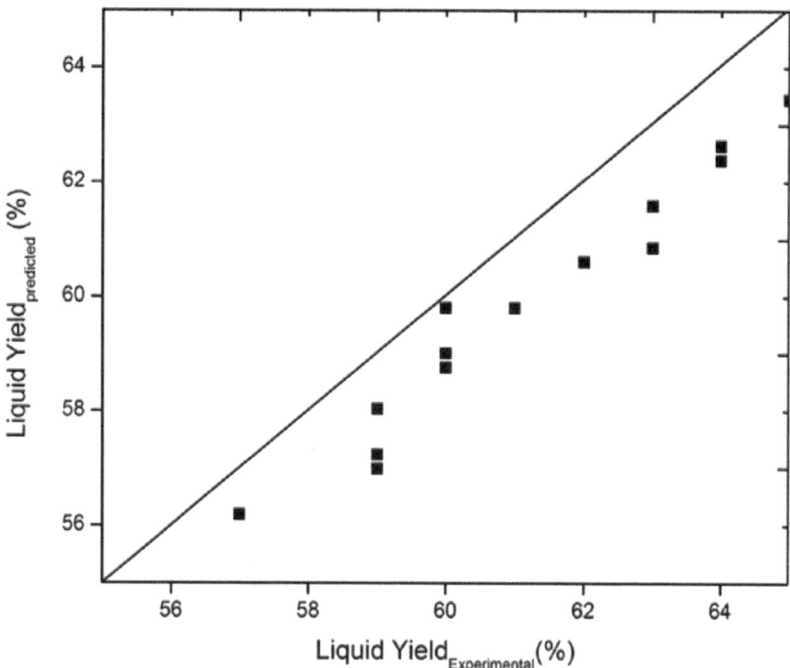

FIGURE 7.3   Plot for the experimentally obtained liquid (OLP) yield and theoretically predicted liquid (OLP) yield.

by p-values. It also indicates that temperature and catalyst weight are significant model terms with p-value lower than 0.05. The weight of catalyst was observed to have more significant effect on bio-oil yield than temperature. Hence, Catalyst weight > Temperature > Tetralin fraction is the order of significance. p-value (probability of error) is below 0.0001 shows model terms are noteworthy. The importance of all coefficients was fixed by p-values. Coefficient of variation value (CV) of 0.8% shows that the outcomes of the fitted model are genuine.

The ANOVA analysis was observed to fit significantly with quadratic model. From the linear, interaction and quadratic terms it was observed that both individual as well as quadratic effect for each parameter were significant whereas only combined effect of catalyst and temperature was significant with p-value lower than 0.05 and a large F value. The weight of catalyst was again more significant than temperature on liquid yield. The quadratic term of catalyst weight has greater significant impact on liquid yield (F value of 33.3) than temperature (F value of 20.66) and tetralin ratio (F value of 12.08). The effect of interaction between temperature and teralin ratio and catalyst weight and tertalin ratio had almost insignificant effect. Correlation coefficient R_2 value of 0.94 indicates excellent correlation between the variables.

TABLE 7.3
Analysis of Variance for Response Surface Model (for Gas, Liquid and Gasoline Range Hydrocarbon)

Source	Sum of Squares	Degree of Freedom	Mean Square	F-value	Prob>F
ANOVA for Gas Yield					
Model	65.05	3	21.68	80.28	<0.0001
X_1 - Temperature	6.66	1	6.66	24.66	0.0003
X_2- Catalyst Weight	57.78	1	57.78	213.92	<0.0001
X_3- Tetralin Ratio	0.61	1	0.61	2.24	0.1584
CV % = 1.31	$R^2 = 0.95$	Adj. $R^2= 0.94$	Predicted $R^2=0.91$		
ANOVA for Liquid Yield					
Model	65.29	3	21.76	95.34	<0.0001
X_1 - Temperature	6.84	1	6.84	29.99	0.0001
X_2- Catalyst Wt	58.32	1	58.32	255.48	<0.0001
X_3- Tetralin Ratio	0.13	1	0.13	0.55	0.4725
CV % = 0.8	$R^2 = 0.96$	Adj. $R^2= 0.94$	Predicted $R^2= 0.93$		
ANOVA for Gasoline range Hydrocarbon Yield					
Model	2338.02	9	259.78	12.25	0.0016
X_1 - Temperature	120.13	1	120.13	5.66	0.0489
X_2 - Catalyst Wt	364.5	1	364.5	17.19	0.0043
X_3 - Tetralin Ratio	300.13	1	300.13	14.15	0.0071
X_1X_2	156.25	1	156.25	7.37	0.03
X_1X_3	9	1	9	0.42	0.5356
X_2X_3	2.25	1	2.25	0.11	0.7542
X_1^2	438.06	1	438.06	20.66	0.0027
X_2^2	706.12	1	706.12	33.3	0.0007
X_3^2	256.17	1	256.17	12.08	0.0103
CV % = 7.07	$R^2 = 0.94$	Adj. $R^2= 0.86$	Predicted $R^2= 0.86$		

7.4.3 RESPONSE SURFACE PLOTS

As discussed earlier the gas yield was observed linearly dependent on the three variables studied in this work. It can be noticed that the impact of catalyst is more significant and as the catalyst weight was increased more gaseous products e.g. CO_2, CO or hydrocarbons with C_1-C_4 were obtained due to availability of increased acid sites whereas no significant change in the gas yield was noticed with change in temperature within in the specified range.

7.4.4 ANOVA ANALYSIS

Response surface plots were obtained for liquid (OLP) yield to study the effect of temperature, catalyst weight and tetralin fraction. It has been reported in literature that increase in catalyst weight enhance the cracking reaction which may lead to formation of small hydrocarbons (hydrocarbons in the range of C_6 to C_9) via secondary cracking reactions.

3D response surface plots for better understanding of gasoline hydrocarbon fraction showing the effect of catalyst weight, temperature and tetralin fraction. Graph plotted for catalyst weight and temperature indicates that the extreme values ($\alpha = \pm 1$) considered in this work do not provide a maxima for maximum yield but rather at $\alpha=0$. This can be concluded that an optimum catalyst weight as well as temperature keeps a balance by providing optimum number of acid sites and temperature which control the secondary cracking reactions. Graph plotted between tetralin fraction and temperature indicates a decrease in the Gasoline range which may be due to increase of tetralin fraction which enhances the possibilities of higher cracking towards uncondensable gas products e.g. CO_2, CO or hydrocarbons with C_1-C_4. Similar results were obtained, which shows that the optimum catalyst weight was 2 g. The optimum conditions for maximum gasoline hydrocarbon fraction were obtained at temperature 425°C with 2g catalyst and absence of tetralin.

7.5 CONCLUSIONS

The application of response surface methodology and BBD for experimental design study of catalytic cracking of jatropha oil was discussed. Model variables investigated in the present study were temperature (X_1), amount of catalyst (X_2) and tetralin oil ratio (X_3). The optimum condition for obtaining gasoline range hydrocarbons was at temperature 425°C with 2g catalyst and without tetralin. Predicted values of yield of gas, liquid (OLP) yield and gasoline hydrocarbons fraction obtained using BBD experimental design were in good agreement with the experimental results. To get a better understanding of the effect of variables, the predicted models can be understood by 3D response surface plots. The study showed that BBD and response surface methodology could be used to get maximum information in a short period of time with minimum number of experiments.

ACKNOWLEDGEMENT

The authors are thankful to Prof. K K Pant, Department of Chemical Engineering, IIT Delhi for his help and support to utilize the facilities.

REFERENCES

1. Huber, G.W., Corma, A.: Synergies between bio- and oil refineries for the production of fuels from biomass. Angew. Chem., Int. Ed., 46, 7184–7201 (2007).
2. AEO2014 Early Release Overview, U.S. Energy Information Administration, Annual Energy Outlook 2013 Early Release Overview, www.fossil.energy.gov/programs/ gasregulation/authorizations/2013applications/scexhibts13116118/Ex.109-AEO2014 EarlyReleaseOvervie.pdf (2014) Accessed on 5 October 2018.

3. Climate Change 2007: Working Group III: Mitigation of Climate Change, Intergovenrment panel on climate change (IPCC), Chapter 5: Transport and its infrastructure, www.ipcc.ch/publications_and_data/ar4/wg3/en/ch5. html (2007) accessed on 30 September 2018.

4. Katikaneni, S.P.R., Adjaye, J.D. & Bakhshi, N.N.: Catalytic conversion of canola oil to fuels and chemicals over various cracking catalysts. Canadian Journal of Chemical Engineering, 73, 484–497 (1995).

5. Taufiqurrahmi, N., & Bhatia, S.: Catalytic cracking of edible and non-edible oils for the production of biofuels. Energy Environment Sciences, 4, 1087–1112 (2011).

6. Twaiq, F., Zabidi, N.A.M., & Bhatia, S.: Catalytic conversion of palm oil to hydrocarbons: performance of various zeolite catalysts. Industrial and Engineering Chemistry Research, 38, 3230–3238 (1999).

7. Alemán-Vázquez, L.O., Cano-Domínguez, J.L. & García-Gutiérrez, J.L.: Effect of tetralin, decalin and naphthalene as hydrogen donors in the upgrading of heavy oils. Procedia Eng., 42, 532–539 (2012).

8. Liu, Y. & Fan, H.: The effect of hydrogen donor additive on the viscosity of heavy oil during steam stimulation. Energy Fuel, 16, 842–846 (2002).

9. Ong, Y.K., & Bhatia, S.: The current status and perspectives of Biofuel production via catalytic cracking of edible and non-edible oils. Energy, 35, 111–119 (2010).

10. Kumar, D., & Pant, K.K.: Biorefinery solid cake waste to biocrude via hydrothermal treatment: optimization of process parameters using statistical approach. Biomass Conversion and Biorefinery, 6(1), 79–85 (2016).

11. Lotfy, W.A., Ghanem, K.M., & El-Helow, E.R.: Citric acid production by a novel Aspergillus niger isolate: II. Optimization of process parameters through statistical experimental designs. Bioresource Technology, 98, 3470–3477 (2007).

12. Chauhan, G., Pant, K.K., & Nigam, K.D.P.: Development of green technology for extraction of nickel from spent catalyst and its optimization using response surface methodology. Green Process Synthesis, 2, 259–271 (2013).

13. Xue, J., Cui, Q., Bai, Y., Wu, Y., Gao, Y., Li, L., & Qiao, N.: Optimization of adsorption conditions for the removal of petroleum compounds from marine environment using modified activated carbon fiber by response surface methodology. Environmental Progress & Sustainable Energy, 35 (5), 1400–1406 (2016).

14. Singh, B., Birla, A., Upadhyay, S. N., Yaakob, Z., & Sharma, Y.C.: Synthesis of biodiesel using potassium fluoride (KF) supported by hydrotalcite and process optimization by Box–Behnken design. Biomass Conversion and Biorefinery, 2, 317–325 (2012).

15. Padhi, S.K., & Singh, R.K.: Non-edible oils as the potential source for the production of biodiesel in India: A review. J. Chem. Pharm. Res. 3(2), 39–49 (2011).

16. Nzikou, J.M., Matos, L., Mbemba, F., Ndangui, C.B., Pambou-Tobi, N.P.G., Kimbonguila, A., Silou, Th., Linder M., & Desobry, S.: Characteristics and composition of Jatropha curcas oils, variety Congo-Brazzaville. Research Journal of Applied Sciences, Engineering and Technology, 1(3), 154–159 (2009).

17. Karaj, Shkelqim, Huaitalla, Roxana Mendozaa, Müller, Joachima, Conference on International Agricultural Research for Development (2008).

18. Masel, Richard I. (2001) Chemical Kinetics and Catalysis. Wiley-Interscience, New York. ISBN 0-471-24197-0.

19. .Laidler, Keith J., & Meiser, John H. (1982). Physical Chemistry. Benjamin/ Cummings. pp. 424–425. ISBN 0-8053-5682-7.

20. Adjaye, J. D., & Bakhshi, N. N.: Catalytic conversion of a biomass-derived oil to fuels and chemicals 2: model compound studies and reaction pathways, Biomass and Bioenergy, 8(4), 26S277 (1995).

21. Biofuels – Second Generation Biofuels". biofuel.org.uk. Retrieved 18 January 2018.

22. Dinh, L. T. T., Guo, Y., & Mannan, M. S. (2009). Sustainability evaluation of biodiesel production using multicriteria decision-making. Environmental Progress & Sustainable Energy, 28, 38–46. doi:10.1002/ep.10335.

23. Riekert L. & Zhou, J-q: Kinetics of cracking of n-decane and n-hexane on zeolites H-ZSM-5 and HY in the temperature range 500 to 780 K. Journal of Catalysis, 137, 437–452 (1992).

24. Kubo, K., Iida, H., Namba, S., & Igarashi, A.: Selective formation of light olefin by n-heptane cracking over HZSM-5 at high temperatures. Microporous and Mesoporous Materials, 149, 126–133 (2012).

25. John D. Adjaye, Sai P.R. Katikaneni, Narendra N. Bakhshi: Catalytic conversion of a biofuel to hydrocarbons: effect of mixtures of HZSM-5 and silica-alumina catalysts on product distribution. Fuel Processing Technology, 48, 115–143 (1996).

.

8 Production of Hydrogen from Waste Biomass

Saptarshi Das[1] Satarupa Pattanayak[2] and Sumit Kumar Jana[1]

[1]Department of Chemical Engineering, Birla Institute of Technology, Mesra, Ranchi, India

[2]Department of Chemistry, Birla Institute of Technology, Mesra, Ranchi, India

CONTENTS

8.1 INTRODUCTION

Hydrogen is emerging as an interesting energy production technology. It can be a promising alternative method for energy production. Hydrogen does not exist freely in nature, rather it is available in the secondary form; thus, it has to be produced. Hydrogen is a clean fuel. It also has a very high energy yield which is 2.75 times higher than hydrocarbon fuels (Kapdan and Kargi, 2006). Hydrogen is becoming significant due to its environmental friendliness and effective burning characteristics. It can be used to generate electricity in fuel cells. The use of hydrogen as a transportation and stationary fuel is attracting a lot of positive attention.

DOI: 10.1201/9781003196358-8

Because of its wide range of applications, hydrogen is increasingly demanding over other conventional fuels (Arregi et al., 2018; Shahbaz et al., 2017). Hydrogen is also a key component in various chemicals such as ethanol, urea, ammonia, etc.

Hydrogen is manufactured from fossil as well as non-fossil fuels, with carbonaceous raw materials accounting for ninety-six percent of total hydrogen production (Bakenne et al., 2016). When hydrogen is being produced from fossil fuels, it leads to carbon dioxide formation. Thus, it contributes to the increase in global warming. So, hydrogen from biomass becomes important. In the production of hydrogen from biomass, carbon dioxide, which is formed during the gasification process (a method used to produce hydrogen from biomass), is the carbon dioxide that has been taken from the atmosphere during the photosynthesis process. In this way, the net amount of carbon dioxide release is reduced.

8.2 BIOMASS

Biomass is referred to as lignocellulose. Lignocellulose's chemical components can be classified into four categories, i.e., cellulose, hemicelluloses, lignin, and extractives (Yaman, 2004). Cellulose, hemicelluloses, and lignin, on average, have large molecular weights and contribute a lot of mass, while extractives have a small molecular size and are less available (Demirbas, 2009). The hardwoods have higher cellulose hemicellulose content (78.8 percent) than the softwoods (70.3 percent). On the other hand, softwoods have higher lignin content (29.2 percent) than hardwoods (21.7 percent) (Balat, 2009). The chemical structure and components of biomass are an important factor in the production (Balat, 2007). Sources of lignocellulosic biomass generally include agri-food by-products, residues from agriculture, forest by-products, etc. Biomass has increased in popularity as a result of its abundant availability and low cost.

There are several advantages in the production of hydrogen from biomass. Firstly, since biomass is readily available in our country so, there would be no dependence on oil imports. Secondly, the net product is remaining in the country. Thirdly, there could be a 30 percent increase in CO_2 balance. Moreover, it has stable pricing. Much research is currently being done on biomass-based energy that is both renewable and environmentally friendly with the aim of replacing traditional fossil fuels. Although there are many advantages, this process is in desperate need of improvement.

Biomass will become more important as a greater emphasis is placed on renewable energy sources and conservation. Biomass having the most resources potential can meet the energy needs to ensure that there will be a sufficient supply of petrol in the future.

8.3 BIOMASS FEEDSTOCKS

To produce hydrogen, the biomass feedstocks can be energy crops, agriculture residues, animal wastes, waste from forestry, industrial and urban wastes, etc. Energy crops may include woody energy crops, farm crops, etc. Some examples of biomass feedstock species corresponding to the type of hydrogen production method are as follows: For pyrolysis, olive husk, crop straw, tea waste are used, for steam gasification bio-nut shell, black liquor is used, for hydrogen production by the microbial fermentation process, waste from pulp and paper and slurry manure are used (Balat, 2008).

8.4 BIOMASS-BASED HYDROGEN PROCESSING METHODS

The two broad classifications of biomass to hydrogen production processes are:

i. thermochemical methods and
ii. biological methods

Thermochemical methods can be as follows:

i. Combustion
ii. Liquefaction
iii. Pyrolysis
iv. Gasification

The biological methods can be as follows:

i. Direct biophotolysis
ii. Indirect biophotolysis
iii. Photo fermentation
iv. Dark fermentation
v. Biological water gas shift reaction

8.4.1 THERMOCHEMICAL METHODS

Thermochemical methods are widely regarded as a cost-effective way of produc- ing hydrogen from biomass. Combustion is a method where the chemical energy of biomass is converted into heat. Here, the biomass is burnt in the presence of air. This might take place in equipment such as a boiler where the steam produced may be further used for conversion into mechanical energy. But due to the poor energy efficiency and pollutant emissions (such as CO_2, NOx, SOx), combustion is not the preferred process for producing hydrogen. Biomass liquefaction is another thermochemical method that produces hydrogen from biomass. In this liquefac- tion, the biomass is heated in the absence of air, at a temperature of 525–600K in water at 5–20 MPa pressure. The process may include the addition of a solvent or a catalyst. Direct liquefaction and hydrothermal liquefaction are the two types of liquefaction processes for biomass. Biomass liquefaction has the disadvantages of being difficult to set up the severe operating conditions and producing less amount of hydrogen. Thus, for producing hydrogen from biomass, pyrolysis and gasifica- tion are the prominent thermochemical methods that are mostly used.

8.4.1.1 Biomass Pyrolysis

Pyrolysis is the process of thermally decomposing biomass in the absence of air to produce charcoal (solid), bio-oil (liquid), and fuel gas. Here, the biomass is heated at temperatures ranging from 280 to 780 degrees Celsius and at low pressure (1–5 bars).

There are two forms of pyrolysis: slow and fast (Naqvi et al., 2019). Slow pyrolysis takes place at temperatures 400–450C. Here, the rate of heating is low (1–5C per second) and it has a longer residence time (four to eight minutes). Fast pyrolysis, on the other hand, maintains a temperature between 450C and 950C. It has a faster heating rate and a shorter residence time of 100–300C per second and one to five seconds, respectively (Naqvi et al., 2015).

Gaseous, liquid, and solid phases may all contain fast pyrolysis products: (i) Gaseous products for pyrolysis are hydrogen gas, methane, carbon monoxide, carbon dioxide and other gases, depending on the organic composition of the biomass. (ii) Liquid materials are tars and oils. (iii) The products at the solid phase can be char and some inorganic materials such as ash and some alkali metals. Although the main purpose of the pyrolysis process is to produce bio-oil, flashing or rapid pyrolysis at high temperatures and a sufficient residence time can also result in the production of hydrogen (Naqvi et al., 2018). The following are the reactions that occur during biomass pyrolysis:

Pyrolysis:

$$\text{Biomass} + \text{heat} \rightarrow \text{hydrocarbons} + \text{tar} + \text{char} + H_2 + CO + CH_4 \qquad (1)$$

Reforming:

$$C_x H_y + H_2O \rightarrow xCO + (x + 0.5y) \, H_2 \qquad (2)$$

Water–gas shift reaction:

$$CO + H_2O \rightarrow CO_2 + H_2 \qquad (3)$$

After pyrolysis, reforming by steam can also be conducted in a catalytic bed reformer. This process causes more production of hydrogen by reforming methane and some hydrocarbon vapours. The carbon monoxide, which is formed in pyrolysis as well as in reforming, can undergo a water gas shift reaction that causes further yield of hydrogen. In addition to the gaseous products, the liquid products can also be used to yield hydrogen. The liquid part, mainly pyrolysis oil, is generally segregated on the basis of water solubility (Arregi et al., 2018). Among the two fractions, the fraction which is soluble in water can be further proceeded to produce hydrogen whereas the water insoluble fraction can be used to make adhesives.

There are several factors on which depend the yield of hydrogen by pyrolysis process. Most significantly, parameters such as temperature, heating rate, the type of catalyst used and residence time are important to consider. Usually higher temperature and heating rate are favourable (Demirbas, 2002a). The residence time

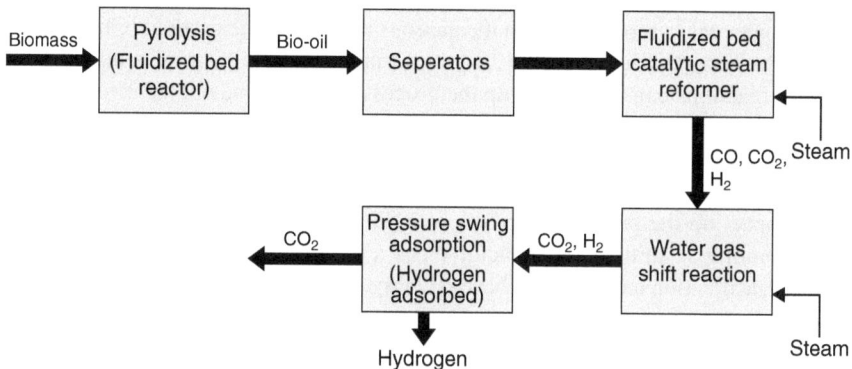

FIGURE 8.1   Hydrogen production by pyrolysis.

of the volatile phase should be long enough for high yield (Demirbas, 2002a). Regarding catalyst type, there had been many research works conducted to find the catalysts which are effective. Many metal oxides such as Al_2O_3, Cr_2O_3 and metals such Ni and Ru have been found to be suitable for using for the process. Among the catalysts, the Ni catalyst is very popular. The type of feedstock used can also play a vital role in hydrogen production. The dependence of reactor type is also an investigating area for the researchers. It has been found that the efficiency of the steam reforming process is degraded by the presence of char and coke, which unfortunately gets deposited on the catalyst surface. The fluidized catalytic beds can reduce this problem and enhance the reforming performance (Bair et al., 2002; Bair et al., 2003). The hydrogen production by pyrolysis process is shown in Figure 8.1.

8.4.1.2 Biomass Gasification

Gasification is a high-temperature thermochemical process that converts biomass into gaseous products. This process may take place in the presence or absence of catalyst (Alipour et al., 2014). It is an effective thermochemical process of biomass conversion with a usable heating/power value. This process gives a gaseous mixture of hydrogen, carbon dioxide, carbon monoxide, methane, light and heavy hydrocarbons, and char. Gasification and pyrolysis are distinguished by the fact that gasification mainly produces gaseous materials while pyrolysis significantly produces bio-oils. Gasification is a better process for producing hydrogen than pyrolysis since it produces mostly gases.

The reaction involved in gasification process is:

$$\text{Biomass} + \text{heat} + \text{steam} \rightarrow H_2 + CO + CO_2 + \\ CH_4 + \text{light and heavy hydrocarbons} + \text{char} \tag{4}$$

Methane, dimethyl ether, diesel, ethanol and methanol are a few of the fuels and chemicals that can be made from the obtained product gas (Shahbaz et al., 2016; Shahbaz et al., 2017).

Hydrogen gas is separated from the gaseous mixture in the product (Ghiat et al., 2020). The gases produced can be steam reformed to create hydrogen, and water–gas shift reactions can help speed up the process. For carrying out this process, the moisture content of biomass should not more than thirty-five percent (Demirbas, 2002b). On the basis of biomass moisture content, gasification is categorized into two types, dry biomass and wet biomass. These two biomass conditions have a major impact on the product distribution in the gaseous mixture produced; therefore, operating conditions and reactor types are highly dependent on them. Dry biomass gasification technologies have recently entered into commercialization for some applications.

A significant engineering problem in biomass gasification is the formation of tar, which can result in forming tar aerosols and polymerizing to a complex structure, both of which are undesirable for hydrogen production via steam reforming. As a consequence, it is crucial to prevent tar from developing. Three approaches are currently used to lower the formation of tar, which are, proper gasifier design, proper process control, and use of additives or catalysts. The hydrogen production by gasification process is shown in Figure 8.2.

Operating temperatures of around 1273C have been shown to increase thermal cracking of tar (Zevenhoven-Onderwater et al., 2001; Inayat et al., 2012). Gasifying agents also play a critical role in forming and decomposing tar and thus must be optimized. Catalysts and other additives have also been shown to decrease tar formation, improve process efficiency and improve product quality. Certain additives such as olivine and dolomite may help to minimize tar in the gasifier (Corella et al., 1999). It is possible to fully remove tar when dolomite is used (Sutton et al., 2001). On the other hand, catalysts lower the amount of tar as well as improve the quality of the gas product. Nickel-based catalysts and alkaline metal oxides are the catalysts used in the gasification process. Many researchers have examined the work that has resulted in a decrease in tar content using catalysts (Shahbaz et al., 2017; Abu El-Rub et al., 2004) as well as other mechanical and thermal methods

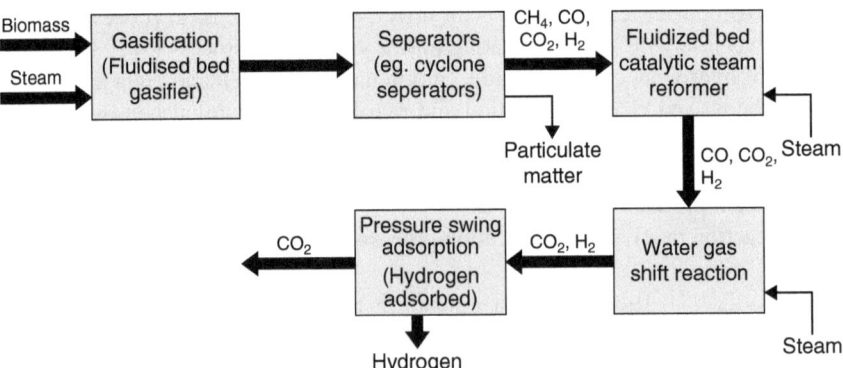

FIGURE 8.2 Hydrogen production by gasification.

(Anis and Zainal, 2011; Devi et al., 2003), both experimentally and through modelling. Tar reduction can also be accomplished by modifying the procedure, such as using two-stage gasification and injecting secondary air into the gasifier (Narvaez et al., 1997).

Another engineering problem is the ash formation which causes fouling, sintering, deposition and agglomeration (Arvelakis and Koukios, 2002; Garcia-Ibanez et al., 2004). Fractionation or leaching can be used to minimize the formation of ash within the reactor to overcome the ash-related problems. Although fractionation is efficient at removing ash, it has the potential to degrade the remaining ash's quality. In contrast, leaching removes the inorganic fraction of biomass while also enhancing the purity of the residual ash (Arvelakis and Koukios, 2002). A fluidized bed gasifier is typically used in this method. If a fluidized bed gasifier and the required catalysts are used, then approximately 60 percent hydrogen can be generated. Gasification of biomass has a high conversion value. Biomass gasification should be a promising option for hydrogen production. There can be several process modifications to optimize the gasification process, some of which are discussed below.

8.4.1.2.1 Supercritical Water Gasification (SCWG)

Many researchers have investigated the aqueous conversion of whole biomass to hydrogen at low temperatures under supercritical conditions. The first thesis on supercritical gasification of wood was published by Modell (1985). A separation technique that uses a supercritical fluid as a solvent is known as supercritical fluid extraction. Every fluid has a critical point, which is the point where it stops flowing. Water becomes a supercritical fluid when it reaches a temperature of 647.2 K and 22.1 MPa pressure. Since water is abundant, simple to obtain, non-toxic, low-cost and environmentally friendly, it has been used as a gasifying medium. It has a significant advantage in that the feed does not need to be pre-dried, saving time, energy, and money. Supercritical water gasification can handle wet biomass without any requirement of drying it first. It also shows high gasification performance at low temperature. Due to the high molar fraction of hydrogen produced by SCWG, the separation of H_2 from other gases is simplified. Very little amount solid residue or char is produced in this process. Compost fruits, vegetables, sea algae, and sewage sludge can all be used as biomass feedstock in this process. Gasafi et al. (2008) reported that sewage sludge could become a potential market for SCWG. Supercritical water gasification can be of two types (i) Non-catalytic supercritical water gasification (ii) catalytic supercritical water gasification. The temperature for non-catalytic supercritical water gasification varies from 500C to 750C. It is also known as steam-only gasification or high-temperature supercritical gasification. The temperature for catalytic supercritical water gasification varies from 350C to 600C. Low-temperature supercritical gasification is another name for it. The catalyst, in combination with a lower reaction temperature, decreases the formation of char and tar.

8.4.1.2.2 Sorbent Enhanced Gasification

This is a process modification. Here, the carbon dioxide produced is being utilized in an excellent strategy. The carbon dioxide gas is absorbed on calcium oxide and calcium carbonate is formed. This absorption process is an exothermic reaction where heat is released. This heat can be used in the gasification process, thus, saving heat energy. But, after a particular time, the calcium oxide used will not absorb any further CO_2. Then the CaO has to be regenerated by heating the $CaCO_3$. The heat required in this step can be obtained by burning the charcoal. The sorbent enhanced gasification process is shown in Figure 8.3.

8.4.1.2.3 The HyPr-RING Method

Lin et al. proposed the Hydrogen Production by Reaction Integrated Novel Gasification or HyPr-RING, which combines water hydrocarbon reaction, water–gas shift reaction, and carbon dioxide and other pollutant absorption in a single reactor under both sub-critical and supercritical water conditions (Lin et al., 2001). The HyPr-RING method produces a high yield of hydrogen at a lower temperature. This method has been found (both theoretically and experimentally) to be an effective process for producing H_2 from biomass since the reaction (involved in producing hydrogen) and the following operation of separating the gases takes place in a single reactor.

8.4.2 BIOLOGICAL METHODS

Biological methods are another way to manufacture hydrogen from biomass. The biological methods mainly are of five kinds, i.e., direct bio photolysis, indirect

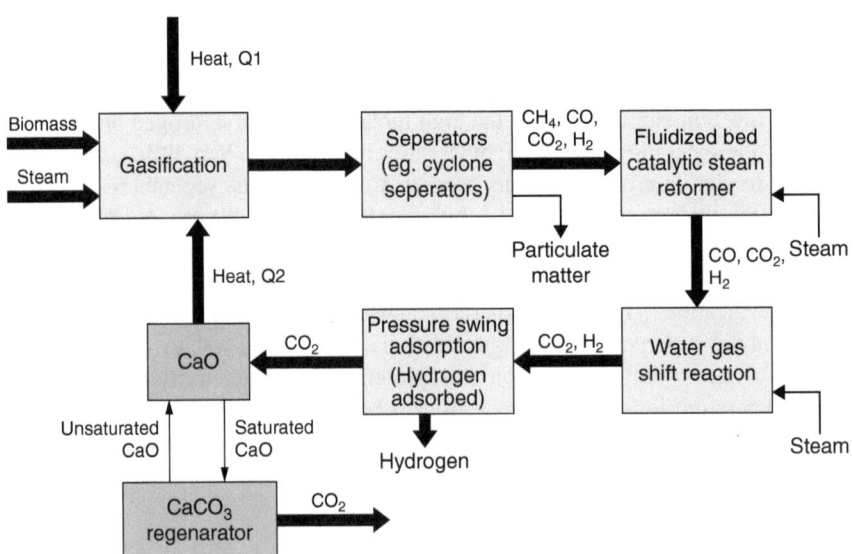

FIGURE 8.3 Sorbent enhanced gasification.

biophotolysis, photo fermentation, dark fermentation and biological water–gas shift reaction. It differs from other methods since there is no CO_2 buildup. Moreover, in these methods, various kinds of waste materials can be used to generate hydrogen. Biological methods are generally conducted at ambient temperature and pressure; thus, it is thought to be less energy-intensive than other techniques (Basak and Das, 2007). In addition, unlike chemical catalysts, which are easily deactivated during thermochemical conversions, the microorganisms which take part in this process get reproduced, therefore, reducing the turnover frequency (Henstra et al., 2007). Hence, it can be seen that biological methods have several advantages over other methods.

8.4.2.1 Direct Biophotolysis

In direct bio photolysis, the photosynthesis system in microalgae converts solar energy into chemical energy, resulting in the generation of H_2 from biomass. The direct biophotolysis process is carried out by photo-autotrophic organisms. The lack of a substrate and the fact that it is a CO_2-neutral method using sunlight are some benefits of this process (Azwar et al., 2014).

$$2H_2O + \text{Solar Energy} \rightarrow 2H_2 + O_2 \qquad (5)$$

The photosynthesis process involves photosynthetic systems of two types i.e., photosystem I (PSI) and photosystem II (PSII). The former produces a carbon dioxide reductant while the latter splits water to produce oxygen. Here, the two photons that have been generated from water may take part in the reduction of the carbon dioxide (carried out by PSI) or, in the presence of the hydrogenase, which may lead to the generation of hydrogen. Basically, photosystem II uses light energy, which results in the formation of electrons. Then these electrons, using the solar energy (which photosystem I has absorbed), get transferred to ferredoxin. Further, the electron from ferredoxin is used by hydrogenase to ultimately produce hydrogen. Due to the lack of this hydrogenase, the green plants, unlike the microalgae, are unable to produce hydrogen, whereas microalgae, like Cyanobacteria, possess hydrogenase and can give hydrogen as a result.

In direct biophotolysis, the oxygen generated shows some negative consequences, such as poisoning hydrogen and disrupting the photosynthetic mechanism, which is the process's primary flaw. For avoiding these problems, species that can absorb oxygen, such as glucose oxidase, can be utilized (Benemann, 1998). An electrochemical photoreactor for spirulina was created to address this issue, and the hydrogen rate increased fourfold. Another analysis conducted to reduce oxygen increase by using four different reducing agents (Marquez-Reyes et al., 2015; Hasnaoui et al., 2020). Several research works have been established for solving this issue, but due to some reasons such as restricted hydrogen production and poor solar energy conversion efficiency, commercial success has yet to be achieved.

8.4.2.2   Indirect Biophotolysis

This process involves the conversion of biomass into hydrogen by Cyanobacteria. The principle of indirect biophotolysis, reported by Gaudernack (1998), entails the following four stages:(i) Photosynthesis for producing biomass, (ii) biomass concentration,(iii) In the algae cell, dark aerobic fermentation produces 4 mol hydrogen/mol glucose, as well as 2 mol acetates, and (iv) 2 mol acetates are converted to hydrogen. The following are the reactions in indirect bio photolysis:

$$12H_2O + 6CO_2 \rightarrow C_6H_{12}O_6 + 6O_2 \qquad\qquad (6)$$

$$12H_2O + C_6H_{12}O_6 \rightarrow 12H_2 + 6CO_2 \qquad\qquad (7)$$

The first reaction is responsible for fixing carbon dioxide from the environment to glucose, and the second is responsible for glucose fermentation to H_2. In both darkness and sunshine, the entire mechanism results in the fixation of carbon dioxide from the atmosphere into carbohydrates, converted to H_2 by hydrogenase. In indirect bio photolysis, the hydrogen production rate depends on the light intensity, i.e., it increases with an increase in light intensity. A report by Troshina et al., where indirect biophotolysis by Cyanobacteria was studied, showed optimum hydrogen production when pH ranging between 6.8 and 8.3 was maintained (Troshina et al., 2002). The hydrogen yield gets doubled when the temperature is increased from 30 to 40°C.

8.4.2.3   Biological Water–Gas Shift (BWGS) Reaction

Photoheterotrophic bacteria such as Rhodospirillum feedstock or Rubrivivax gelatinosus (non-sulphur, purple photosynthetic bacteria) are used in this technology to produce hydrogen. At 25C and 4 atm, they can conduct the water-gas shift reaction (Wolfrum et al., 2003). CO_2 and H_2 are the critical by-products of this process.

$$CO + H_2O \rightarrow CO_2 + H_2 \ldots\ldots (8) \; \Delta H = \tfrac{1}{4}\ 154kJ$$

On a laboratory scale, there have been few studies on the biological water–gas shift reaction. All these studies had the same goal, i.e., to find suitable microorganisms with high carbon monoxide uptake as well as find out the rate at which H_2 is produced. The BWGS is less costly than the thermochemical water-gas shift process because it needs fewer equipment items. Several researchers are interested in this hydrogen production process because of its low cost. Moreover, this reaction takes place at lower pressure and temperature. But this process has some problems, such as carbon monoxide degrades the performance of the bacteria, thus has to be limited. Another issue that has to be considered is the mass transfer of the gas to the liquid.

8.4.2.4 Photo Fermentation

Photo fermentation is a process in which hydrogen is produced from biomass using nitrogenase enzymes in the presence of sunlight. The microorganisms for photo-fermentations are purple non-sulphur (PNS) bacteria. The bacteria are anoxygenic, i.e., they absorb light energy and store the energy in the form of ATP. Here, photo-system I (PSI) is involved thus, preventing oxygen production and allowing for long-term H_2 output. The purple non-sulphur bacteria can convert organic acids to clean H_2 that can be used in PEM fuel cells (Redwood,2007). These bacteria produce hydrogen by the action of nitrogenase.

Photo fermentation is advantageous as higher concentration of hydrogen is produced. Furthermore, organic acids being the substrate, is available at ample amount such as, in wastewater of many industries. Unfortunately, there are some disadvantages of the process (Fedorov et al.,1998). Firstly, the efficiency of solar energy conversion is quite low. Secondly, for the activation of the nitrogenase enzyme, high energy is required. Thirdly, huge anaerobic photobioreactors are needed, which occupies a large area. Photo fermentation has not been scaled up to the commercial level, and many research works are being carried out on this process.

8.4.2.5 Dark Fermentation

In dark fermentation, anaerobic bacteria or microalgae produce hydrogen from carbohydrate-rich substrates under anoxic conditions and in the absence of light. These carbohydrates are degraded anaerobically by heterotrophic microorganisms. Many bacteria depend on the energy generated by organic matter synthesized by other species. Bio photolysis produces only H_2 (depending on the substrate), while dark fermentation produces a mixture of H_2 and CO_2, as well as other gases such as CH_4 or H_2S. When using glucose as the model substrate, 4 mol of H_2 per mole glucose is produced when acetic acid is the end product, and 2 moles of H_2 is produced when butyrate is the end product.

(acetate as end product)

$$C_6H_{12}O_6 + 2H_2O \rightarrow 2CH_3COOH + 2CO_2 + 4H_2 \qquad (10)$$

(butyrate as an end product)

$$C_6H_{12}O_6 + 2H_2O \rightarrow CH_3CH_2CH_2COOH + 2CO_2 + 2H_2 \qquad (11)$$

The performance of the dark fermentation can be affected by the pH value. The pH significantly impacts the activity of hydrogenase as well as the metabolic pathway of the bacteria to produce hydrogen (Khanal et al., 2004). In order to know the effect of pH, several experiments were carried out examining the initial pH. The experiment conducted on mixed inocula showed a pH range of 4.2 to 5, giving the maximal amount of hydrogen production, whereas Citrobacter CDN1 favoured a pH value of 5 (Wang and Wan, 2009; Sinha and Pandey, 2011). Thus, it can be concluded that a slightly acidic environment is favourable for this process. Another significant factor affecting hydrogen production is HRT (hydraulic retention time). A report by Ueno et al. (1996) on hydrogen production by an anaerobic microflora showed that, as the hydraulic retention time was increased, the rate of production of H_2 decreased, whereas the efficiency of decomposition of the carbohydrates increased. In hydrogen production, the partial pressure of hydrogen gas is also significant (Niel et al., 2003). The fermentative biological process is also influenced by nitrogen, phosphate, metal ions, and magnesium. Since dark fermentation does not require solar energy, the weather conditions cannot affect the process. Also, this process does not require huge land. Due to these advantages, dark fermentation is gaining much importance.

8.4.2.5.1 Hybrid Systems

It can be noted that the moles of hydrogen produced from the dark fermentation are maximally four moles per mole of glucose. In addition to this limitation, there is also a thermodynamic limitation. The complete oxidation of glucose is not possible since the Gibb's energy value turns out to be positive. In order to surpass this limit, a two-stage process can be set up. This two-step process is known as a hybrid system. In this case, a hybrid system comprising dark fermentation followed by photo fermentation can be established. The organic acids produced in the dark fermentation can be used as feed for the photo fermentation process. Thus, the following reactions can take place in this system.

$$C_6H_{12}O_6 + 2H_2O \rightarrow 2CH_3COOH + 2CO_2 + 4H_2 \ldots\ldots\ldots \text{(dark fermentation)}$$

$$2CH_3COOH + 4H_2O \rightarrow CO_2 + 8H_2 \ldots\ldots\ldots \text{(photo fermentation)}$$

Considering the above reactions, we can see that in totality, 12 moles of hydrogen are produced per mole glucose. Therefore, a higher amount of hydrogen is obtained.

There had been many strategies adopted for the choice of the microorganisms to be used for the two processes involved in this system. But the most important concern is the bridging of the two processes. Researchers have tried to establish a co-culture setup where the different microorganisms would perform simultaneously, but this approach has proven to be unfeasible due to the incompatibility of the organisms in such an arrangement.

An alternative approach to setting up sequential reactors can be followed. These sequential reactors arrangement would provide different optimal conditions for the different microorganisms. It is evident that a transportation process is required between the two reactors mainly to convey the organic acids from the first reactor conducting dark fermentation to the second carrying out photo fermentation. For carrying out the continuous process, strategies such as using continuous centrifugation and using membranes can be followed. But unfortunately, these methods are costly.

A novel method of electrodialysis was conducted by Dr. Mark Redwood (Redwood, 2007). In this method, three membranes (i.e., cation-selective membrane, anion-selective membrane and bipolar membrane) were used. This process successfully separates the organic acids and delivers them to the reactor conducting photo fermentation. There is also an extra advantage of this process, i.e., additional generation of hydrogen in the cathode. Thus, there is a high amount of hydrogen produced if a hybrid system, including electrodialysis, is used. Although the experimental results have proven this process to be successful, still some improvements are required to make this method economically viable.

8.5 SEPARATION OF HYDROGEN PRODUCED

To obtain pure hydrogen from the product mixture generated from the reactions is also an important aspect. There are various strategies followed for the separation process, such as membrane systems, adsorption, etc. Adsorption techniques such as pressure swing adsorption, temperature swing adsorption can be used. The separation of hydrogen by membranes is also evolving as a promising technology that can be used. Membrane separation technology has certain benefits such as simple in operating, low cost and environment friendly. The membranes used for this purpose can be dense metal membranes or polymeric membranes, which are chosen on the basis of selectivity and permeability, i.e., polymeric membranes are used in the case of higher permeability (but low selectivity), whereas dense metal membranes are utilized for high selectivity (but low permeability) (Shahbaz et al., 2020).

8.6 SUMMARY

Biomass, which is both sustainable and environmentally friendly, is being used to replace conventional fossil fuel-based hydrogen production. Thermochemical and biological methods are used for producing hydrogen from biomass. Thermochemical processes are mainly of four kinds, combustion, liquefaction, pyrolysis and gasification. Among these methods, the first two, i.e. combustion and liquefaction, are not viable for producing hydrogen at commercial level, whereas pyrolysis and gasification are suitable and are mostly used for hydrogen production. For more yield of hydrogen steam reforming of hydrocarbons and water–gas shift reaction had been carried out. Some issues such as tar formation and ash formation have also been encountered. The gasification processes have been further modified for better results. For example, supercritical water gasification, sorbent enhanced gasification had improved the gasification process. The HyPr-Ring method had integrated many processes, such as the water gas shift reaction, water hydrocarbon reaction as well as carbon dioxide absorption in a single reactor. Direct bio photolysis, indirect bio photolysis, photo fermentation, dark fermentation, and biological water gas shift reaction are the most common biological methods. There is an absence of CO_2 accumulation in biological methods, thus, it has drawn the attention of many researchers. In dark fermentation process, the number of moles of hydrogen produced per mole of glucose can be increased by establishing a hybrid system which comprises a two-stage process starting with dark fermentation followed by photo fermentation. But presently, most of these methods are at a pilot scale. In the hydrogen production processes, the hydrogen produced from the reactions has to be separated from the product mixture. This can be done by various techniques such as adsorption, utilization of membrane separation systems, etc.

REFERENCES

Abu El-Rub Z, Bramer EA, Brem G. Review of catalysts for tar elimination in biomass gasification processes. Ind Eng Chem Res. 43 (2004) 6911–19.

Alipour Moghadam R, Yusup S, Azlina W, Nehzati S, Tavasoli A. Investigation on syngas production via biomass conversion through the integration of pyrolysis and air steam gasification processes. Energy Convers Manag. 87 (2014) 670–5.

Anis S, Zainal ZA. Tar reduction in biomass producer gas via mechanical, catalytic and thermal methods: a review. Renew Sustain Energy Rev. 15 (2011) 2355–77.

Arregi A, Amutio M, Lopez G, Bilbao J, Olazar M. Evaluation of thermochemical routes for hydrogen production from biomass: a review. Energy Convers Manag. 165 (2018) 696–719.

Arvelakis S, Koukios EG. Physicochemical upgrading of agroresidues as feedstocks for energy production via thermochemical conversion methods. Biomass and Bioenergy 22 (2002) 331.

Azwar MY, Hussain MA, Abdul-Wahab AK. Development of biohydrogen production by photobiological, fermentation and electrochemical processes: a review. Renew Sustain Energy Rev. 31 (2014) 158–73.

Bair KAM, Czernik S, French R, Chornet E. Fluidizable catalysts for hydrogen production from biomass pyrolysis/steam reforming, FY 2003 Progress Report, National Renewable Energy Laboratory (2003).

Bair KAM, Czernik S, French R, Parent Y, Ritland M, Chornet E. Fluidizable catalysts for producing hydrogen by steam reforming biomass pyrolysis liquids, Proceedings of the 2002 U.S. DOE Hydrogen Program Review, NREL/CP-610-32405, National Renewable Energy Laboratory (2002).

Bakenne A, Nuttall W, Kazantzis N. Sankey-Diagram-based insights into the hydrogen economy of today. Int J Hydrogen Energy 41 (2016) 7744–53.

Balat M. Gasification of biomass to produce gaseous products. Energy Sources Part A 31 (2009) 516–26.

Balat M. Hydrogen in fueled systems and the Significance of hydrogen in vehicular transportation. Energy Sources Part B 2 (2007) 49–61.

Balat M. Hydrogen-rich gas production from biomass via pyrolysis and gasification processes and effects of catalyst on hydrogen yield. Energy Sources Part A 30 (2008) 552–64.

Basak N, Das D. Microbial biohydrogen production by Rhodobactersphaeroides OU 001 in photobioreactor. In: Proceedings of World Congress on Engineering and Computer Science, San Francisco, USA (2007).

Benemann JR. Process analysis and economics of biophotolysis of water. IEA Hydrogen Program Paris (1998).

Corella J., Aznar MP., Gil J., Caballero MA. Biomass gasification in fluidised bed: where to locate the dolomite to improve gasification? Energy and Fuels 13 (1999) 1122.

Demirbas A., Gaseous products from biomass by pyrolysis and gasification: effects of catalyst on hydrogen yield, Energy Conversion and Management 43 (2002a) 897.

Demirbas A., Hydrogen production from biomass by the gasification process. Energy Sources 24 (2002b) 59.

Demirbas A. Pyrolysis of biomass for fuels and chemicals. Energy Sources Part A 31 (2009) 1028–37.

Devi L, Ptasinski KJ, Janssen FJJG. A review of the primary measures for tar elimination in biomass gasification processes. Biomass Bioenergy. 24 (2003) 125–40.

Fedorov AS, Tsygankov AA, Rao KK, Hall DO. Hydrogenphotoproduction by Rhodo bactersphaeroidesimmobilised on polyurethane foam, Biotechnology Letters 20 (1998) 1007.

Garcia-Ibanez P, Cabanillas A, Sanchez JM. Gasification of leached orujillo (olive oil waste) in a pilot plant circulating fluidized bed reactor. Preliminary results, Biomass and Bioenergy 27 (2004) 183.

Gasafi E, Reinecke M-Y, Kruse A, Schebek L. Economic analysis of sewage sludge gasification in supercritical water for hydrogen production. Biomass Bioenergy 32 (2008) 1085–96.

Gaudernack B., Photoproduction of hydrogen, IEA Agreement on the Production and Utilization of Hydrogen Annual Report, IEA/H2/AR-98 (1998).

Ghiat I, AlNouss A, McKay G, Al-Ansari T. Biomass-based integrated gasification combined cycle with postcombustion CO_2 recovery by potassium carbonate: technoeconomic and environmental analysis. Comput Chem Eng (2020) 106758.

Hasnaoui S, Pauss A, Abdi N, Grib H, Mameri N. Enhancement of bio-hydrogen generation by spirulina via an electrochemical photo-bioreactor (EPBR). Int J Hydrogen Energy. 45 (2020) 6231–42.

Henstra AM., Sipma J, Rinzema A, Stams AJ. Microbiology of synthesis gas fermentation for biofuel production. Curr. Opin. Biotechnol. 18 (2007) 200–6.

Inayat A, Ahmad MM, Mutalib MIA, Yusup S. Process modeling for parametric study on oil palm empty fruit bunch steam gasification for hydrogen production. Fuel Process Technol 93 (2012) 26–34.

Kapdan IK, Kargi F. Bio-hydrogen production from waste materials. Enzym Micro Tech 38 (2006) 569–82.

Khanal, SK, et al., Biological hydrogen production: effects of pH and intermediate products. International Journal of Hydrogen Energy, 29 (2004) 1123–131.

Lin, S-Y, et al., Hydrogen Production from Hydrocarbon by Integration of Water–Carbon Reaction and Carbon Dioxide Removal (HyPr–RING Method). Energy & Fuels, 15 (2001) 339–43.

Marquez-Reyes MLA, Sanchez-Saavedra MdP, Valdez-Vazquez I. Improvement of hydrogen production by reduction of the photosynthetic oxygen in microalgae cultures of Chlamydomonas gloeopara and Scenedesmus obliquus. Int J Hydrogen Energy 40 (2015) 7291–300.

Naqvi SR, Prabhakara HM, Bramer EA, Dierkes W, Akkerman R, Brem G. A critical review on recycling of end-of-life carbon fibre/glass fibre reinforced composites waste using pyrolysis towards a circular economy. ResourConservRecycl. 136 (2018) 118–29.

Naqvi SR, Tariq R, Hameed Z, Ali I, Naqvi M, Chen W-H, et al. Pyrolysis of high ash sewage sludge: kinetics and thermodynamic analysis using Coats-Redfern method. Renew Energy 131 (2019) 854–60.

Naqvi SR, Uemura Y, Yusup S, Sugiura Y, Nishiyama N. In situ catalytic fast pyrolysis of paddy husk pyrolysis vapors over MCM-22 and ITQ-2 zeolites. J Anal Appl Pyrolysis. 114 (2015) 32–9.

Narvaez I., Corella J, Orio A. Fresh tar (from a biomass gasifier) elimination over a commercial steam-reforming catalyst. Kinetics and effect of different variables of operation. Industrial and Engineering Chemistry Research 36 (1997) 317.

Niel EWJV, Claassen PAM, Stams A.J.M. Substrate and production inhibition of hydrogen production by the extreme thermophile Caldicellulosiruptor saccharolyticus, Biotechnology and Bioengineering 81 (2003) 255.

Redwood, M.D., Bio-hydrogen production and biomass-supported palladium catalyst for energy production and waste minimization, in PhD thesis, School of Biosciences, University of Birmingham, UK (2007).

Shahbaz M, Yusup S, Inayat A, Ammar M, Patrick DO, Pratama A, et al. Syngas production from steam gasification of palm kernel shell with subsequent CO_2 capture using CaO sorbent: an aspen plus modeling. Energy Fuel. 31 (2017) 12350–7.

Shahbaz M, Yusup S, Inayat A, Patrick D, Partama A. System analysis of poly-generation of SNG, power and district heating from biomass gasification system. Chem Eng Trans. 52 (2016) 781–6.

Sinha, P, Pandey A., An evaluative report and challenges for fermentative biohydrogen production. International Journal of Hydrogen Energy, 36 (2011) 7460–78.

Sutton D., Kelleher B., Ross J. Review of literature on catalysts for biomass gasification. Fuel Processing Technology 73 (2001) 155.

Troshina O, Serebryakova L, Sheremetieva M, Lindblad P. Production of H2 by the unicellular Cyanobacterium gloeocapsa alpicola CALU 743 during fermentation. International Journal of Hydrogen Energy 27 (2002) 1283.

Ueno Y, Otsuka S, Morimoto M. Hydrogen production from industrial wastewater by anaerobic microflora in chemostat culture. Journal of Fermentation and Bioengineering 82 (1996) 194.

Wang, J., Wan W. Factors influencing fermentative hydrogen production: A review. International Journal of Hydrogen Energy, 34 (2009) 799–811.

Wolfrum EJ, Vanzin G, Huang J, Watt AS, Smolinski S, Maness P-C. Biological water gas shift development. DOE hydrogen, fuel cell, and infrastructure technologies program review (2003).

Yaman S. Pyrolysis of biomass to produce fuels and chemical feedstocks. Energy Convers Manage. 45 (2004) 651–71.

Zevenhoven-Onderwater M, Backman R, Skrifvars B-J, Hupa M. The ash chemistry in fluidised bed gasification of biomass fuels. Part I: predicting the chemistry of melting ashes and ash-bedmaterial interaction. Fuel. 80 (2001) 1489–502.

9 Microbial Mediated Waste Management and Bioenergy Production

Janeeshma E.[1], Gurudatta Singh[2] and Jos T. Puthur[1]

[1] Plant Physiology and Biochemistry Division, Department of Botany, University of Calicut, C.U. Campus P.O., Kerala – 673635, India

[2] Institute of Environment and Sustainable Development, Banaras Hindu University, Varanasi, India – 221005

CONTENTS

DOI: 10.1201/9781003196358-9

9.1 INTRODUCTION

An alternative energy source for the fossil fuels is a demand of this hour and one of the best options is the utilization of renewable energy. There are different sources of renewable energy such as sunlight, wind, water, geothermal heat and bioenergy. Of the different forms of renewable energy, bioenergy considered as the most sustainable form due to the availability of substrate, i.e. waste. Wastes management is another crisis of this decade due to the uncontrolled anthropogenic activities. So it is very important to find a sustainable paradigm for waste management and fuel production (Kwok 2019). Microbes have a potential role in biomass production and the transformation of biomass to bioenergy, and thus microbial mediated biocatalytic reactions are utilized in waste management and bioenergy production (Ge et al. 2016). Different microbes showed tolerance towards extreme conditions with an ability to utilize hazardous materials as the substrate or carbon source. Bacteria, fungi, and algae showed the potential of bioenergy production and these organisms are also capable to utilize the energy in the waste materials (Timmers et al. 2010). The enzymes produced by the bacteria, fungus and algae are helpful in imparting xenobiotic tolerance and its transformation to useful products.

Different wastes such as municipal solid waste, wastewater from landfills, and lignocellulosic biomass are utilized by the microbes as their source of energy (Pagliano et al. 2017). However, the microbial species richness and diversity varies with the medium or the substrate which they dependent for the nutrition, where the optimal growth in different xenobiotics determining the tolerance level of these organisms. Therefore, the tolerance level of different microbes towards different wastes is varying. It is essential to know the sensitivity of different microbes towards different xenobiotic substrates and this article have explored this phenomenon too.

The microbial fuel cell (MFC) is a recent technology, which utilizes microbes to release electrons and protons from the organic or inorganic substrates for the production of bioelectricity. MFCs are highly exploited for wastewater treatment associated with electricity production (Ieropoulos et al. 2005). Thus, it is important to improve the efficiency of MFCs by incorporating microbes with high degradation and biotransformation potential.

Microbial mediated biomass production, bioconversion of inorganic xenobiotics and degradation of organic wastes have high impact in bioenergy production. The quality of the waste material used as the substrate also have to be evaluated and selected appropriately. This work intends to find efficient microbial candidates and different waste materials with high bioenergy production capacity.

9.1.1 MICROBIAL MEDIATED BIOMASS PRODUCTION

Manufacturing of bioenergy via ecological source have been deliberated comprehensively because of the lessening remnant energy assets. Carbon-based trashes remain superlative as well as reasonable substrate intended for the bacterial oil manufacture through oleaginous microorganisms. Micro-algae remain considered by means of utmost encouraging feed stock intended for biofuel manufacture because of the benefits of the advanced oil contents, advanced level of photo-synthesis, no unswerving struggle for the agrarian lands as well as relaxed farming. Usage of the carbon-based trashes as the substrate for micro-algae farming for the production of microbial oil was studied (Sibbi 2018). Four categories of the microorganisms were utilized for the production biomass: microbes, yeast, molds as well as algae's. The selection of the microorganism be contingent on several standards, the utmost significant of this is the nature of the raw material accessible.

The biomass stockpiled energies remain consumed for the synthesis of the renewable electrical or high temperature energies. Ignition or the gasification progressions intended for desiccated biomass or biogas (methane) which is underneath the control of anaerobic microbial hydrolysis can be utilized for the production of bio-power. Undeviating heating (at high temperature and pressure in presence of oxygen) of the lesser as well as compacted masses of the woods, residues of the woods after trimming, as well as additional desiccated biomass quantities were get exploited intended for the construction of the thermal powers intended for the heating as well as conserving purpose at individuals at the trading scales (Srivastava 2019).

9.1.2 Micro-algae Biomass

Micro-algae are the single celled plants biomass through varied natural surroundings, displaying the numerous probables intended for the improvement of the liquescent transference fuels. Micro-algal organisms be able to cultivate into the freshwater as well as inundated brackish (or both) through proficiently use of carbon di-oxide from the atmosphere by additional involvement into the worldwide carbon fascination (up to 40%). These biomass of the algae have capacity to produce very quickly (doublings periods 6–24 h) dependence of genre or class by natural capability for producing energy rich oil into its entire dehydrated cell biomass. The achievement of the bio-diesel productions, this is unique types of biofuel, on or after the micro-algae depending upon the contented of TAG (triglyceride), this has been composed of more than that of 70% of the phosphatide contents (Bigogno et al. 2002) as well as the biomass efficiency. Diverse species algae produces diverse varieties of the bioenergy have been deliberated in Table 9.1. Algae's has been reconnoitered intended for its exclusive potentials to yield the varieties of the biofuels concurrently by the generations of the value added products as well as phyco-remediation of wastewater has been obtainable in Figure 9.1.

9.2 CONTRIBUTIONS OF MICROBES IN WASTE MANAGEMENT

Wastes are generally classified as biodegradable and non-biodegradable, and its excess accumulation in the ecosystem is a growing concern of human population. Urban or municipal wastes, industrial wastes, commercial wastes, electronic wastes, and agricultural wastes are the common types of wastes, which may be

TABLE 9.1
Different Algal Species and Different Types of Bioenergy

Bioenergy	Role of microbial system	Sources of bioenergy	References
Algal biodiesel/ bioethanol	Biomass of *S.* obliquus. contains almost 489 mg/l lipids and 99.8% theoretical yield for ethanol	Biomass of *Scenedesmus obliquus* contains about 50.39% w/w of total carbohydrates	Ho et al. (2013a)
Bio-butanol and biohydrogen	Sequential alkali pretreatment and acid hydrolysis is used for production of butanol titer (13.1g/L) and productivity (90.58 mol/ mol sugar). Hydrolysate from pretreated with NaOH (1%) followed by H2SO4 (3%) without inhibitors for fermentation. Hydrogen has reported the yield (0.39 mol/ mol sugar) and productivity (104.2 g.L^{-1} h^{-1})	*Chlorella vulgaris* JSC-6 Biomass	Knápek et al. (2015)
Hydrogen, ethanol / lipid production for biodiesel	*C. reinhardtii* biomass (total carbohydrate of 59.7% w/w) for carbon-based fuels and chemical precursors (alkane, ethylene), as well as gas (hydrogen) or (lipid-based) triacylglycerols (TAGs) for biodiesel application. It has 0.24g ethanol g^{-1} DCW and liquefaction of algal biomass by α-amylase and sacharification by amyloglucosidase with Saccharomyces cerevisiae S288C.	Metabolic pathways involved in biofuel production from Chlamydomonas reinhardtii	Dubini (2011); Choi et al. (2010)
High quality biodiesel production from algae	From biomass of microalga Scenedesmus abundans, lipid extraction as well as transesterification process are used for fatty acids with carbon chain length of C16 and C18 containing biodiesel	CHU-13 medium has reported highest biomass (1.13 g.L^{-1}) and lipid yield (0.49 g.L^{-1})	Ho et al. (2013b)

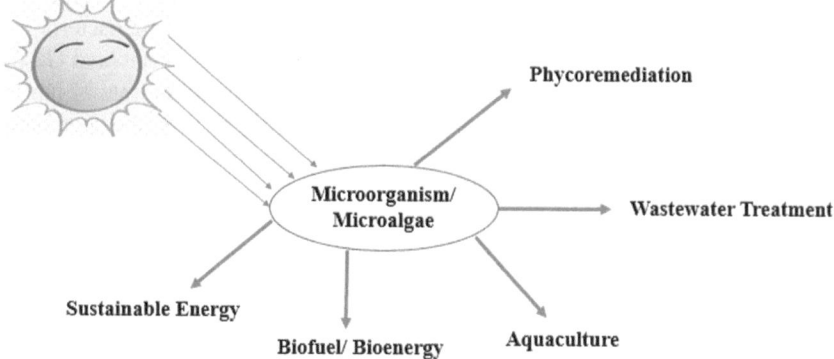

FIGURE 9.1 Production of bioenergy and different uses of microorganism/microalgae.

hazardous to the life and health of man. Recently, different approaches were experimented to utilize microbes in waste degradation and transformation. The role of microbes in the recycling of sewage sludge (Shalaby 2011), composting and decomposing of solid waste (Rastogi et al. 2020), recycling of electronic wastes (Kwok 2019) was elaborated by many researchers. These studies showed that the action of different microbes on different types of wastes are varying. Improvement in the biodegradation of organic waste and conversion of inorganic wastes to its inactive form using microbes is a sustainable strategy in waste management and it is important to understand the interaction between different microbes and waste materials to improve the degradation process.

9.2.1 Microbes in Organic Waste Management

Different microbes such as bacteria, fungi, and algae are involved in waste management. Of these, bacterial mediated waste management is a relevant strategy used to clean the contaminated environment. Furthermore the organic waste decomposition yields the vermicompost with the aid of bacteria and earthworms, which act as a nutrient rich substrate for the growth of the plants as well as other microbes (Pathma and Sakthivel 2012). The augmentation in the hydrolysis and acidogenesis processes of municipal solid wastes using exogenous aerobic bacteria mixture was pointing out the importance of the microorganisms in the waste management (Ge et al. 2016). Some of the mesophilic bacteria *Escherichia, Klebsiella, Aeromonas*, and *Alcaligenes, Enterococcus* and *Bacillus* contribute significantly to the initial stages of composting process, whereas the thermotolerants *B. stearothermophilus, B. acidocaldarius*, and *B. schleglii* were involved in the later stages of composting process and this may be due to the substrate selectivity of microbial systems (Ghazifard et al. 2001). *B. cereus* PCM 2849 is a keratinolytic bacterium and it is utilized in the composting of pig bristles and it accelerated the mineralization of the same (Choińska-Pulit et al. 2019). Heavy metals, ammonia, and dissolved organic content are the major toxic compounds leached from the landfills which always increase the chances of ground water contamination and luminescent bacteria were identified as an indicator for the presence of the contaminants (Pivato and Gaspari 2006). Treatment of chicken manure with bacteria augmented the rate of biodegradation of the waste materials (Mazza et al. 2020). Acceleration in the degradation of lignin was achieved with the inoculation of *Pseudomonas*, *Phenylobacterium*, and *Caulobacter* which are potentially producing laccase enzymes (Jiang et al. 2020). *Bacillus* sp. NKSP-7 showed improved keratinolytic activity which was utilized to increase the degradation of poultry feathers waste (Haq et al. 2020).

Fungus also plays a key role in the natural decomposition of the wastes and this was examined using different types of white-rot fungi (*Phanerochaete chrysosporium* Burds., *Trametes versicolor* and *Fomes fomentarius*), and a consortium of these fungi (Voběrková et al. 2017). Application of theses microbes enhanced the rate of degradation of municipal solid waste by accelerating the activity of dehydrogenase and protease (Voběrková et al. 2017). When municipal landfill leachate was treated with *Dichomitus squalens,* it improved the degradation process and further enhancement in degradation was observed with application of microbial originated extracellular ligninolytic enzymes (Kalčíková et al. 2014). The food waste hydrolysis via solid state fermentation (SSF) using *Aspergillus awamori* Nakaz. was found o improve the waste biotransformation as well as fungal enzyme production and thus this technique was exploited for the production of glucoamylase enzyme (Lam et al. 2013). Some of the basidiomycetes significantly contributes to the degradation of persistent organic pollutants by the biosynthesis of different enzymes (Kües 2015). The potential of *Aspergillus flavus* Link. and *Aspergillus fumigatus* Fresenius., *Penicillium roqueforti* Thom., *Rhizomucor pusillus* (Lindt) Schipper, *Thermoascus crustaceus* Miehe and *Thermomyces lanuginosus* Tsiklinsky in organic waste digestion was evaluated (Awasthi et al. 2014). Of these *Talaromyces* and *Paecilomyces* species was more tolerant to the mesophilic anaerobic digestion (Schnürer and Schnürer, 2006). The fungal consortium containing *T. viride*, *A. niger* and *A. flavus* improved the degradation of organic fraction of municipal solid waste and aid in the early maturity of compost (Awasthi et al. 2014). *Aspergillus awamori* Nakaz. and *Aspergillus oryzae* (Ahlburg) E. Cohn were used in food waste treatment process and the fungal hydrolysis process helped to degraded 80–90% of waste within 36–48 h, converting to glucose, free amino nitrogen and phosphate (Pleissner et al. 2014). Another study examined the possibility of using textile waste as feedstock for cellulase production through solid state fermentation. *Aspergillus niger* CKB with 0.43 ± 0.01 FPU g^{-1} of cellulase activity was selected for this process and it potentially involved in the hydrolysis of another important category of waste, i.e. food waste (Hu et al. 2018).

Photosynthetic green algae were also found to contribute towards the waste management and *Chlorella* sp. was used for the municipal wastewater treatment (Wang et al. 2010). These organisms improved the degradation of the waste material and at the same time enhanced its growth rate (Wang et al. 2010). Algae also helps to degrade lipid-rich fat, oil, and grease waste and it indicate the potential of these organism to utilize different waste material as their substrates (Park and Li 2012).

9.2.2 Microbes in Inorganic Waste Management

In this scenario, bacterial assisted phytoremediation of inorganic xenobiotics such as metals, metalloids and radionuclides has achieved considerable attention due to the persistent nature of the xenobiotics. Heavy metals, such as zinc (Zn), copper (Cu), chromium (Cr), mercury (Hg), and lead (Pb) are increasing in the environment which are potentially toxic to human life. Degradation, immobilization, mobilization, volatilization, and accumulation of inorganic pollutants are the different strategies of microbes and the associated host plants. The plant-bacteria interaction improved the bioaccumulation potential of plants as well as supported the establishment of the plant in a contaminated environment (Ahmad et al. 2017).

Different researchers reported the potential of the microbes in the recovery of heavy metals from electronic wastes. A mesophillic bacteria *Acidithiobacillus thiooxidans* was used to prevent the contamination of copper that leached from the electronic wastes by altering the bioleaching properties of the copper (Hong and Valix 2014). The enhancement in the acidification of soil with the aid of sulphur-oxidising bacteria and a mixed culture of biosurfactant-producing bacteria and sulphur-oxidising bacteria helps to improve the bioleaching properties of zinc, copper, lead, nickel, cadmium and chromium (Karwowska et al. 2014). The bacteria *Geobacter metallireducens* improved the degradation rate of toluene with the co-dissolution of arsenic (Lee et al. 2012). Similarly, Bacillus sp. isolated from *Hymeniacidon heliophila* Parker sponge cells exhibited bioleaching properties and thus could be exploited in the recovery of copper from electronic wastes (Rozas et al. 2017). The elimination of polystyrene, acrylonitrile butadiene styrene, and polypropylene from the bioleaching of Cu and Ni using *Acidithiobacillus ferrooxidans* was practiced in e-waste contaminated sites (Arshadi et al. 2019). Presently contamination of the environment with microplastic is increasing and the effect of bacterial community on the microplastic degradation was analysed at Guiyu, an e-waste dismantling area in Guangdong Province, China. Further the study proved that microorganisms played a role in the degradation of microplastic in e-waste zone (Chai et al. 2020). In the same place, the changes in the bacterial species composition and species richness was reported and the presence of *Clostridium* sp. *Massilia* sp., β-*proteobacteria* and *Firmicutes* was observed (Zhang et al. 2010). The potential of *Escherichia coli, Bacillus,* P*seudomonas, Flavobacterium* and *Alcaligenes* to tolerate different metal ions like Zn, Cu, Cr, Hg, and Pb was proved in the samples collected from Common Effluent Treatment Plant (CETP), Pallavaram, Chennai (Selvi et al. 2012).

Different fungal members are also contributing towards the degradation or the transformation of inorganic xenobiotics. More than the degradation, the reuse of metals can also be achieved by the utilization of microbes. It was proved that the precious metals from printed circuit boards could be mobilized through *Aspergillus niger* DDNS1 mediated bioleaching and this specific fungus was isolated from the sludge samples collected from the Premier Mills Textile Pvt, Ltd., Hosur, Krishnagiri District, Tamil Nadu, India (Narayanasamy et al. 2018). Similar to this, *A. niger* was used to leach out copper from a polymer contaminated site and this organism potentially aid to degrade the plastic wastes also (Kapoor and Viraraghavan, 1998). In an advanced study, the importance of the immobilization of fungal biomass in a solid polymer matrix was explained and it helps for the easy separation of fungal biomass from the treated water. When *A. niger* biomass was immobilized in a polysulfone matrix, the metal absorption was increased to 3.60, 2.89, 10.05, and 1.08 mg/g per unit weight of beads for cadmium, copper, lead and nickel respectively (Kapoor and Viraraghavan 1998).

More than this, algae also helps to transform the inorganic pollutants like plastics and heavy metals. The potential of microalgae to use plastic wastes as the source of carbon was exploited in the waste management (Chia et al. 2020). The ability of different species such as *Chlorella kessleri, Chlorella ellipsoidea, Chlorella vulgaris* Beijerinck, *Scenedesmus bijuga, Scenedesmus bijugatus, Scenedesmus obliquus, Anabaena laxa, Anabaena subcylindrica, Nostoc muscoum, Oscillatoria angusta* and *Nitzschia perminuta* in the degradation of monoazo and diazo dyes was investigated. Finally, it was found that *Scenedesmus bijugatus* and *Nostoc muscoum* showed maximum decolorization and degradation of these dyes (Omar 2008). A brown alga named *Ecklonia radiata* helps in the degradation of organic arsenic to inorganic arsenic. The degradation of high density and low density polyethylene was boosted with the help of indigenous microalgae native to Malaysia and furthermore this organism helps to transform plastic into CO_2, H_2O, and minerals (Bhuyar et al. 2018). *Chlorella* sp. showed the potential to clear Al, Ca, Fe, Mg, and Mn from the contaminated water (Wang et al. 2010).

9.2.2.1 Mechanisms in Microbial Mediated Waste Management

The major mechanism in the algal autotrophs for the utilization of waste material is photosynthesis, where algae fix solar energy in the form of chemical energy using carbon from the waste material (Walker et al. 1990). The nitrogen enriched waste is utilizing as a source of inorganic nitrogen for the biosynthesis of organic nitrogen. This nitrogen assimilation aid to clear waste water enriched with nitrogen (Li et al. 2005).

9.3   MICROBIAL MEDIATED BIOENERGY PRODUCTION

Three broad categories of bioenergy are available: solid biomass (e.g., wood, harvesting residues), liquid biofuels (e.g., bioethanol, biodiesel), and gaseous biofuels (e.g., biogas). Recently the application of biofuel is increasing and anaerobic digestion is emerging as a technique to meet the global energy need by

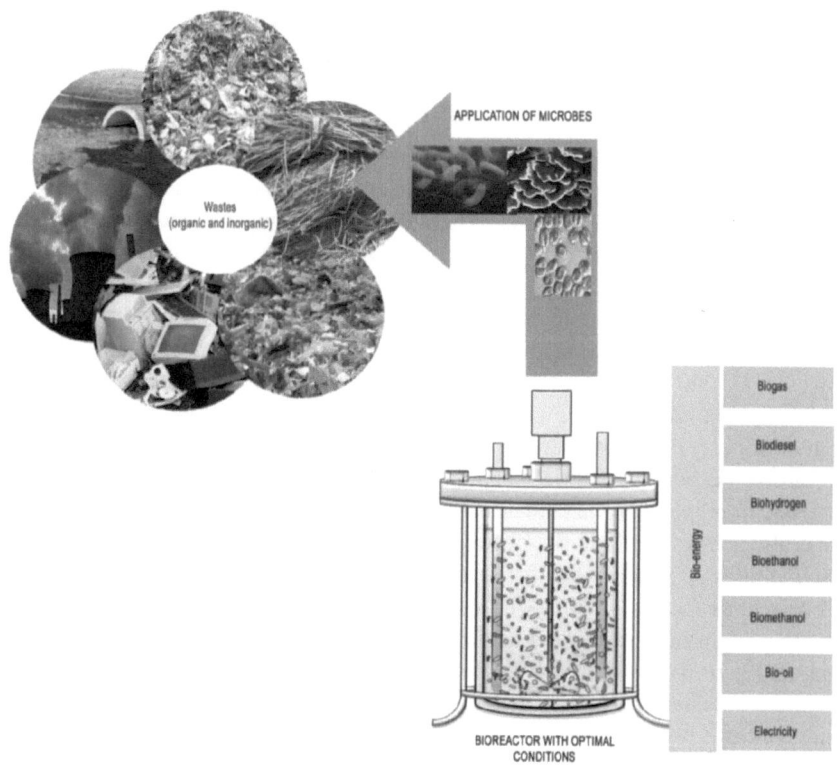

FIGURE 9.2   A general overview of microbial mediated bioenergy production.

the production of bioenergy where the microbial system helps the synthesis of biomass as well as aid in the degradation of wastes (Timmers et al. 2010; Cerrillo et al. 2016). Utilization of different organic and inorganic substrates for the metabolic processes of microorganism helps to generate bioenergy. The desirable substrate selection and the efficiency of different microbial candidate have to be evaluated before the execution of the bioenergy production process. The microbial cell metabolisms transforms the different substrates into different products such as ethanol, butanol, methanol, alkanes, biohydrogen, and bioelecricity. Bioelectrochemical cells (BEC) was the technology used to produce bioenergy from organic biomass and wastewaters. The extensive utilization of microbial fuel cells (MFCs) and microbial electrolysis cells (MECs) for the generation of bioelectricity and biohydrogen was reported in recent studies (Gajda et al. 2018). Moreover, utilization of different waste material as the substrate in the bioenergy production protect the food production sector and facilitate an alternative for fossil fuels (Timmers et al. 2010). A general overview of microbial mediated bioenergy production is illustrated in Figure 9.2.

9.4 CONTRIBUTIONS OF MICROBES TO WASTE MANAGEMENT AND BIOENERGY PRODUCTION

9.4.1 Role of Bacteria in Bioenergy Production

Bacteria are utilized for the production of bioenergy by degrading different substrates like food waste, municipal solid waste, agricultural waste etc. Exogenous aerobic bacteria mixture is being used to make methane which aid to generate electricity and heat (Ge et al. 2016). This technique is known as biomethanization and was produced as the result of anaerobic digestion (Voběrková et al. 2017). Lignocellulose is the most abundant waste produced in this world and the microbial cellulosome aid in the conversion of these lignocellulose materials to ethanol (Bayer et al. 2007). *Alcaligenes eutrophus, Alcaligenes latus, Azotobacter vinelandii* Lipman, *Azotobacter chroococcum* Beijerinck, *Azotobacter beijerincki*, methylotrophs, *Pseudomonas* spp., *Bacillus* spp., *Rhizobium* spp., *Nocardia* spp. are some potent bacterial candidates which can transforme different waste materials to polyhydroxyalkanoates (Pagliano et al. 2017). Bioelectricity production is also a very important perspective of microbe mediated bioenergy production. *Lysinibacillus sphaericus* can generate a maximum power density of 85 mW/m^2 and current density of $\approx$270 mA/m^2 using graphite felt as electrode and this strain can use protein rich wastewater as the substrate for electricity production (Nandy et al. 2013, p. 5). Similarly, beta-*Proteobacteria*, nitrobacteria and denitrifying bacteria associated with constructed wetlands aid in the bioelecricity production from synthetic wastewater (Wang et al. 2016). *Clostridial* species are considered as efficient producers of butanol and this species was used to convert rice straw and sugarcane bagasse to butanol. Moreover, cellulases originating from *Pseudomonas* sp. CL3 also could transform agricultural wastes into butanol (Cheng et al. 2012). The butanol production from the agricultural waste was explained in the Figure 9.3, which was illustrates based on the study of Cheng et al. (2012).

Biological hydrogen production by the metabolic process of purple non-sulfur (PNS) bacteria from different waste materials was also reported and this biohydrogen can be utilized as an efficient biofuel. For biohydrogen production the tofu (soybean product) wastewater, palm oil mill effluent, olive mill wastewater, and brewery wastewater was used as the substrate and the source of energy. Genetically transformed *G. thermoglucosidasius* divert their fermentative carbon flux that can aid in the production of ethanol (Novik et al. 2018). Dairy industry wastewater was used as the substrate for the production of hydrogen and methane by the action of *Enterobacter aerogens* and methanogenic bacteria isolated from cow dung (Kothari et al. 2017). The presence of *Archaea* was reported in the petroleum waste and this organism have a role in methanogenesis (de Rezende et al. 2020). Importance of temperature in bacterial mediated bioenergy production is a topic of interest and Kovalovszki and his coworkers (2020) developed dynamic microbial temperature-dependence function for modeling which could probably improve the biodegradation.

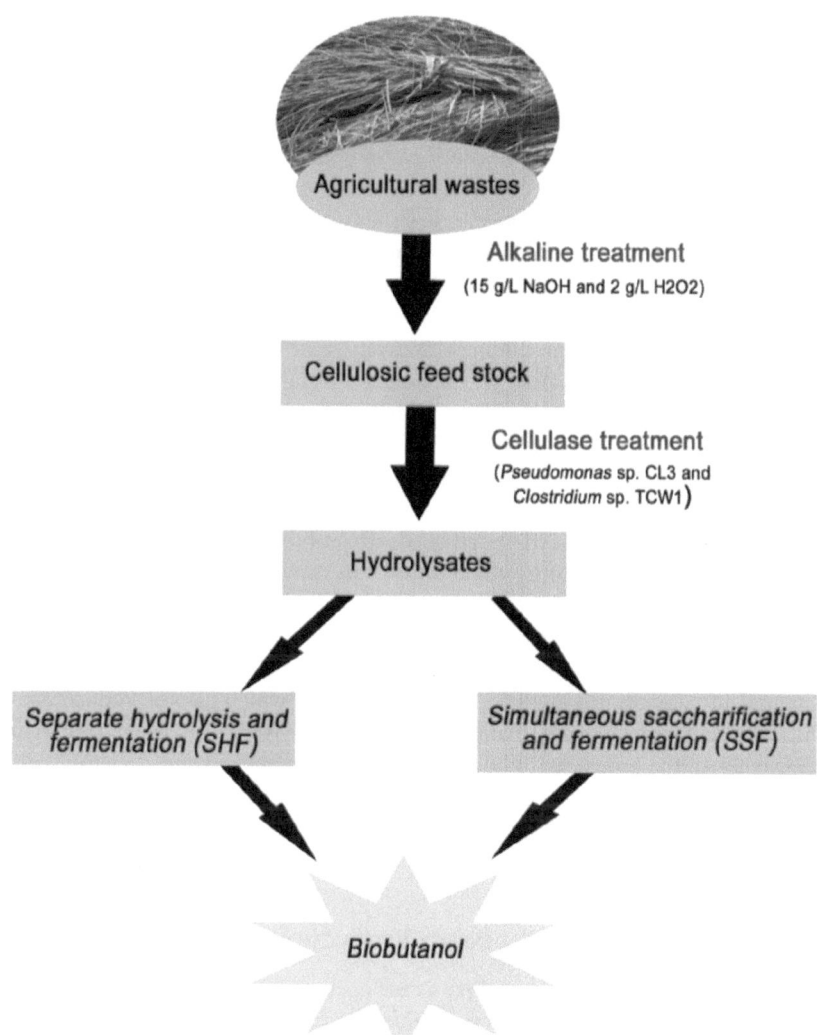

FIGURE 9.3   Microbial mediated butanol production from the agricultural wastes.

Source: Cheng et al. (2012).

9.4.2   ROLE OF FUNGUS IN BIOENERGY PRODUCTION

Different fungal strains showed the potential to transform wastes to bioenergy. Agrowastes can be hydrolysed with the help of different fungal strain, for improved degradation of the lignocellulosic waste these fungal strains have to produce different enzymes. In a recent study, *Aspergillus niger* CBS 110.42, *Phanerochaete chrysosporium* CBS 246.84, *Oidiodendron echinulatum* CBS 113.65 and

Bjerkandera adusta CBS 143380 showed high degrading activity due to the production of cellulase, laccase and peroxidase (Mbareche et al. 2018; Ming et al. 2019). In the same study, importance of the application of fungal consortium was also explained. *Co-culturing of A. niger* CBS 554.65 with *Trichoderma reesei* CBS 383.78 and *A. niger* CBS 110.42 with *P. chrysosporium* CBS 246.84 resulted in best degradation activity and this could be further utilized in the bioenergy production. Yeast is a model organism and novel researches were conducted to improve the waste degradation potential of yeast. Yeast cell surface engineering is one of the strategy to improve the oxidation of waste where different hydrolytic enzymes were genetically immobilized on yeast cell surface which aid to increase the production of ethanol from biomass (Fukuda et al. 2009). Moreover, yeast was shown to have the potential to transform waste glycerol to value-added metabolic products (Chatzifragkou et al. 2011). The *Trichoderma viride* EMCC 107 and baker's yeast were used to convert agrowastes to bioethanol, and *T. viride* EMCC 107 showed better activity (El-Tayeb et al. 2012). Solid-state anaerobic digestion is a good strategy for the conversion of albizia biomass (a forestry waste) to biogas with the help of Ceriporiopsis subvermispora (Ge et al. 2014). When dextrose broth medium and corncob (the central core of an ear of corn) waste liquor was used as the substrate, *Aspergillus.* transformed it into biodiesel (Subhash and Mohan 2011). And it was observed that the biodiesel produced from corncob waste liquor was with better qualities than the biodiesel produced from synthetic medium. The fungal genus *Phlebia* showed the potential to degrade non-pretreated lignocelluloses which aid to produce ethanol with different organic acids (Mattila et al. 2017). White rot fungi significantly contributes to the bioenergy production and waste management. Anaerobic digestion of lignocellulosic wastes with the help of *Trametes Versicolor* MES 11914 and *Pleurotus Sajor-Caju* MES 03464 resulted in the production of fermentative volatile fatty acids enriched with acetate and propionate (Fang et al. 2018).

9.4.3 Role of Algae in Bioenergy Production

Biofuel production from microalgae is an upcoming technique to overcome the issues created by the hazardous contaminants. High-rate algal pond (HRAP) is considered as a new strategy in wastewater treatment, which can also generate algal biomass. Further these algal biomass could be transformed to bioenergy with the help of anaerobic digestion (Park et al. 2011). Microbial fuel cell (MFC) could be modified into an integrated photobioelectrochemical (IPB) system by the incorporation of algal bioreactor (Xiao et al. 2012). Glycerol from biodiesel waste was used as the substrate for the production of biofuel using MFCs and algae (Sharma et al. 2011). The evaluation of bioenergy production potential of *Chlorella vulgaris, Chlorella pyrenoidosa, Chroococcus* sp.1 and *Chroococcus* sp. 2 was carried for the production of methane. In this study cattle dung was used as the substrate and methane production was found to be maximum with *Chroococcus* sp.1 (Prajapati et al. 2014). Dairy manure could also act as a good source for bioenergy production (Chowdhury and Freire 2015). Anaerobic digestion and the associated algal biodiesel production are the major processes involved in the bioenergy production, where pyrolysis and enzymatic hydrolysis also have significant roles. *Chlorella pyrenoidosa* was used for the treatment of dairy wastewater and helps to produce bioenergy (Kothari et al. 2012). *Coelastrum* showed best biotransformation of waste molasses as compared to *Chlamydomonas reinhardtii* CC503 and *Chlorella vulgaris* and *these species could be utilized in the bioenergy production* (Chen et al. 2019). Waste molasses was used in another study where *Chlorella protothecoides* transformed these wastes to biodiesel (Yan et al. 2011). *Botryococcus* sp. showed the potential to tolerate domestic sewage and produced 3.2 g/L of dry biomass which was further utilized in the production of biofuel (Ashokkumar et al. 2019a). Photoautotrophic as well as heterotrophic algal cultivations got attention as these are the commonly used technique. At the same time, these techniques have some disadvantages, which could overcome with the help of mixotrophic cultivation and advanced research works are progressing in this aspect (Zhan et al. 2017).

However, microalgal mediated bioenergy production is not an economically feasible according to the current informations (Prajapati et al. 2014). New approaches are being developed to reduce the cost of the growing medium. For this, *Synechocystis* PCC6803 was adapted under laboratory conditions using municipal wastewater for biodiesel production (Ashokkumar et al. 2019b).

9.5 MICROBIAL FUEL CELL (MFC)

A microbial fuel cell (MFC) is a device in which microbes can convert chemical energy to electrical current by the oxidation of organic or inorganic substrates. Generally, MFC have two compartments: anode and cathode compartments (Mishra et al. 2019). These two compartments are separated with a cationic membrane. Microbes are growing in the anode compartment and they transform organic compounds such as glucose, which potentially donate electron. Electrons and protons generated in the biotransformation process of microbes have different paths. Electrons are transferred to the anode surface and then the electrons move to cathode through a conductive material. At the same time, the protons move through the electrolyte and then pass through the cationic membrane placed in the centre. And finally the electrons and protons move to the cathode by the reduction of soluble electron acceptors (Ieropoulos et al. 2005). The most common electron acceptor for the cathode reaction is oxygen (O_2). The collection of electrical current was achieved by placing a load between the anode and cathode compartments (Chaturvedi and Verma 2016).

Three different generations of MFC are available (MFC). Generation-I utilized synthetic redox mediators combined with *Escherichia coli*. However, the sulphate/sulphide and the sulphate reducing bacteria *Desulfovibrio desulfuricans* Beijerinck were used in Generation-II. In Generation-III the anodophillic species *Geobacter sulfurreducens* was used and it lack a soluble mediator. Different substrates can be used in the MFCs and according to 'substrate to power conversion efficiency', Generation-II MFCs were more efficient as compared to the other two (Ieropoulos et al. 2005). Extensive research in the innovations on MFCs proved the potential of different substrates in the production of electricity. Moreover, the organic and inorganic xenobiotics were also used as the electron donors. Hexavalent chromium, agro wastes, azo dyes, selenite, and nitrate were the common waste materials experimented in the MFCs development (Chaturvedi and Verma 2016). Hexavalent chromium contamination in the environment is a growing concern due to the uncontrolled discharge of this metal from different fields such as leather tanning, metallurgy, electroplating, and in wood preservation. The major toxic form of chromium is hexavalent chromium [Cr(VI)] (Humphries et al. 2004). Therefore, there is a need for detoxification of hexavalent chromium [Cr(VI)] by converting it to non-toxic trivalent chromium [Cr(III)]. The potential of acidic hexavalent chromium [Cr(VI)] to perform as cathodic electron acceptor in the MFC was proved in recent research findings (Chaturvedi and Verma 2016). *Trichococcus pasteurii* and *Pseudomonas aeruginosa* were used as the reducers of hexavalent chromium [Cr(VI)] in MFC for the generation of electricity (Tandukar et al. 2009). This experiment was coupled with the wastewater treatment and efficiently generated bioelectricity.

The wastes from farming and different agro industries are increasing and these agro-wastes could be utilized for the generation of electricity. The major concern on the degradation of cassava wastewater is the presence of the toxic biomolecule cyanoglycosides, at the same time, it contains large amount of starch (Kaewkannetra et al. 2009). Therefore, the utilization cassava wastewater in MFC is a good way to solve the waste management problem and it helps to synthesize bioenergy. A N_2-fixing bacterium, *Azotobacter vinelandii* TISTR 1094 have the capability to grow in the cassava wastewater and it aid in to removal of the toxic cyanide content (Kaewkannetra et al. 2009). Another important agro-waste cellulose also helps to generate electricity using MFC (Rismani-Yazdi et al. 2007). Many scientists have investigated the possibility of exploitation of rumen microbes in the MFC for the production of electricity. When cellulose act as the electron donor and ferricyanide act as the catholyte in the cathode of a MFC, electricity was generated. For this experiment, *Clostridium* spp. and *Comamonas* spp. abounded consortia was inoculated in the medium of anode chamber and the final power density reached up to 55 mW/m^2 (Rismani-Yazdi et al. 2007).

Utilization of *Clostridium cellulolyticum* and *Geobacter sulfurreducens* in MFC for the biotransformation of carboxymethyl cellulose aided in the production of 143 mW/m^2 electric current (Ren et al. 2008). *Enterobacter cloacae* (Rezaei et al. 2009), *Nocardiopsis* sp. KNU and *Streptomyces enissocaesilis* KNU (Hassan et al. 2012) were also used in the biotransformation process.

Many reports have showed that the bioelectricity could also be generated using azo dyes. The presence of N = N bond in all the azo dyes aid to perform bioelectricity generation in MFC as an electron acceptor. When *Klebsiella pneumoniae* L17 was employed in the degradation of methyl orange I, orange II, and methyl orange, these azo dyes were used as cathode oxidants (Liu et al. 2009). The detail of this

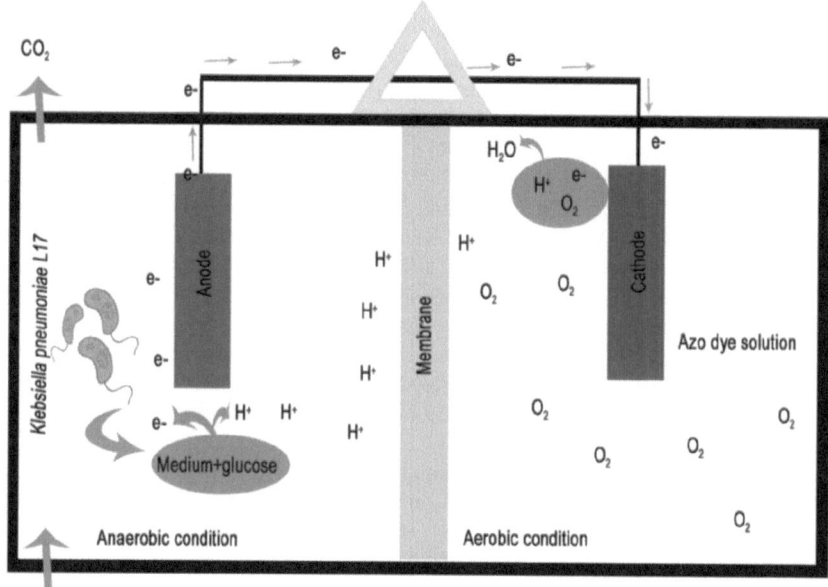

FIGURE 9.4   An illustration of a microbial fuel cell.

Source: Liu et al. (2009).

study was illustrated in Figure 9.4. Wastewater from glass and electronics manufacturing contains high amounts of selenium and its derivatives. When the application of nitrate-based fertilizers was increased, an exponential increase was observed in the nitrite contamination of water. The nitrate ions and selenium could act as the oxidant in MFC to generate electricity (Chaturvedi and Verma 2016).

However, the industrial application of MFCs have some drawbacks, such as the high cost of different materials and the low buffering capacity of domestic wastewater.

9.6 CONCLUSION

Different microbes showed high tolerance towards xenobiotic substrates and they efficiently degraded the wastes. Moreover, the energy released by the oxidation of the materials are used for the bioenergy production. Therefore, microbial mediated organic and inorganic waste management and associated bioenergy production is an environmentally sustainable technology. Execution of efficient MFCs for waste management and bioenergy production is a need of the hour. For the production of bioenergy potential bacterial, fungal and algal candidate were pointed in this study with suitable substrates.

REFERENCES

Ahmad, I., Imran, M., Hussain, M.B. and Hussain, S., 2017. Remediation of organic and inorganic pollutants from soil: The role of plant-bacteria partnership. In Anjum, N.A. (ed.) *Chemical Pollution Control with Microorganisms*, New York, USA, 197–244.

Arshadi, M., Yaghmaei, S. and Mousavi, S.M., 2019. Study of plastics elimination in bioleaching of electronic waste using Acidithiobacillus ferrooxidans. *International Journal of Environmental Science and Technology*, 16 (11), 7113–7126.

Ashokkumar, V., Chen, W.H., Ala'a, H., Kumar, G., Sathishkumar, P., Pandian, S., Ani, F.N. and Ngamcharussrivichai, C., 2019a. Bioenergy production and metallic iron (Fe) conversion from Botryococcus sp. cultivated in domestic wastewater: Algal biorefinery concept. *Energy Conversion and Management*, 196, 1326–1334.

Ashokkumar, V., Chen, W.H., Ngamcharussrivichai, C., Agila, E. and Ani, F.N., 2019b. Potential of sustainable bioenergy production from Synechocystis sp. cultivated in wastewater at large scale–A low cost biorefinery approach. *Energy Conversion and Management*, 186, 188–199.

Awasthi, M.K., Pandey, A.K., Khan, J., Bundela, P.S., Wong, J.W. and Selvam, A., 2014. Evaluation of thermophilic fungal consortium for organic municipal solid waste composting. *Bioresource Technology*, 168, 214–221.

Bigogno, C., Khozin-Goldberg, I., Boussiba, S., Vonshak, A. and Cohen, Z., 2002. Lipid and fatty acid composition of the green oleaginous alga Parietochloris incisa, the richest plant source of arachidonic acid. *Phytochemistry*, 60(5), 497–503.

Bayer, E.A., Lamed, R. and Himmel, M.E., 2007. The potential of cellulases and cellulosomes for cellulosic waste management. *Current Opinion in Biotechnology*, 18(3), 237–245.

Bhuyar, P., Muniyasamy, S. and Govindan, N., 2018, August. Green revolution to protect environment–An identification of potential micro algae for the biodegradation of plastic waste in Malaysia. *In World Congress on BIOPOLYMERS AND BIOPLASTICS & RECYCLING Expert Opin Environ Biol* (Vol. 7).

Cerrillo, M., Viñas, M. and Bonmatí, A., 2016. Removal of volatile fatty acids and ammonia recovery from unstable anaerobic digesters with a microbial electrolysis cell. *Bioresource Technology*, 219, 348–356.

Chai, B., Li, X., Liu, H., Lu, G., Dang, Z. and Yin, H., 2020. Bacterial communities on soil microplastic at Guiyu, an E-Waste dismantling zone of China. *Ecotoxicology and Environmental Safety*, 195, 110521.

Chaturvedi, V. and Verma, P., 2016. Microbial fuel cell: a green approach for the utilization of waste for the generation of bioelectricity. *Bioresources and Bioprocessing*, 3(1), 38.

Chatzifragkou, A., Makri, A., Belka, A., Bellou, S., Mavrou, M., Mastoridou, M., Mystrioti, P., Onjaro, G., Aggelis, G. and Papanikolaou, S., 2011. Biotechnological conversions of biodiesel derived waste glycerol by yeast and fungal species. *Energy*, 36(2), 1097–1108.

Cheng, C.L., Che, P.Y., Chen, B.Y., Lee, W.J., Lin, C.Y. and Chang, J.S., 2012. Biobutanol production from agricultural waste by an acclimated mixed bacterial microflora. *Applied Energy*, 100, 3–9.

Chen, M., Li, Y., Li, P., Wang, W., Qi, L., Li, P. and Li, S., 2019. A novel native bioenergy green alga can stably grow on waste molasses under variable temperature conditions. *Energy Conversion and Management*, 196, 751–758.

Chia, W.Y., Ying Tang, D.Y., Khoo, K.S., Kay Lup, A.N., Chew, K.W., 2020. Nature's fight against plastic pollution: Algae for plastic biodegradation and bioplastics production. *Environmental Science and Ecotechnology,* 4, 100065.

Choi, S.P., Nguyen, M.T. and Sim, S.J., 2010. Enzymatic pretreatment of Chlamydomonas reinhardtii biomass for ethanol production. *Bioresource Technology,* 101(14), 5330–5336.

Choińska-Pulit, A., Łaba, W. and Rodziewicz, A., 2019. Enhancement of pig bristles waste bioconversion by inoculum of keratinolytic bacteria during composting. *Waste Management,* 84, 269–276.

Chowdhury, R., Freire, F., 2015. Bioenergy production from algae using dairy manure as a nutrient source: Life cycle energy and greenhouse gas emission analysis. *Applied Energy,* 154, 1112–1121.

de Rezende, J.R., Oldenburg, T.B., Korin, T., Richardson, W.D., Fustic, M., Aitken, C.M., Bowler, B.F., Sherry, A., Grigoryan, A., Voordouw, G. and Larter, S.R., 2020. Anaerobic microbial communities and their potential for bioenergy production in heavily biodegraded petroleum reservoirs. *Environmental Microbiology,* 22(8), 3049–3065.

Dubini, A., 2011. Green energy from green algae: Biofuel production from Chlamydomonas reinhardtii. *The Biochemist,* 33(2), 20–23.

El-Tayeb, T.S., Abdelhafez, A.A., Ali, S.H., Ramadan, E.M., 2012. Effect of acid hydrolysis and fungal biotreatment on agro-industrial wastes for obtainment of free sugars for bioethanol production. *Brazilian Journal of Microbiology,* 43, 1523–1535.

Fang, W., Zhang, P., Zhang, X., Zhu, X., van Lier, J.B., Spanjers, H., 2018. White rot fungi pretreatment to advance volatile fatty acid production from solid-state fermentation of solid digestate: *Efficiency and mechanisms. Energy,* 162, 534–541.

Fukuda, H., Kondo, A. and Tamalampudi, S., 2009. Bioenergy: Sustainable fuels from biomass by yeast and fungal whole-cell biocatalysts. *Biochemical Engineering Journal,* 44, 2–12.

Gajda, I., Greenman, J. and Ieropoulos, I.A., 2018. Recent advancements in real-world microbial fuel cell applications. *Current Opinion in Electrochemistry,* 11, 78–83.

Ge, X., Matsumoto, T., Keith, L. and Li, Y., 2015. Fungal pretreatment of albizia chips for enhanced biogas production by solid-state anaerobic digestion. *Energy & Fuels,* 29(1), 200–204.

Ghazifard, A., Kasra-Kermanshahi, R. and Far, Z.E., 2001. Identification of thermophilic and mesophilic bacteria and fungi in Esfahan (Iran) municipal solid waste compost. *Waste Management & Research,* 19(3), 257–261.

Hassan, S.H., Kim, Y.S. and Oh, S.E., 2012. Power generation from cellulose using mixed and pure cultures of cellulose-degrading bacteria in a microbial fuel cell. *Enzyme and Microbial Technology,* 51(5), 269–273.

Haq, I.U., Akram, F. and Jabbar, Z., 2020. Keratinolytic enzyme-mediated biodegradation of recalcitrant poultry feathers waste by newly isolated Bacillus sp. NKSP-7 under submerged fermentation. *Folia Microbiologica,* 65, 823–834.

Ho, S.H., Huang, S.W., Chen, C.Y., Hasunuma, T., Kondo, A. and Chang, J.S., 2013a. Bioethanol production using carbohydrate-rich microalgae biomass as feedstock. *Bioresource Technology,* 135, 191–198.

Ho, S.H., Li, P.J., Liu, C.C. and Chang, J.S., 2013b. Bioprocess development on microalgae-based CO2 fixation and bioethanol production using Scenedesmus obliquus CNW-N. *Bioresource Technology,* 145, 142–149.

Hong, Y. and Valix, M., 2014. Bioleaching of electronic waste using acidophilic sulfur oxidising bacteria. *Journal of Cleaner Production,* 65, 465–472.

Hu, Y., Du, C., Leu, S.-Y., Jing, H., Li, X., Lin, C.S.K., 2018. Valorisation of textile waste by fungal solid state fermentation: An example of circular waste-based biorefinery. *Resources, Conservation and Recycling* 129, 27–35.

Humphries, A.C., Nott, K.P., Hall, L.D. and Macaskie, L.E., 2004. Continuous removal of Cr (VI) from aqueous solution catalysed by palladised biomass of Desulfovibrio vulgaris. *Biotechnology Letters,* 26(19), 1529–1532.

Ieropoulos, I.A., Greenman, J., Melhuish, C. and Hart, J., 2005. Comparative study of three types of microbial fuel cell. *Enzyme and Microbial Technology,* 37(2), 238–245.

Jiang, Z., Li, X., Li, M., Zhu, Q., Li, G., Ma, C., Li, Q., Meng, J., Liu, Y. and Li, Q., 2020. Impacts of red mud on lignin depolymerization and humic substance formation mediated by laccase-producing bacterial community during composting. *Journal of Hazardous Materials,* 124557.

Kaewkannetra, P., Imai, T., Garcia-Garcia, F.J. and Chiu, T.Y., 2009. Cyanide removal from cassava mill wastewater using Azotobactor vinelandii TISTR 1094 with mixed microorganisms in activated sludge treatment system. *Journal of Hazardous Materials,* 172(1), 224–228.

Kalčíková, G., Babič, J., Pavko, A. and Gotvajn, A.Ž., 2014. Fungal and enzymatic treatment of mature municipal landfill leachate. *Waste Management,* 34(4), 798–803.

Kapoor, A. and Viraraghavan, T., 1998. Removal of heavy metals from aqueous solutions using immobilized fungal biomass in continuous mode. *Water Research,* 32(6), 1968–1977.

Karwowska, E., Andrzejewska-Morzuch, D., Łebkowska, M., Tabernacka, A., Wojtkowska, M., Telepko, A. and Konarzewska, A., 2014. Bioleaching of metals from printed circuit boards supported with surfactant-producing bacteria. *Journal of Hazardous Materials,* 264, 203–210.

Knápek, J., Králík, T., Valentová, M. and Voříšek, T., 2015. Effectiveness of biomass for energy purposes: a fuel cycle approach. Wiley Interdisciplinary Reviews: *Energy and Environment,* 4(6), 575–586.

Kothari, R., Kumar, V., Pathak, V.V. and Tyagi, V.V., 2017. Sequential hydrogen and methane production with simultaneous treatment of dairy industry wastewater: bioenergy profit approach. *International Journal of Hydrogen Energy,* 42(8), 4870–4879.

Kothari, R., Pathak, V.V., Kumar, V. and Singh, D.P., 2012. Experimental study for growth potential of unicellular alga *Chlorella pyrenoidosa* on dairy wastewater: an integrated approach for treatment and biofuel production. *Bioresource Technology,* 116, 466–470.

Kovalovszki, A., Treu, L., Ellegaard, L., Luo, G. and Angelidaki, I., 2020. Modeling temperature response in bioenergy production: Novel solution to a common challenge of anaerobic digestion. *Applied Energy,* 263, 114646.

Kües, U., 2015. Fungal enzymes for environmental management. Current Opinion in Biotechnology, Environmental biotechnology. *Energy Biotechnology,* 33, 268–278.

Kwok, R., 2019. Inner Workings: How bacteria could help recycle electronic waste. *Proceedings of the National Academy of Sciences,* 116(3), 711–713.

Lam, W.C., Pleissner, D. and Lin, C.S.K., 2013. Production of fungal glucoamylase for glucose production from food waste. *Biomolecules,* 3(3), 651–661.

Lee, K.Y., Bosch, J. and Meckenstock, R.U., 2012. Use of metal-reducing bacteria for bioremediation of soil contaminated with mixed organic and inorganic pollutants. *Environmental Geochemistry and Health*, 34(1), 135–142.

Li, X.K., Ji, W.J., Zhao, J., Wang, S.J. and Au, C.T., 2005. Ammonia decomposition over Ru and Ni catalysts supported on fumed SiO2, MCM-41, and SBA-15. *Journal of Catalysis*, 236(2), 181–189.

Mattila, H., Kuuskeri, J. and Lundell, T., 2017. Single-step, single-organism bioethanol production and bioconversion of lignocellulose waste materials by phlebioid fungal species. *Bioresource Technology*, 225, 254–261.

Liu, L., Li, F.B., Feng, C.H. and Li, X.Z., 2009. Microbial fuel cell with an azo-dye-feeding cathode. *Applied Microbiology and Biotechnology*, 85(1), 175.

Mazza, L., Xiao, X., Ur Rehman, K., Cai, M., Zhang, D., Fasulo, S., Tomberlin, J.K., Zheng, L., Soomro, A.A., Yu, Z. and Zhang, J., 2020. Management of chicken manure using black soldier fly (Diptera: Stratiomyidae) larvae assisted by companion bacteria. *Waste Management*, 102, 312–318.

Mbareche, H., Veillette, M., Dubuis, M.È., Bakhiyi, B., Marchand, G., Zayed, J., Lavoie, J., Bilodeau, G.J. and Duchaine, C., 2018. Fungal bioaerosols in biomethanization facilities. *Journal of the Air & Waste Management Association*, 68(11), 1198–1210.

Ming, C., Dilokpimol, A., Zou, C., Liao, W., Zhao, L., Wang, M., De Vries, R.P. and Kang, Y., 2019. The quest for fungal strains and their co-culture potential to improve enzymatic degradation of Chinese distillers' grain and other agricultural wastes. *International Biodeterioration & Biodegradation*, 144, 104765.

Mishra, S., Roy, M. and Mohanty, K., 2019. Microalgal bioenergy production under zero-waste biorefinery approach: Recent advances and future perspectives. *Bioresource Technology*, 292, 122008.

Nandy, A., Kumar, V. and Kundu, P.P., 2013. Utilization of proteinaceous materials for power generation in a mediatorless microbial fuel cell by a new electrogenic bacteria Lysinibacillus sphaericus VA5. *Enzyme and Microbial Technology*, 53(5), 339–344.

Narayanasamy, M., Dhanasekaran, D., Vinothini, G. and Thajuddin, N., 2018. Extraction and recovery of precious metals from electronic waste printed circuit boards by bioleaching acidophilic fungi. *International Journal of Environmental Science and Technology*, 15(1), 119–132.

Novik, G., Savich, V., Meerovskaya, O., 2018. Geobacillus Bacteria: Potential Commercial Applications in Industry, Bioremediation, and Bioenergy Production. Growing and Handling of Bacterial Cultures. https://doi.org/10.5772/intechopen.76053

Omar, H.H., 2008. Algal decolorization and degradation of monoazo and diazo dyes. *Pakistan Journal of Biological Sciences*, 11(10), 1310–1316.

Pagliano, G., Ventorino, V., Panico, A. and Pepe, O., 2017. Integrated systems for biopolymers and bioenergy production from organic waste and by-products: a review of microbial processes. *Biotechnology for Biofuels*, 10(1), 113.

Park, J.B.K., Craggs, R.J., Shilton, A.N., 2011. Wastewater treatment high rate algal ponds for biofuel production. Bioresource Technology, Special Issue: Biofuels - II: *Algal Biofuels and Microbial Fuel Cells* 102, 35–42.

Park, S. and Li, Y., 2012. Evaluation of methane production and macronutrient degradation in the anaerobic co-digestion of algae biomass residue and lipid waste. *Bioresource Technology*, 111, 42–48.

Pathma, J. and Sakthivel, N., 2012. Microbial diversity of vermicompost bacteria that exhibit useful agricultural traits and waste management potential. *SpringerPlus*, 1(1), 26.

Pivato, A. and Gaspari, L., 2006. Acute toxicity test of leachates from traditional and sustainable landfills using luminescent bacteria. *Waste Management*, 26(10), 1148–1155.

Pleissner, D., Kwan, T.H. and Lin, C.S.K., 2014. Fungal hydrolysis in submerged fermentation for food waste treatment and fermentation feedstock preparation. *Bioresource Technology*, 158, 48–54.

Prajapati, S.K., Choudhary, P., Malik, A., Vijay, V.K., 2014. Algae mediated treatment and bioenergy generation process for handling liquid and solid waste from dairy cattle farm. *Bioresource Technology*, 167, 260–268.

Rastogi, M., Nandal, M. and Khosla, B., 2020. Microbes as vital additives for solid waste composting. *Heliyon*, 6(2), 03343.

Ren, Z., Steinberg, L.M. and Regan, J.M., 2008. Electricity production and microbial biofilm characterization in cellulose-fed microbial fuel cells. *Water Science and Technology*, 58(3), 617–622.

Rezaei, F., Xing, D., Wagner, R., Regan, J.M., Richard, T.L. and Logan, B.E., 2009. Simultaneous cellulose degradation and electricity production by Enterobacter cloacae in a microbial fuel cell. *Applied and Environmental Microbiology*, 75(11), 3673–3678.

Rismani-Yazdi, H., Christy, A.D., Dehority, B.A., Morrison, M., Yu, Z. and Tuovinen, O.H., 2007. Electricity generation from cellulose by rumen microorganisms in microbial fuel cells. *Biotechnology and Bioengineering*, 97(6), 1398–1407.

Rozas, E.E., Mendes, M.A., Nascimento, C.A., Espinosa, D.C., Oliveira, R., Oliveira, G. and Custodio, M.R., 2017. Bioleaching of electronic waste using bacteria isolated from the marine sponge Hymeniacidon heliophila (Porifera). *Journal of Hazardous Materials*, 329, 120–130.

Schnürer, A. and Schnürer, J., 2006. Fungal survival during anaerobic digestion of organic household waste. *Waste Management*, 26(11), 1205–1211.

Selvi, A.T., Anjugam, E., Devi, R.A., Madhan, B., Kannappan, S. and Chandrasekaran, B., 2012. Isolation and characterization of bacteria from tannery effluent treatment plant and their tolerance to heavy metals and antibiotics. *Asian Journal of Experimental Biological Sciences*, 3(1), 34–41.

Shalaby, E.A., 2011. Prospects of effective microorganisms technology in wastes treatment in Egypt. *Asian Pacific Journal of Tropical Biomedicine*, 1(3), pp.243–248.

Sharma, Y., Parnas, R. and Li, B., 2011. Bioenergy production from glycerol in hydrogen producing bioreactors (HPBs) and microbial fuel cells (MFCs). *International Journal of Hydrogen Energy*, 36(6), 3853–3861.

Sibi, G., 2018. Bioenergy Production from Wastes by Microalgae as Sustainable Approach for Waste Management and to Reduce Resources Depletion. *International Journal of Environmental Sciences & Natural Resources*, 13(3), 77–80.

Srivastava, R.K., 2019. Bio-Energy production by contribution of effective and suitable microbial system. *Materials Science for Energy Technologies*, 2(2), 308–318.

Subhash, G.V. and Mohan, S.V., 2011. Biodiesel production from isolated oleaginous fungi Aspergillus sp. using corncob waste liquor as a substrate. *Bioresource Technology*, 102(19), 9286–9290.

Tandukar, M., Huber, S.J., Onodera, T. and Pavlostathis, S.G., 2009. Biological chromium (VI) reduction in the cathode of a microbial fuel cell. *Environmental science & technology*, 43(21), 8159–8165.

Timmers, R.A., Strik, D.P., Hamelers, H.V. and Buisman, C.J., 2010. Long-term performance of a plant microbial fuel cell with Spartina anglica. *Applied Microbiology and Biotechnology*, 86(3), 973–981.

Voběrková, S., Vaverková, M.D., Burešová, A., Adamcová, D., Vršanská, M., Kynický, J., Brtnický, M. and Adam, V., 2017. Effect of inoculation with white-rot fungi and fungal consortium on the composting efficiency of municipal solid waste. *Waste Management*, 61, 157–164.

Walker, S.E., Mostaghimi, S., Dillaha, T.A. and Woeste, R.E., 1990. Modeling animal waste management practices: impacts on bacteria levels in runoff from agricultural lands. *Transactions of the ASAE*, 33(3), 807–0817.

Wang, J., Song, X., Wang, Y., Abayneh, B., Li, Y., Yan, D. and Bai, J., 2016. Nitrate removal and bioenergy production in constructed wetland coupled with microbial fuel cell: establishment of electrochemically active bacteria community on anode. *Bioresource Technology*, 221, 358–365.

Wang, L., Min, M., Li, Y., Chen, P., Chen, Y., Liu, Y., Wang, Y. and Ruan, R., 2010. Cultivation of green algae Chlorella sp. in different wastewaters from municipal wastewater treatment plant. *Applied Biochemistry and Biotechnology*, 162(4), 1174–1186.

Xiao, L., Young, E.B., Berges, J.A. and He, Z., 2012. Integrated photo-bioelectrochemical system for contaminants removal and bioenergy production. *Environmental Science & Technology*, 46(20), 11459–11466.

Yan, D., Lu, Y., Chen, Y.F. and Wu, Q., 2011. Waste molasses alone displaces glucose-based medium for microalgal fermentation towards cost-saving biodiesel production. *Bioresource technology*, 102(11), 6487–6493

Zhan, J., Rong, J. and Wang, Q., 2017. Mixotrophic cultivation, a preferable microalgae cultivation mode for biomass/bioenergy production, and bioremediation, advances and prospect. *International Journal of Hydrogen Energy*, 42(12), 8505–8517.

Zhang, W., Wang, H., Zhang, R., Yu, X.Z., Qian, P.Y. and Wong, M.H., 2010. Bacterial communities in PAH contaminated soils at an electronic-waste processing center in China. *Ecotoxicology*, 19(1), 96–104.

10 Use of Waste Biomass as Remediator for Environmental Pollution

Amit Kumar Tiwari, Nirupama Prasad and Dan Bahadur Pal
Department of Chemical Engineering, Birla Institute of Technology, Mesra, Ranchi – 835215, Jharkhand, India

CONTENTS

10.1 INTRODUCTION

Agricultural and forest waste have been considered equitable, environment-friendly, sustainable, and affordable natural resources for many years. In the past decades, global use of these resources has been increased; agricultural and food wastes and emissions have grown in line with the growing population. A country like India is the most dynamic country with respect to urban and industrial transformation; therefore, there is tremendous growth in these waste biomass resources and their use. Due to the shortage of fossil fuels to achieve the goals of sustainable development, more focus is now switched to the use of agricultural and food wastes as an alternative fuels. According to the World Investment Report of UNCTAD (2014), the global annual investment required to achieve the goals of sustainable development is around \$5–7 trillion; whereas, India needs to spend \$960 billion to achieve this goal with a gap of \$560 billion (Bhamra et al., 2015). Use of agricultural, food and kitchen wastes biomass for sustainable development through clean energy production may be the best solutions for many developing countries of the world including India. Agricultural, food, and kitchen wastes biomass are mainly source of good amount of lignin, cellulose and hemicellulose. The term "biomass" therefore used with these waste received from these resources because it covers a range of organic materials produced by plants and animals. The biomass can be collected and utilized for conversion into useful clean and bioenergy. Agricultural, food and kitchen waste includes crop residues, forest and wood residues, kitchen residues (except plastics and non-organic materials) etc. A considerable and meaningful amount of biomass is produced by the Asian countries (Eastern region), including China, Bangladesh, India, Thailand, Indonesia, Philippines, and Vietnam. Using updated technologies of bioenergy production, the potential of bioenergy production from lignocellulosic biomass is in the last few decades been evaluated in India (Mandade et al., 2015). It is estimated that the availability of biomass in India is around 915 million metric tons (Ghosh, 2016). Waste biomass is been an significant source of energy for many countries including India, because it offers several benefits for generation of power, production of cooking gas, and conversion into other value-added products. It is renewable and widely available natural source in the rural and urban areas. Biomass is also capable to provide firm energy, around 32% of the total basic energy use in India is mostly derived from the biomass and >70% population of the country is depends on it for their energy requirement. In India, several programs have been started to promote new and improved technologies for the use of waste biomass to ensure better utilization and maximum benefits by the Ministry of New and Renewable Energy (Raghuwanshi and Arya, 2019). In India, around 40% of the total food produced is wasted due to different reasons; the value of this wastage is around Rs 920 trillion (\$12.2 trillion) per year (Luthra et al., 2021). Several programmes have been started to promote new and improved technologies for the use of waste biomass to ensure better utilization and maximum benefits by the Ministry of New and Renewable Energy (Raghuwanshi and Arya, 2019). Power generation from biomass is an industry in India that attracts good investment (above Rs 600 crores) per year, with this much of investment we are generating electricity (around 5,000 M units/year) and providing rural employment of around 10 million man-days. For the effective utilization of waste biomass, bagasse-based cogeneration and biomass power generation process are adopted. (Kolisetty and Jose, 2018; Narnaware et al., 2015). Bioethanol production using lignocellulosic rich waste biomass is a complicated technique; but, it has been noticed that broad research and development efforts are being conducted by several industrial and academic institutions to make it more technically sound and economically viable in the coming future. More efforts are required to develop resource databases, process and data evaluation, development of the novel pretreatment techniques, logistic system, designing of reactors, development of microorganism cultures and other engineering aspects. Presently, due to increased understanding, adaptability of these energy sources by energy consumers and government enterprises, the use of renewable resources is increasing for power generation in many countries including India. India will become one of the major green energy creators very soon among the numerous developed countries. The yield of kitchen waste is significantly increasing due to the growth of the hotel and restaurant industries worldwide. Most of the collected kitchen waste is used as animal feed (especially for pigs). The direct use of kitchen waste as feedstuff is not permitted in many countries due to the chances of foot and mouth diseases, especially in China (Chen et. al., 2014)). Around 30 million tons or more quantity of kitchen waste is produced every year by China alone (Yangyang et. al., 2016). As India has the second largest population in the world, with good agricultural production, it has a good amount of agricultural and kitchen waste biomass, which is a good raw material for value addition. Due to the availability of an enormous amount of waste biomass, the Indian government has shown a strong desire to achieve 40% electric power production capacity from these sources by 2030. Through the establishment of 1,00,000 domestic biogas plants, energy plants and biomass gasifier, the power production from waste biomass is targeted at around 10,000 MW by the Indian government. According to Mandade et al. (2015), the bioenergy production potential of lignocellulosic material has been investigated in the last few decades in India by using updated and technologically proven processes. Nowadays, global warming is a big concern because of the emission of the huge amount of gases and radiations during the production of energy. Increased energy production is a result of the increased world's population; therefore, the world scientific community has urged to find out the renewable, suitable, and sustainable way out. Waste biomass is one of the best suitable alternatives since it also includes the production of biofuels and biomaterials; this biomass includes agricultural waste and residues, food industry wastes, industrial effluents etc. Apart from these resources, kitchen waste is also a rich source of lignocellulosic components that can be used for energy production. After pre-treatment of biomass, it can be converted into biofuels such as biodiesel, CH_4, H_2, C_2H_5OH, etc. (Chun-Min Liu et al., 2016). The possible by-products and co-products that are produced in the agro-processing industry during the production of main products can be encouraged for biofuel purposes which will also help to reduce the cost of processing. As per Carriquiry et al. (2011), it is found that biofuel production cost is comparatively two to three times more than petroleum fuels. To minimize the production cost, several challenges are involved in the conversion of lignocellulosic materials into biofuels and other chemicals because it requires biotechnological and chemical platforms together (Hoekman, 2009; Menon and Rao, 2012; Luo et al., 2010). These hurdles are related to the areas of-

   i.   production of feedstock;
   ii.   transportation of feedstock;
   iii. development of advance and effective pre-treatment and treatment techniques (enzymatic hydrolysis and microbial fermentation);
   iv.   development of by-product and co-product;

v.   controlling the environmental impacts of the production process
vi.   setting-up of standards for biochemical and biofuel;
vii. acceptance in between consumer community and society; and the
viii. distribution and marketing of biofuels.

To minimize all of these challenges some special expertise required from the field of agriculture, transport, biotechnology, chemical engineering, chemistry, genetics, microbiology, economics, mechanical engineering, and environmental science. Almost all over the world especially in India, there is a very good scope of utilization of this waste biomass by using economic and ecofriendly processes. (Kumar et al., 2015). According to Greenwell et al., 2012, there are several benefits of biofuels and biochemical production by using non-edible raw materials (lignocellulosic biomass) such as the following:

i.   It may share in improved local economic development.
ii.   It has minimum emission of carbon dioxide responsible for global warming.
iii. It is a renewable and sustainable resource.
iv.   It has the ability to reduce air pollution (if burning or rotting is to be avoided).
v.   It may provide energy security for countries involved in the import of oil.
vi.   It can create technical jobs for agriculture graduates, engineers, microbiologists, mechanical engineers etc.

Other potential uses of agricultural waste biomass and food residues are:

a.   feed for animal,
b.   mulching of crops,
c.   building a roof,
d.   fuels for boilers in processing industries, and
e.   the raw material for cement plant, brick kiln, paper and power plants based on biomass.

Kitchen waste (salt, fat etc.) is rich in moisture content, therefore burning it is not a proper solution because it releases different toxic contaminants such as CO, CO_2, and dioxins in the environment. This kitchen waste can be utilized in three ways i) after sterilization as animal feed, ii) as fertilizer after composting and iii) conversion into bioenergy through the microbial fermentation process.

During the Second World War, the need to improve production and distribution of food gained importance, in order to guarantee the adequate nutrition and health of people worldwide. After seven years of war, governments prioritized the restoration of the food system with the aim to solve hunger and malnutrition problems. In 1945, the Food and Agricultural Organization (FAO) was created as a specialized body of the United Nations, which at that time was composed of 44 nations (FAO, 2015). Since then, agriculture has become a major resource to reduce poverty and contribute to improving the standard of living of world's population, and the FAO has 149 members (FAO, 2017). The recovery of world agriculture after the war was gradual. The fluctuations in production in different regions during the early postwar years, was due to the negative impacts that were more serious than in the First World War. For 8,000 years, cereals such as maize, wheat, and rice have been the staple food supply for humans and animals worldwide (Curtis, 2021). The major cereal producers are the United States, China, India, Brazil, and the European Union. The Green Revolution has been one of the major contributions of the FAO to transform global agriculture (FAO, 2017), the important increase of high yield cereal crop varieties (wheat and rice) in developing countries such as India and Mexico between 1960 and 1970 (Eliazer Nelson et al., 2019), contributed to millions of people overcoming poverty.

Future projections predict higher growth in agricultural production, wherein end purposes are not only limited to food for the global population (FAO, 2017), but also use as animal food and industrial needs. The rapid growth of bioenergy production from biofuel (Hazell and Pachauri, 2006) is an example of agricultural crop diversification in recent years, especially those crops with a high content of starch and cellulose. For example, the cereal starch, mainly derived from maize and wheat, is mostly used as raw material to produce ethanol. Agricultural vegetal wastes, known as biomass, have an important potential to produce sustainable energy from renewable fuels (FAO and UNEP, 2010). In 2006, bioenergy represented 10% of world energy (Hazell and Pachauri, 2006).

Between 2000 and 2015, biofuel production significantly influenced increased crop demand. In 2009, some experts recognized the close link between energy and agriculture and warned about a possible risk of food production for direct human consumption (FAO, 2009). At that time, it was foreseen that the total demand for cereals would depend on the increase in biofuel demand and the expansion of technology designed to transform agricultural biomass into bioproduct. In 2018, cereal production increased compared with 2017, and Asia became the major producer (FAO, 2018). This increase in biofuel production was mainly due to some countries introducing compulsory regulation to mix this bioproduct with traditional fuel. The United States, Brazil, China, and the European Union are among the major ethanol producers. Agricultural production has increased more than three times over the last 50 years because of: the expansion of soils for agricultural use; the technological contribution of the green revolution which influenced productivity; and the accelerated growth of population (OECD/FAO, 2019). Agriculture produces an average of 23.7 million food tons per day worldwide.

This growth in worldwide production has created greater pressure on the environment, up to the point of causing negative impacts on soil, air and water resources, with subsequent influences population health and the sustainability of ecosystems put at risk. Agriculture is responsible for 21% of greenhouse gases emissions (Jantke et al., 2020). According to FAO (2016), in recent years, this new situation has driven a model with more sustainable development, which implies important changes in the current agricultural production systems. For more than 30 years, the Report of the World Commission on environment and development has alerted us to the profound crisis of the environment due to accelerated population growth, which has increased the demand for natural resources, in order to improve economic development. At the same time, nations have been urged to articulate and coordinate political actions through a global program for change that would allow them to assume the responsibility of managing natural resources to ensure durable human progress (Halisçelik and Soytas, 2019). From the Rio Declaration on Environment and Development (1992), the approach of a new model of sustainable development was consolidated, which implies the integration of economic, social and environmental factors to guarantee the welfare and improvement in the quality of life of the populations.

These new evolutionary guidelines for governments have been oriented since then by global development objectives such as the Millennium Development Goals, defined in 2010, and the new Agenda 2030 for Sustainable Development, approved in 2015, with 17 Sustainable Development Goals (SDGs) and 169 targets that reflect the desire and compromise of world leaders to implement strategies and policies oriented towards the preservation of natural resources in order to guarantee environmental sustainability (FAO, 2017). These new global goals pose the need for real structural reform in the world economy (Fukuda-Parr, 2016) to reduce the environmental degradation that threatens the future of humanity. Thus, according to the recent report of United Nations Sustainable Development Goals, the measures adopted until now have not been enough to make satisfactory progress in achieving these SDGs in the next 10 years (Fukuda-Parr, 2016). More ambitious and rapid measures are required from governments, focused on the efficient use of natural resources and the integration of sustainability criteria in all sectors of the economy, to ensure a real social and economic transformation. The priority of world leaders is not only to mitigate the impacts already caused, but also to respond to the need to produce more food and energy for a population that will exceed 10 billion people by 2050 (Fróna et al., 2019). All this must be achieved with less fossil fuel, lower emissions of polluting gases and zero solid waste (UN, 2019). This challenge has started to materialize through some regulatory and management instruments, such as the Europe 2020 Strategy of the European Union (EC, 2020), which aims to ensure smart, sustainable and inclusive growth for Europe. This general framework established the base line for the implementation of initiatives oriented towards sustainable production and consumption, among them, the strategy of Circular Economy and Bioeconomy. More than 40 countries have developed and adopted national policies, policy instruments and strategies related to this new economic model in the last decade (Dietz et al., 2018; Valenti et al., 2019), which is mainly based on an efficient resource management system in which the priority is to extend the useful life of materials and products and prevent their loss of value by incorporating their waste into production processes (Molina-Moreno et al., 2017).

Agriculture is one of the largest biological sectors with the highest biomass production, which becomes an essential input for the bio-economy, (Bracco et al., 2018). This represents a great opportunity, not only because its use and exploitation favours the reduction of fossil fuel use and greenhouse gas emissions, but also because it contributes to the development of new green markets and jobs by promoting the conversion of vegetable waste into value-added products (by-products), such as food, feed, bioproducts and bioenergy (Scarlat et al., 2015). "Bioenergy", or biomass energy, is generally produced by living organisms. Currently, three different forms (heat, fuels, and electrical energy) of this bioenergy are existing with present adapted technologies. India's farmers are in a good position to utilize this bioenergy because they already know the biomass, production of biomass and its utilization. Well-trained and educated farming communities who are also energy consumers are able to generate and utilize bioenergy at their own site.

Heat is a form of bioenergy that has been produced for more than 1,000 years, it provides the best example to plan for its uses in agriculture and farming. The burning of agricultural waste biomass or kitchen waste is known as direct combustion. Direct combustion of waste material for specific purposes like to create heat, steam, etc. is an efficient means of the use of bioenergy. Due to its minimum process requirement and diversity of waste biomass, no need for specific equipment and comparatively high recovery of energy it can be used as feedstock for bioenergy production. Biomass gasification is also a practical bioenergy technology, which can also be used as an on-farm bioenergy source, for which waste of oil crops can be utilized. Agricultural waste biomass such as residues of starchy crops, cellulosic crops, fat-containing crops, sugar crops, etc. are the best sources of energy production; therefore, the production of different energy producing products like biodiesel and bioethanol can be done on-farm.

10.2   A BRIEF INTRODUCTION OF WASTE BIOMASS

Biomass can be obtained from the various agricultural and animal husbandry waste. Waste biomass is imperative resources which are renewable and abundantly available (Abdeen, 2008). The economies of developing countries like India rely mainly on agriculture. Agricultural waste is consisting plant waste biomass having a significant amount of lignin, cellulose and hemicelluloses, these agricultural wastes are grouped into various groups like leaves, grasses, fruit and vegetable waste, wood residues, crop residues, and food wastes from kitchen and industries (Qi et al., 2005; Roig et al., 2006; Rodríguez et al., 2008). Thousands of different crops are produced each year, leaving behind large amount of residues. However, most of the agricultural wastes are still not being utilized properly. These residues if gets rotten it emits methane and if open burned to clean lands it will generates carbon dioxide and various pollutants. Therefore, it is very important to utilize these agricultural wastes to lower burden on our environment. With respect to material and energy recovery, these agricultural wastes have high significance (Sriram and Mohammad, 2005). Agricultural wastes can be utilized for the generation of heat, biogas, compost, and energy. Also, these biomasses can be converted into charcoal, methanol, ethanol, and bio diesel. Nowadays, these biomasses considered as one of the potential substitute for the generation of electricity (Donald, 1998).

In Kampala City, almost 1,000 metric tons of agricultural and animal wastes are accumulated daily. Only 30% of the wastes are removed and dumped into a dump fill. These organic wastes contents good amount of fertilizing nutrients such as nitrogen, phosphorous and potassium. These nutrients are very important for the crop's growth and yields. Utilization of these free of cost fertilizers for agricultural production can able to enhance food security and helps in the reduction of environmental pollution. Thus, awareness among the public and farmers is required for proper management and utilization of agricultural wastes in order to protect our environment (Westerman and Bicudo, 2005). It is very important to set up institution that can focus research and development on the maximum utilization of these agricultural wastes in energy production. With the use of appropriate conversion technologies, agricultural and animal's wastes can be converted into useful resource.

10.3   ROLE OF BIOMASS IN ENVIRONMENTAL POLLUTION

The impact of agricultural waste on the environmental pollution not only depends on the amount of waste generated but it also depends on the methods employed for its disposal. Few disposal techniques pollute the environment such as burning of agricultural wastes (Sabiiti et al., 2004; Tumuhairwe et al., 2009). Burning is the most common technique used for the disposal of residues. But, burning of the agricultural wastes releases number of pollutants such as CO, NO_2, NO_x and smoke particles (Ezcurra et al., 2001). These pollutants further accompanied for the formation of ozone and nitric acid (Hegg et al., 1987). The formation of ozone and nitric acid poses risk to the living organisms and ecological systems (Lacaux et al., 1992). Environmental pollution resulting from animal waste is also one of the major global concerns. This is acute and serious problem to those countries who have limited land for manure disposal. Animal wastes can be excreted in solid, liquid, and gaseous forms. Respiration and fermentation gases discharged to the environment soon after being produced by the animal. These wastes have serious impact on our environment, water quality, soil quality and air pollution. Odour generated by these wastes another major problem among the farmers. Solid and liquid waste can be subjected to anaerobic microbial conversion. This converts organic substrates into microbial biomass as well as soluble and gaseous products (Katongole, 2009). These organic wastes can be used as a fertilizer and soil conditioner. However, excess use of these wastes can leads to the surface run-off and leaching problems which may contaminate ground or surface waters. Nitrate leaching into the surface water causing major nitrogen (N) pollution (Mackie et al., 1998). Leaching of phosphorus (P) in water bodies stimulates the growth of algae and other aquatic plants and their decomposition increases the demand of oxygen which interferes the welfare of fish.

Another issue associated with animal waste is the decomposition of the manure, which generates greenhouse gases such as methane (CH_4), ammonia (NH_3) and nitrogen oxides, etc. Ammonia volatilization causes acid deposition, which results in acid precipitation (Wang et al., 2014; Sabiiti, 2011). Nitrification-denitrification cycle emits nitrous oxide (N_2O) which causes ozone depletion (Khalil et al.,2004). Over the past few decades, preserving ecological system is one of the major concerns for the mankind. Pollutants when enters into the environments it causes undesirable changes in the physical, chemical and biological characteristics of air, water and soil. These changes have serious risk on the all the living beings.

Different overviews of current research and case studies with respect to heavy metals and synthetic chemicals (organic and inorganic) were reviewed and it was noticed that these materials are the major soil contaminants. These contaminants also come from the agricultural residues left in the field after harvesting. When these materials are decomposed due to natural processes, it provides the platform to microbial communities for their growth and development and also contains a variety of biological soil and water contaminants such as pathogens, parasites, and worms. This microbial community is responsible for various adverse impacts on human and animal health.

10.3.1 AIR

Outdoor air pollution is generally resulted from the combustion of combustible materials. Automobiles, process industries, and power stations generate energy by the combustion of petroleum products. However, in a few countries, energy in the process industries and power stations is generated by the combustion of wood and agricultural waste. Industrial processes discharge dust and harmful gases to the environment (Massawe et al., 2016). Apart from the process industries, indoor sources such as kitchen wastes also contribute to the outdoor air pollution. A heavily populated area generates high levels of outdoor air pollution. A kitchen waste emits greenhouse gases such as CO_2 and CH_4 while its decomposition in landfills. It is also important to consider the amount of greenhouse gases that emitted during production, transportation and landfills. According to one of the study in the Finland it was found that the yearly pollution from kitchen waste is almost equivalent to the pollution generated by 100,000 cars in the same year. A US study shows landfill gases emitted by kitchen waste is responsible for almost 17% of USA CH_4 emissions. Furthermore, emission of greenhouses through kitchen wastes is keeps on growing. Over the last 50 years, kitchen wastes have led to an almost 300% increase in greenhouse gas emissions and the figure is expected to increase another 400% by mid-century. Therefore, it is necessary to control the greenhouse gases that are generated by the kitchen wastes (Massawe et al., 2016).

10.3.2 WATER

Ground water is polluting day-by-day due to excess generation of kitchen wastes and fertilizers which are used to improve yield of the crops. Kitchen wastes dumped in landfills and constituents of fertilizers that leached and run-off into the ground water are the source of nutrient and pathogen pollution in water bodies. This polluted groundwater has serious impact on the land and aquatic life.

10.3.3 SOIL

Soil can be polluted either by natural means and/or by human activities such as discharge of industrial effluent, use of pesticides and synthetic fertilizers in agriculture. Adiningsih et al. (1998) carried out a rigorous study on a lowland rice cultivation area of West Java. They found that low level of lead (10–43 ppm) and cadmium (0.19–0.49 ppm) were present in the soil. These heavy metals may be accumulated due to the use of high amount of phosphorous containing fertilizer. Heavy rainfall countries like Indonesia, essentially requires phosphate fertilizer to improve crops yields. But due to heavy rainfall, leaching and run-off of these fertilizers into the ground is also rapid. Alloway (2013), reported 35–255 g/mt of cadmium is present in the soil of Indonesia. Rapid leaching and run-off also results in low soil pH and increase iron oxide and aluminium oxide in the soil. These leads to immobilization of phosphorus content in the soil solution which hinders its uptake by the plants. Kasno et al. (2000) observed two districts of West Java have intensive lowland rice area. This study was based on the content of lead and cadmium in the soil. This intensive lowland rice area can be further divided into highly polluted soil, medium polluted soil and unpolluted soil. Study showed only 7% of the total lowland areas were polluted due to lead content and about 4% were polluted by the cadmium content. These studies concluded that even after 30–40 years of phosphate fertilizer application, productivity of these soils can be sustain. Sofyan et al. (1997) studied the effect of air pollution arises by the automobiles on the soil quality. They conducted their study in tea plantations area of West Java. This area is important for agro forestry and tourism. They observed that the lead content in soil is high near the main road. Combustion of diesel and petrol in automobiles is the measure source that releases lead content. However, cadmium content is not influenced the automobiles discharge. Thus, it can be concluded that the lead can be resulted from the application of phosphate fertilizer as well as use of automobiles, whereas, cadmium is only resulted due to phosphate fertilizer in these areas.

10.4 OVERALL IMPACT ON HUMAN HEALTH

The pollutants that emits from the combustion have serious impact on the human health such as cardiovascular morbidity and amplified respiratory and death (Brunekreef and Holgate, 2002). To calculate health effects in a huge and unprotected population, it is essential to perform an epidemiological analysis. In 1952 in London, one disease outbreak occurred and around 4,000 people died prematurely within a week, followed by 8,000 more deaths in the next few months. The reason of this disease outbreak was air pollution (Bell and Davis, 2001). By doing long-term studies it was found that the increase in cardiovascular and respiratory related mortality was related with exposure to particulate matters; a 16-year follow-up study was also conducted for a group of 500,000 people residing in US cities. After completion of study they found that there was strong association of peoples were with PM 2.5. Many researchers also found that the most of the deaths were associated with the lung's cancer due to air pollution. For ecological studies of small areas another approach was used. This approach was based on the data of census, air pollution and health events and with adjustments for potential confounding factors, including socioeconomic status (Scoggins et al., 2004). A study using this approach indicates increase in mortality for every 10 μg/m^3 of PM 2.5. In the cities of developed countries, average annual PM 2.5 levels was 10 to 30 g/m^3. Many urban areas of developing countries, air pollutants result in rise in significant respiratory morbidity. For example, Romieu et al. (1996) report an exacerbation of asthma among children in Mexico City. Xu and Wang (1993) reported an increasing risk of respiratory symptoms in middle-aged nonsmokers in Beijing. Wang et al. (1997) reported increasing risk of low birth weight among the very young people due to PM exposure and SO$_2$ exposure in Beijing. Pereira et al. (1998) reported in intrauterine mortality in São Paulo due to air pollution. Other impact of air pollutants on health are post-neonatal mortality and mortality that results from acute respiratory infections, children's lung function, cardiovascular and respiratory problems and functional damage of the heart muscle. Asthma is another common disease that caused by air pollutants in urban areas is (Sabine et al., 2004). Exposure to ozone is the major cause of asthma attacks. Ozone exposure as a trigger of asthma attacks is of particular concern. The mechanism behind an air pollution and asthma link is not fully known, but early childhood NO$_2$ exposure may be important (Ponsonby et al., 2000).

In urban areas, leaded petrol used in automobiles causes release of lead in the atmosphere which risk primarily the brain in young children which leads to the behavioural aberrations and reduced or delayed intellectual or motoric ability development. Lead exposure implicated hypertension in adults. Other pollutants that is to be considered is the carcinogenic volatile organic compounds which causes increase in lung cancer (Nyberg et al., 2000).

A chemical pollutant that lies on the surface of the water bodies can create health issues, because that water is used as drinking water source. Also, this water is required for washing and cleaning, fishing and fish farming and for recreation. Groundwater which is another major source of drinking water, these waters contents low concentrations of pathogens. Moreover, toxic metals such as arsenic and fluoride can easily get dissolved in groundwater through run-off of the soil or rock layers. If industrial effluents do not treated properly prior to its discharge can contaminates the surface and groundwater. In 1980s in the United States, the government started the Superfund Program investigation and clean-up program to deal with such sites (U.S. Environmental Protection Agency, 2000). Pollution near sea coastal causes health hazards due to the consumption of contaminant sea foods. For example, eating of mercury-contaminated fish led to the infamous Minamata disease outbreak in Japan in 1956. Small concentration of polychlorinated biphenyls (PCBs) and dioxins in sea water can cause.

10.5 SCOPE FOR UTILIZATION OF BIOMASS

Due to rapid growth in population and changes and improvements in living standards, a rapid increase in the types and volume of kitchen and agricultural waste biomass is noticed. This intensive growth of the waste generated from the agriculture sector and the domestic sector is going to becoming a big problem because the rotten waste biomass releases pollutants like leachate and methane, the open burning of these waste biomass also generates CO_2 and several other pollutants. Hence, the improper disposal and management of agricultural waste biomass and kitchen waste biomass are contributing to environmental pollution, climate change, water contamination, soil contamination and air pollution. Whereas, these waste biomass are of high valued items with respect to value addition and energy production. Presently, a special drive is going on to proper use of agricultural and kitchen waste biomass and to convert these materials into an energy resource, sufficient efforts are being done by the Governments and other private organizations of different countries; but still, few major gaps are there, which needs to be filled. These gaps are due to the lack of awareness and the lack of capacity to convert all the waste biomass into value-added material or energy. The conversion of waste into valuable material and energy could not only reduce the costs of waste disposal but would also generate revenue from the sale of the produced items and energy. Optimal utilization of agricultural and kitchen waste biomass could be achieved by following ways:

- production of biogas,
- production of biofuels, and
- production of energy by co-production

The production and utilization of renewable energy from waste biomass, for both the national and global concern is a necessity due to the combined effect of climate change and disposal problems. The prices of fossil fuel is rising continuously, providing the main reason to explore renewable energy resources to ensure increased production of energy. The use of kitchen and agricultural waste biomass for the production of energy will also ensure the security of the energy supply to the population. Production of biogas is an eco-friendly process that can be achieved by using anaerobic digestion of organic waste. Agricultural, kitchen food wastes, and crop residues can be converted into biogas by this technique. An important advantage of this process is that biogas is produced as a main renewable energy resource, while the residue (by-product) can be used as manure (Ward et al., 2008). The efficiency of the biogas production process mostly depends on the properties of the raw material (waste) and the activity of the microbes involved in the process (Batstone et al., 2002). The entire process can be divided into three steps:

  i. hydrolysis,
 ii. formation of acid and
iii. production of methane (Yong et al., 2015; Merlin et al., 2014; Chasnyk et al., 2015; Abdeshahian et al., 2016).

10.6 CONCLUSION

With the assumptions of lack of availability of land, intensive agriculture provides significant production of food products; apart from this it also offers other benefits like energy supply, development of economic structure and ecological benefits. Agriculture of any country helps the nation by providing several benefits such as in the earning and savings of foreign money, improved and secure energy supply and enhancement of socio-economic status. The waste or residues like leaves, stems, fruits, flowers, straw, etc. of the agricultural crops have good potential to produce value-added items and energy. Kitchen wastes like food waste, vegetable residues like stalks, leaves, flowers and pulps, etc. have a good amount of starch, fat, cellulosic components and other carbohydrates that can be used to produce useful energy either by gasification, microbial digestion (production of biogas). As fossil fuels are too much costly now and a limited source of energy; therefore, efforts to be done to reduce fossil energy uses and to upgrade and promote clean and green energy production using agricultural waste and kitchen waste as biomass sources. Our society is hesitating about the reuse of these wastes, therefore more awareness should be created among the peoples of the society regarding the adoption of clean, green and sustainable approaches so that they may explore the solution to the energy crisis. The conversion of waste biomass into energy will be the key factor to reducing the CO_2 level in the environment because CO_2 is a major contributor to global warming. More research is to be done to find out the other alternative approaches for energy production.

ACKNOWLEDGEMENT

The authors are very thankful to all the researchers whose work is cited in the article; the authors also giving thanks to their family members for continuous support and patience during the writing of this article. The authors also showing gratitude to their institute (BIT Mesra) and the management of the institute for providing valuable support.

REFERENCES

Abdeen, M.O. (2008). Energy, environment and sustain-able development. Renewable and Sustainable Energy Reviews, 12: 2265–2300.

Abdeshahian, P., Lim, J.S., Ho, W.S., Hashim, H., Lee, C.T. (2016). Potential of biogas production from farm animal waste in Malaysia. Renewable and Sustainable Energy Reviews, 60: 714–723.

Adiningsih, J.S., J. Soejitno, dan Subowo. (1998). Ameliorasi pencemaran agrokimia pada lahan sawah intensifikasi Jalur Pantura, Jawa Barat. Laporan Akhir RUT, Kantor Negara Riset dan Teknologi, BPPN, DRN, LIPI, BPPT. (In Bahasia Indonesia).

Alloway, B.J. (2013). Sources of Heavy Metals and Metalloids in Soils. In: Alloway B. (eds) Heavy Metals in Soils. Environmental Pollution, vol 22. Springer, Dordrecht. https://doi.org/10.1007/978-94-007-4470-7_2

Batstone, D.J., Keller, J., Angelidaki, I., Kalyuzhnyi, S.V., Pavlostathis, S.G., Rozzi, A., Sanders, W.T.M., Siegrist, H., and Vavilin, V.A. (2002). The IWA Anaerobic Digestion Model No 1 (ADM 1). Water Science and Technology, 45(10): 65–73.

Bell, M. L. and Davis, D. I. (2001). Reassessment of the lethal London fog of 1952: Novel indicators of acute and chronic consequences of acute exposure to air pollution. Environmental Health Perspectives, 109 (Suppl. 3): 389–394.

Bhamra, A., Shanker, H., and Niazi, Z. (2015). Achieving the Sustainable Development Goals in India - A Study of Financial Requirements and Gaps, Technology and Action for Rural Advancement, New Delhi.

Bracco, S., Calicioglu, O., Gomez San Juan, M., and Flammini, A. (2018). Assessing the contribution of bioeconomy to the total economy: a review of national frameworks. Sustainability, 10: 1698.

Brunekreef, B. and Holgate, S. T. (2002). Air Pollution and Health. Lancet, 360: 1233–1242.

Carriquiry, M. A., Du X., and Timilsina, G. R. (2011). Second generation biofuels: economics and policies. Energy Policy, 39(7): 4222–4234.

Chasnyk, O., Sołowski, G., and O. Shkarupa (2015). Historical, technical and economic aspects of biogas development: Case of Poland and Ukraine. Renewable and Sustainable Energy Reviews, 52: 227–239.

Chen, T., Jin, Y., Qiu, X., and X. Chen (2014). A hybrid fuzzy evaluation method for safety assessment of food-waste feed based on entropy and the analytic hierarchy process methods. Expert Systems with Applications, 41: 7328–7337.

Chun-Min Liu Shu-Yii Wu. (2016). From biomass waste to biofuels and biomaterial building blocks. Renewable Energy, 96: 1056–1062.

Curtis, B.C. Wheat in the World. Online: www.fao.org/3/y4011e/y4011e04.htm (accessed on March 2021).

Dietz, T., Börner, J., Förster, J.J. and Braun, J.V. (2018). Governance of the bioeconomy: a global comparative study of national bioeconomy strategies. Sustainability, 10(9): 3190.

Eliazer Nelson, A.R.L., Ravichandran, K. and Antony, U. (2019). The impact of the Green Revolution on indigenous crops of India. J. Ethn. Food, 6(8). https://doi.org/10.1186/s42779-019-0011-9

European Commission (2020). The Post-2020 Common Agricultural Policy: Environmental Benefits and Simplification. https://ec.europa.eu/info/sites/info/files/food-farming-fisheries/key_policies/documents/cap-post-2020-environ-benefits-simplification_en.pdf (accessed on March 2021)

Ezcurra AI, Ortiz de Zarate, Pham Vhan Dhin and JP Lacaux (2001) Cereal waste burning pollution observed in the town of Vitoria (northern Spain). Atmospheric Environment, 35: 1377–1386.

Food and Agriculture Organization of the United Nations (FAO) and Unite Nations Environment Programme (UNEP), (2010). A Decision Support Tool for Sustainable Bioenergy. Online: www.fao.org/docrep/013/am237e/am237e00.pdf, (accessed on March 2021)

Food and Agriculture Organization of the United Nations (FAO), (2018). Crop Prospects and Food Situation #4, December 2018 www.fao.org/3/CA2726EN/ca2726en.pdf, (accessed on March 2021)

Food and Agriculture Organization of the United Nations (FAO), (2016). Crop Prospects and Food Situation www.fao.org/3/a-i6558e.pdf, (accessed on March 2021)

Food and Agriculture Organization of the United Nations (FAO), 2015. 1945–70 Years of FAO – 2015, www.fao.org/3/i5142e/i5142e.pdf

Food and Agriculture Organization of the United Nations (FAO), 2017. Towards Zero Hunger 1945–2030. www.fao.org/3/a-i6196e.pdf.

Food and Agriculture Organization of the United Nations (FAO),(2009). Meeting Report on How to Feed the World in 2050. www.fao.org/3/ak542e/ak542e19.pdf, (accessed on March 2021)

Fróna, D., Szenderák, J., and Harangi-Rákos, M. (2019). The Challenge of Feeding the World. *Sustainability*, 11(20), 5816. doi:10.3390/su11205816

Ghosh, S.K. (2016) Biomass and Bio-waste Supply Chain Sustainability for Bio-energy and Bio-fuel Production. Procedia Environmental Sciences, 31: 31–39.

Greenwell, H. C., Loyd-Evans M., and Wenner C. (2012). Biofuels, science and society. Interface Focus, 3, pp. 1–4.

Halisçelik, E. and Soytas, M.A. (2019). Sustainable development from millennium 2015 to sustainable development goals 2030, Sustainable Development, 27, pp. 545–572.

Hazell, Peter and Pachauri, J. (2006). Overview: bioenergy and agriculture promises and challenges. International Food Policy Research Institute (IFPRI), 2020 vision briefs.

Hegg, D.A, Radke L.F., Hobbs P.V., Brock, C.A. and Riggan, P. J. (1987). Nitrogen and Sulphur emissions from the burning of forest products near large urban areas. Journal of Geophysical Research, 92: 14701–14709.

Hoekman S. K. (2009) "Biofuels in the U.S.—challenges and opportunities," Renewable Energy, 34(1): 14–22. www.chintan-india.org/sites/default/files/2019-09/Food%20waste%20in%20India.pdf (visited on 17.03.2021)

Jantke, K., Hartmann, M. J., Rasche, L., Blanz, B., and Schneider, U. A. (2020). Agricultural Greenhouse Gas Emissions: Knowledge and Positions of German Farmers. Land, 9(5), 130. doi:10.3390/land9050130

Kasno, A., Sri Adiningsih, Sulaeman dan Subowo. (2000). Status pencemaran Lead and Cadmium pada padi sawah intensifikasi jalur Pantura Jawa Barat. Jurnal Ilmu Tanah dan Lingkungan 3(2): 25–32. (In Bahasia Indonesia).

Katongole, C.B. (2009). Developing rations for meat goats based on some urban market crop wastes PhD. Thesis, Makerere University, Uganda.

Khalil, K., Bruno, M. and Pierre Renault. (2004). Nitrous oxide production by nitrification and denitrification in aggregates as affected by O_2 concentration. Soil Biology and Biochemistry, 36: 687–699.

Klass, D.L. *An Introduction to Biomass Energy a Renewable Resource*, 1998, (www.beral. org).

Kolisetty, D. and Jose, B. B. (2018). Indian Progress in the Renewable Technologies: A Review on Present Status, Policies, and Barriers. International Journal of Renewable Energy Research, 8(2): 805–819.

Kumar, Anil, Kumar Nitin, Prashant Baredar Prashant, Ashish Shukla Ashish. (2015). A review on biomass energy resources, potential, conversion and policy in India. Renewable and Sustainable Energy Reviews, 45: 530–539.

Lacaux, J.P., Loemba-Ndembi, J., Lefeivre, B., Cros, B. and Delmas, R. (1992). Biogenic emissions and biomass burning influences on the chemistry of the fogwater and stratiform precipitations in the African equatorial forest. Atmospheric Environment, 26 (A/4): 541–551.

Lamarque, J.F., Bond, T.C., Eyring, V., Granier, C., Heil, A., Klimont, Z., Lee, D., Liousse, C., Mieville, A., Owen, B. and Schultz, M.G., (2010). Historical (1850–2000) gridded anthropogenic and biomass burning emissions of reactive gases and aerosols: methodology and application. Atmospheric Chemistry and Physics, 10(15), 7017–7039.

Luo, L. van der Voet, E. and Huppes, G. (2010). Bio refining of lignocellulosic feedstock—technical, economic and environmental considerations. Bioresource Technology, 101(13): 5023–5032.

Luthra, A., Chaturvedi, B., and Mukhopadhyay, S. (2021). Air pollution, waste management and livelihoods: Patterns of cooking fuel use among waste picker households in Delhi. Geographical Review, https://doi.org/10.1080/00167428.2021.1941016

Mackie, R.I., Stroot, P.G. and Varel, VH. (1998). Biochemical identification and biological origin of key odour components in livestock waste. Journal of Animal. Science, 76: 1331–1342.

Mandade, P., Bakshi, B. R., & Yadav, G. D. (2015). Ethanol from Indian agro-industrial lignocellulosic biomass—a life cycle evaluation of energy, greenhouse gases, land and water. The International Journal of Life Cycle Assessment, 20(12): 1649–1658.

Massawe, S.B., Olorunnisola, A.O., Adenikinju, A. (2016). The Environmental Challenges of Biomass Utilisation for Combined Heat and Power Generation in a Paper Mill in Tanzania. J Fundam Renewable Energy Appl. 6: 202. doi:10.4172/2090-4541.1000202

Menon, V. and Rao, M. (2012) Trends in Bioconversion of Lignocelluloses: Biofuel, Platform Chemicals and Biorefinery Concept. Progress in Energy and Combustion Science, 38: 522–550.

Merlin Christy, P., Gopinath, L.R., Divya, D., (2014). A review on anaerobic decomposition and enhancement of biogas production through enzymes and microorganisms. Renew. Sust. Energy Rev. 34: 167–173.

Molina-Moreno, V., Leyva-Díaz, J., Llorens-Montes, F., Cortés-García, F. (2017). Design of indicators of circular economy as instruments for the evaluation of sustainability and efficiency in wastewater from pig farming industry. Water, 9: 653.

Nyberg F., Gustavsson P., Jarup L., Bellander T., Berglind N., Jacobsson R. (2000). Urban Air Pollution and Lung Cancer in Stockholm. Epidemiology. 11:487–495.

OECD/FAO (2019), *Background Notes on Sustainable, Productive and Resilient Agro-Food Systems: Value Chains, Human Capital, and the 2030 Agenda*, OECD Publishing, Paris/FAO, Rome, https://doi.org/10.1787/dca82200-en.

Pereira L. A., Loomis D., Conceição G. M., Braga A. L., Arcas R. M., Kishi K. S. (1998). Association between Air Pollution and Intrauterine Mortality in São Paulo, Brazil. Environmental Health Perspectives. 106: 325–329.

Ponsonby A. L., Couper D., Dwyer T., Carmichael A., Kemp A., Cochrane J. (2000). The Relation between Infant Indoor Environment and Subsequent Asthma. Epidemiology. 11: 128–135.

Preeti H. Narnaware, Ramesh G. Surose, Swati V. Gaikwad. (2015). Current Status and The Future Potentials Of Renewable Energy. In India - A Review. International Journal of Advances in Science Engineering and Technology, ISSN: 2321-9009 (1): 33–38.

Qi, B.C, Aldrich, C., Lorenzen, L., and Wolfaardt, G.W. (2005). Acidogenic fermentation of lignocellulosic substrate with activated sludge. Chem. Eng. Communications, 192(9): 1221–1242. doi:10.1080/009864490515676.

Rodríguez, G., Lama, A., Rodríguez, R., Jiménez, A., Guillén, R., and Fernández-Bolaños, J. (2008). Olive stone an attractive source of bioactive and valuable compounds. Bioresource Technology, 99 (13): 5261–5269. https://doi.org/10.1016/j.biortech.2007.11.027

Roig, A., Cayuela, M. L., and Sánchez-Monedero, M. A. (2006). An overview on olive mill wastes and their valorisation methods. Waste Management (New York, N.Y.), 26(9), 960–969. https://doi.org/10.1016/j.wasman.2005.07.024

Romieu I., Meneses F., Ruiz S., Sienra J. J., Huerta J., White M. C., Etzel R. A. (1996). Effects of Air Pollution on the Respiratory Health of Asthmatic Children Living in Mexico City. American Journal of Respiratory Critical Care Medicine. 154, 300–307.

Sabiiti E.N., Bareeba, F., Sporndly, E., Tenywa, J.S., Ledin, S., Ottabong, E., Kyamanywa, S., Ekbom, B., Mugisha, J. and Drake, L. (2004). Urban market garbage: A hidden resource for sustainable urban/peri-urban agriculture and the environment in Uganda. The Uganda Journal, 50: 102–109.

Sabiiti, E. (2011). Utilising agricultural waste to enhance food security and conserve the environment. African Journal of Food, Agriculture, Nutrition and Development. 11: 1–9.

Sabine, C.L., Feely, R.A., Gruber, N., Key, R.M., Lee, K., Bullister, J.L., Wanninkhof, R., Wong, C.S.L., Wallace, D.W., Tilbrook, B. and Millero, F.J., (2004). The oceanic sink for anthropogenic CO2. Science, 305(5682), 367–371.

Sakiko Fukuda-Parr (2016) From the Millennium Development Goals to the Sustainable Development Goals: shifts in purpose, concept, and politics of global goal setting for development. Gender and Development, 24(1): 43–52.

Santosh Singh Raghuwanshi and Rajesh Arya (2019) Renewable energy potential in India and future agenda of research. International Journal of Sustainable Engineering, 12(5): 291–302.

Scarlat, N., Dallemand, J.F., Monforti –Ferrario, F. and Nita, V. (2015). The role of biomass and bioenergy in a future bioeconomy: policies and facts. Environ. Dev., 15: 3–34.

Scoggins A., Kjellstrom T., Fisher G., Connor J., Gimson N. (2004). Spatial Analysis of Annual Air Pollution and Mortality. Science of the Total Environment, 321: 71–85.

Sofyan, A., Murjaya, dan Subowo. (1997). Identifikasi status dan jangkauan pencemaran Lead dan Cadmium dalam tanah dan tanaman the. Laporan Akhir P5SL. (In Bahasia Indonesia).

Sriram, N., & Shahidehpour, M. (2005, June). Renewable biomass energy. In IEEE Power Engineering Society General Meeting, IEEE, 612–617.

Tumuhairwe J.B., Tenywa J.S., Otabbong E. and Ledin, S. (2009). Comparison of four low-technology composting methods for market crop wastes. Waste Management, 29: 2274–2281.

U.S. Environmental Protection Agency, (2000). Superfund: 20 Years of Protecting Human Health and the Environment. EPA 540-R-00-007. Washington, DC: U.S. Environmental Protection Agency. www .epa.gov/superfund.

UNCTAD Annual Report 2014, United Nations Publication. https://unctad.org/system/files/official-document/dom2015d1_en.pdf

United Nations, (2019). The Sustainable Development Goals Report 2019. United Nations Publications, New York, USA. https://unstats.un.org/sdgs/report/2019/The-Sustainable-Development-Goals-Report-2019.pdf, (accessed on March 2021).

Valenti, F., Porto, S.M.C., Selvaggi, R. and Pecorino, B. (2019). Co-digestion of by-products and agricultural residues: a bioeconomy perspective for a Mediterranean feedstock mixture. Sci. Total Environ., 700: 134440.

Wang X., Ding H., Ryan L., Xu X. (1997). Association between Air Pollution and Low Birth Weight: A Community-Based Study. Environmental Health Perspectives. 105: 514–520.

Wang, J., Hu, Z., Xu, X., Jiang, X., Zheng, B., Liu, X., Pan, X., and Kardol, P. (2014). Emissions of ammonia and greenhouse gases during combined pre-composting and vermicomposting of duck manure. Waste Management, 34(8): 1546–1552.

Ward, A.J., Hobbs, P.J., Holliman, P.J., and Jones, D.L. (2008). Optimisation of the anaerobic digestion of agricultural resources. Bioresour. Technol. 99(17): 7928–7940.

Westerman, P.W. and Bicudo, J.R. (2005). Management considerations for organic waste use in agriculture. Bioresource Technology, 96: 215–221.

Xu X., Wang L. (1993). Association of Indoor and Outdoor Particulate Level with Chronic Respiratory Disease. American Review of Respiratory Diseases. 148: 1516–1522.

Yangyang Li, Yiying Jina, Jinhui Lia, Yixing Chena, Yingyi Gongc, Yuezhong Lid and Jinfeng Zhangd (2016). Current situation and development of kitchen waste treatment in China. Procedia Environmental Sciences, 31: 40–49.

Yong, Z., Dong, Y., Zhang, X., Tan, T. (2015). Anaerobic codigestion of food waste and straw for biogas production. Renew. Energ. 78; 527–530.

11 Recent Trends in Biomass Conservation and Management

Amarendra Kumar Dash[1], Vivek Kumar[2], Dipti Sahoo[3], Jay Mant Jha[4], Balgovind Tiwari[5] and Rahul[6]

[1]Department of English, Rajiv Gandhi University of Knowledge Technologies, Nuzvid, AP – 521202, India

[2]Department of Chemical Engineering, Rajiv Gandhi University of Knowledge Technologies, RK Valley, AP – 516330, India

[3]Department of Management, Rajiv Gandhi University of Knowledge Technologies, Nuzvid, AP – 521202, India

[4]Department of Chemical Engineering, Maulana Azad National Institute of Technology, Bhopal, MP – 462003, India

[5]Department of Physics, Rajiv Gandhi University of Knowledge Technologies, RK Valley, AP – 516330, India

[6]Department of Chemical Engineering, Jaipur National University, Jaipur, RJ – 302017, India

CONTENTS

DOI: 10.1201/9781003196358-11

11.1   STATEMENT OF PROBLEM AND OBJECTIVES

The global demand for energy is continuously increasing due to population growth and large-scale economic developments. Industrial expansions are power-hungry. Between 2005 and 2030, energy needs are estimated to increase by 55%, from 11.4 billion toe (tons of oil equivalents) to 17.7 billion toe (WEO, 2007). Fossilized, renewable, and nuclear resources are the three major components of the world's energy continuum. Although fossil fuel remains the major energy front, efforts are being made to generate 80% of the energy from renewable sources. At present, hydroelectricity remains the most widely used renewable energy whereas an investment in solar energy is going to add significant stimulus to the energy demands worldwide. Besides, biomass, in the form of natural, agricultural, industrial, and human waste, is also going to be a major source of the renewable energy chain.

There are two major challenges with the use of biomass:

1 Although biomass is a renewable resource, the ever-increasing demand for energy, in the long run, will bring in a scenario where any amount of biomass will also be inadequate to fulfill the demand. This provokes the scientific community to optimize the efficiency of the biomass.
2 Increased use of biomass for electric power generation will rely on the qualitative and quantitative supply of biomass as well as the scale of sustainability of the entire fuel cycle and its impact on the natural ecosystem.

Against this backdrop, this article offers an overview of the recent trends in scientific biomass conservation and management.

11.2   BIOMASS POTENTIALS AND USAGE

According to Balat and Ayar (2005), "The only other naturally-occurring, energy-containing carbon resource known that is large enough to be used as a substitute for fossil fuels is biomass. Biomass is all non-fossil organic materials that have intrinsic chemical energy content." Biomass is available in the form of forest, agricultural, industrial, and urban waste. People get timber, food, feed, fibre, and energy from biomass. Energy is produced from wood wastes, agricultural and forest thinning residues, food processing wastes, and fibre or energy crops. Around 75% of the world's biomass are logging residues and waste by-products (Miles and Miles, 1992), which if not recycled or reused, would contaminate the natural ecosystem either being dumped in landfills or burnt.

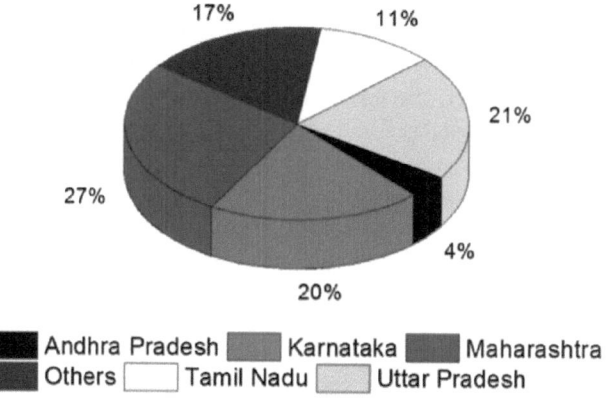

FIGURE 11.1 Installed capacity of grid-connected biomass in India.

Source: Rodl and Partner (2020).

India's vast population, economic expansion, rapid industrialization, and urbanization have created a thriving market for fuel and energy. The nation is blessed with large agricultural fields. The volume of biomass from plantations, forests, industrial byproducts, urban waste, and aquatic remains provide a readymade solution to its energy needs. The Government of India has expressed its determination to install the facility to generate 175 GW of renewable power by the end of 2022 that includes 10 GW worth of bioenergy capacity (Rödl and Partner, 2020), as shown in Figure 11.1.

According to the Council on Energy, Environment and Water, India could create 160,000 jobs and reduce its dependence on crude oil imports by using green methanol for the manufacturing of petrochemicals (Joshi, 2019). Biomass-based green methanol is considered to reduce India's carbon footprint, contribute to its grand vision of import substitution, create new green jobs, and bring down the soaring prices of fuel and energy.

Biomass is subjected to direct combustion to generate the steam that runs the turbine. The turbine drives a generator that produces electricity. Specific biomass materials that release less ash are used for direct combustion. Biomass is converted to combustible biogas with the help of gasifiers.

After this step, the biogas is utilized in driving a combined cycle, high-efficiency gas turbine. Devices such as micro-turbines, turbines, fuel cells, and reciprocating engines are used to convert biogas into energy. Simply, biogas is produced in an anaerobic digestion through the decomposition of organic matter, e.g., domestic wastes or farm residue. The biogas thus produced is a mixture of some gases, however, it is used for cooking, space heating, or generate electricity. A digester tank has relatively high temperature due to the exothermic nature of reactions. Hence, it kills many parasites, disease-producing species of organisms. Thus it is safe to use the digested matter as manure or fish feed.

A smaller scale biogas plant provides sufficient biogas so as to fulfil the requirement of energy for cooking and organic fertilizer for agriculture and horticulture use. On a bigger scale, biogas needs to be scrubbed for H_2S (hydrogen sulphide) removal, at first. Then it can be further utilized in street lighting, electricity generation through biogas-based generators at telecom towers or at community level. Even bigger scale provides the opportunity to compress the biogas and then it can be used in transportation applications.

Chemical conversion of biomass into pyrolysis oil is made possible by using heat. Storage and transportation of oil is more convenient than the solid biomass. It also can generate electricity. Phenol oil, a chemical adhesive, is also extracted from the biomass through pyrolysis.

A major part of the industrial biomass wastes is used for process steam, combined heat and power (CHP) generation and residential heating. Biomass energy conversion for power generation and the demand for the same are uneven in India. For example, urban and industrial clusters need a lot of energy but the supply of biomass is less and the cost of production is high in those places. On the contrary, the availability of biomass is high in the rural and agrarian zones with not much demand for energy.

11.3 ELECTRICITY, ETHANOL, AND HYDROGEN FROM BIOMASS

In a steam power plant, there are two fluids at work. The energy is produced by burning the biogas or biomass and then passed to water. In the coming years, biomass for the generation of electric power would depend mainly on the technologies for integrated gasification or gas turbine technology. These technologies have high conversion efficiencies for the overall energy. Biomass run power plants use such technology that is similar to that of coal-fired plants. Thus, biomass plants have efficiencies of 25% for steam-turbine generation and fuel delivery. The average size of a biomass power plant (BPP) is approx. 20 MW, although some of the wood-fired plants are constructed for 40–50 MW size and running with gas turbine/steam combined cycle. When burnt in the boiler, biomass produces steam. The steam runs a turbine which in turn propels a generator that produces electricity.

Pyroligneous oils are produced by pyrolysing the lignocellulosic biomass and these oils are used to produce hydrogen. The syngas thus formed is put for reforming into hydrogen. This way two problems can be solved together: (1) first the disposal of municipal waste, and (2) second is renewable source for hydrogen production to run vehicles on hydrogen (Veziroglu, 2001). Mechanism of hydrogen production from biomass can be given as the following:

$$Municipal\ solid\ waste + Air = CO + H_2$$

$$Biomass + H_2O + Air = H_2 + CO_2$$

$$Cellulose + H_2O + Air = H_2 + CO + CH_4$$

To produce ethanol we can use conventional high starch/sugar food crops, for example, corn, sorghum and sugarcane etc. for the fermentation which produces ethanol. Ethanol is a clean-burning, fuel which has high-energy compared to biogas. Choosing a particular biomass for ethanol production is very important for cost analysis.

Sugar can be extracted from any feedstock after suitable physical pretreatment steps followed by enzymatic hydrolysis. Then the extract is added with water and yeast and warmed in a fermenter. The yeast converts the sugars into alcohols. A distillation process can filter water as well as impure particles from the fermented alcohol. Typically it has 10–15% of ethanol. Then ethanol can be concentrated up to 95% by volume via a distillation step.

11.4   BIOMASS, ENERGY, AND STAKEHOLDERS IN INDIA

India's national biomass policy of the 1970s emerged out of its rural and renewable energy missions. The aim was to improve the efficiency of traditional biomass use through technological innovation and institutional support. Subsequently, the National Biogas and Manure Management Program (NBMMP) was founded to set up domestic biogas plants that would provide cooking fuel as well as organic manures for the rural households.

Currently, the Ministry of New and Renewable Energy (MNRE) and the Ministry of Environment and Forests (MoEF) are the key promoters of the bioenergy programs. Bioenergy plantation has been adopted as a new task under the national employment creation plan. The Integrated Energy Policy (2003) of the Planning Commission of India has undertaken several initiatives to promote the supply and use of bioenergy (Planning Commission, 2003) and there has been a stress upon one more time in the Integrated Energy Policy (2006) over bio-ethanol and diesel; plantation of bio-mass and gasification of wood; bio-gas plants (Planning Commission, 2003). Figure 11.2 captures the interdependent web of stakeholders dedicated to India's bioenergy expansion programs.

The existing bioenergy programmes and policies in India have multiple missions such as direct combustion and cogeneration to advance technologies for the optimal use of biomass reserves for power generation. Additional actions include the promotion of biomass gasifiers for electricity and biogas-based distributed/grid power, the development of fuel-efficient cookstoves, and the plantation of oil crops like Jatropha and Karanja.

11.5   MAJOR PROBLEMS IN BIOMASS CONSERVATION AND MANAGEMENT

The major challenges to biomass conservation and management are: 1) Identification and preservation of a sustainable biomass inventory; 2) Managing the pre-production and post-production supply chain; 3) Efficient technology for biomass conversion; 4) Effective environmental life-cycle management; and 5) Commercial viability at each stage of the process. The environmental side effects of biomass

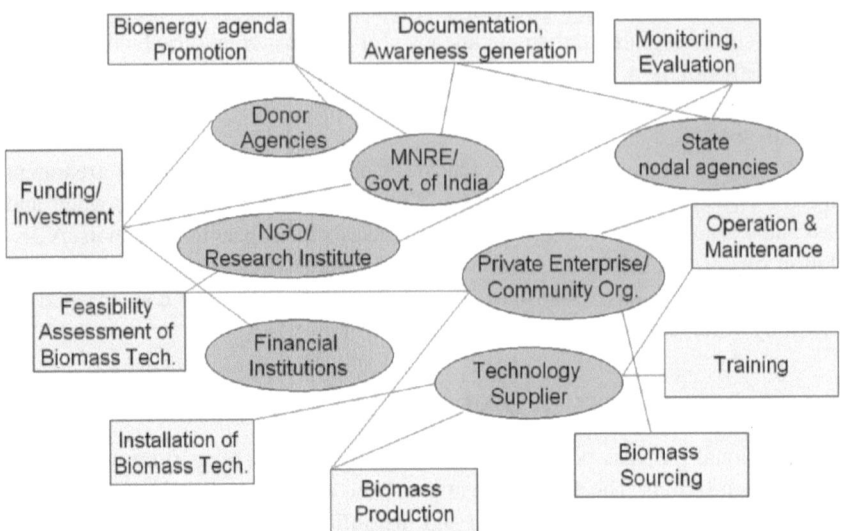

FIGURE 11.2   Bioenergy stakeholders in India.

Source: TERI (2010, p. 21).

production and conversion can be traced to their impact on agriculture and wildlife on one side and land, water, and air on the other.

Agricultural residues are usually available in huge amounts. However, their cost of collection is prohibitory. The physical and chemical properties of agri-residues are far less energy efficient compared to wood. High level of alkali metals like potassium and sodium in agri-residues leads to the heavy deposit of salts on the interiors of boilers. Agri-residues also need the additional mechanism to control sulphur dioxide emissions. One such mechanism is sulphur dioxide absorption using limestone in the circulating or bubbling beds (Miles and Miles, 1992).

Increased use of agricultural land for commercial energy crops will put pressure on agriculture for food. The collection of forest residues involves increased human activities in wildlife habitats. It also involves the erosion of grassland and the food supply chain for birds and animals. The former problem can be overcome by focusing upon the utilization of wasteland and non-agricultural land. The latter issue can be addressed by proper planning of the biomass collection from wildlife habitats. It is to be ensured that only abandoned or wastelands are diverted towards biomass fibre or energy crops and new habitats for birds and animals are created.

The supply of water for biomass conversion plants is a matter of concern in the drier and arid parts of India, especially the South and Western states. This issue can be addressed by developing rainwater storage systems, watershed management, and protection of the riparian areas. Table 11.1 presents an overview of environmental impact of biomass production and conversion activities.

TABLE 11.1
Environmental impact of biomass production and conversion

Biomass production	• **Soils:** Soil properties affected due to felling as well as yarding work. Nitrogen loss and site productivity gets affected due to residue removal. • **Air Quality:** Effect on emissions due to residue removal and combustion of wood.
Biomass conversion	• **Solid Waste Disposal:** Disposal and recycling of Wood ash. Issues related to energy requirements at such solid waste management sites. • **Air Quality:** Impact on particulate emission due to residential wood combustion.

Sources: Miles & Miles, 1992, Page 14.

Steam-turbine power plants demand a continuous supply of large quantity of premium quality water into their boilers. Although, water is usually recycled over and again, but a large amount of water gets evaporated in the process. Gas turbines do not have boilers, but they also need a lot of water for cooling to avert the formation of NOx. Disposal of waste-water is a permanent challenge because there is no final solution to this. Therefore, efficient use of water catering to the principle of "zero-discharge" is steadily becoming the new normal.

The availability of fuel-bound nitrogen in biomass is minimal. Nowadays, non-catalytic methods for the control of emission like ammonia injection, are preferably being followed to regulate any type of thermal NO_x formed during combustion. Dust or fugitive emissions are regulated by spraying water on the fuel stacks. The fly ash particulates can be either be kept covered or wet. At a later stage, they can be ploughed into the soil to enhance the quality of the land.

11.6 RECENT TRENDS IN BIOMASS CONSERVATION

The converted form of biomass is more energy-efficient than its natural form. Conversion technologies are in place for (a) direct use, (b) transformation, (c) electricity generation, (d) gasification, (e) pyrolysis, (f) charcoal production, and (g) modular systems (Intelligent Energy Europe, p. 11). The abundant resources of biomass are used in several ways, however, the major portion is utilized in following sectors:

11.6.1 THERMAL ENERGY AND THERMAL POWER

It has been suggested that direct burning of wood is not as efficient as gasification (Biagini et al., 2014). Gasification can be utilized in heating as well as electric power generation. In the second case of power generation, the overall efficiency has been found to increase (Situmorang et al., 2020). Hence, other than electricity production, the recovery and generation of heat from flue gases contributes to overall power and heat production and this is an energy-efficient pattern (Marchenko et al., 2020). Such power plants are called combined heat and power (or CHP) plants. For decentralized systems, small CHP plants work best and utilize the locally available fuel. In such case, transportation is not expensive and distribution of energy is at local level (Patuzzi et al., 2016). Such plants are suitable for low density of electricity and heat demand.

As waste wood need not be grown separately, it puts no effect of any kind on land, production of crops, and price of wood based materials. In most cases, it can be used as fuel in comparison to be used for material recovery or putting to pretreatment for further processing (Vis et al., 2016). However, it is found that when waste wood is utilized as fuel in the boiler systems of the thermal power plants this practice has a tendency to lead towards deterioration of the internal surfaces inside and in turn causes fouling and corrosion. This affects the efficiency and working life of plant. It has been studied that waste wood contains such matter which, during combustion, forms various deposits besides alkali chlorides (Jenkins et al., 1998). Due to this, the boilers of biomass based cogeneration power plants have greater chances of slag deposition, fouling and corrosion (Åmand et al., 2006). However, such ash-related problems can be reduced by mixing mineral and sulphur additives before the combustion step (Wang et al., 2012).

In a detailed study over different type of biomass based energy production plants at four different countries by Corona et al. (2020) it is observed that any biomass based energy generation plant would have following major areas responsible for creating an impact on environment: low efficiency for exergy, fuel transportation distance and flue gas emission for given energy production. However, disposal of ashes does not seem to be a problem as it can be recycled. Flue gas cleaning agents (ammonia, urea or sodium hydroxide) pose a concern as the processes of production of these agents have the factors responsible for climate change, eutrophication, and cumulative energy demand etc. Low dust removal efficiency adds to increased acidification and particulate matter content. In most of the cases, the use of additives reduced the harmful impacts by 3–12% in various categories.

11.6.2 BIOGAS

In the wake of sharp increase in fossil fuel prices the biomass based as well as other forms of renewable energies are in a good position to compete in the energy sector and claim to solve the energy scarcity issues in the remote areas. Among these alternatives, anaerobic digestion is most sustainable technology which is able to provide green energy using organic matter including waste food as raw material (Wall et al., 2017; Dahunsi et al., 2018; 2019; Dahunsi, 2019; Voelklein et al., 2019).

More than 1.3 billion tons of food is wasted globally which is approximately 30% of world food production (FAO, 2013). Besides, 15–63% of wasted food is ending up in municipal waste in developed and developing countries (Zhang et al.,

2014; Kim et al., 2016). It has been found that waste food is among top 3 (out of 15) bioresources with attractive investment opportunities (Dobbs et al., 2011).

An innovative study by Dahunsi (2020) for the electricity generation from the blends of food waste and spent animal beddings via anaerobic digestion route presents a way to utilize the fermentation liquids to run fuel cell by physically separating the fermentation and methanogenesis steps. Food waste have tendency to be easily biodegraded leading to accumulation of volatile fatty acids in the system which are converted to acids in further conversion steps. After sufficient time, in the environment of low pH, thus created, the microbes find their activities inhibited and methane production step is continuously slowed down and may even go for a complete halt.

In another study, a thorough review was conducted to compare the electricity generation potential and costs among food waste (FW)-based bioenergy, solar, and wind power. It concluded that promotion of the FW-based bioenergy has been helpful in various aspects and would be more competitive in developing countries (Thi et al., 2016).

11.6.3 Pellet and Briquette Manufacturing

Normally, biomass has low bulk density, inhomogeneous structure and low energy density. This makes biomass-based energy production less competitive. However, worldwide efforts are being made to improve on these parameters. Briquette and pellet making is a great solution to such case.

Briquette and pellet are the compacted and combinational biomass. Raw materials for briquette can be from wood chips, waste from textiles, agricultural and food waste to residues from furniture industry including industrial as well as urban biomass residues (Sawadogo et al., 2018) and charcoal (Lohri et al., 2016). Briquette has a greater density, less than 12% moisture, better space-to-weight ratio compared to its constituents making transportation more efficient. Due to their superior density and energy density, low moisture and good combustion properties biomass briquettes make themselves an attractive biofuel for electricity production and heating applications (Gangil, 2015; Yank et al., 2016). They are produced with diameter from 2–20 cm and length from 15–50 cm. Pellets, smaller in size, are similar to briquettes. Diameter from 6–7.25 mm and length from 10–36 mm. Effectively, they are a granulated form of biomass (Garcia et al., 2016; Spirchez et al., 2018).

In 2016, more than 800 sites produced above 29 million tonnes pellets globally (FAO, 2017). Last decade has witnessed an approximate 20% annual growth in biomass pelletization industry (WBA, 2014) mainly led by the policies made at global level (Dwivedi et al., 2014). Pellets and briquettes are useful in wood stoves, external combustion engines and in gasification and pyrolysis applications. Pellets offer better energy density and combustion efficiency which leads to reduction in ash production. Thus, it can be used in boilers with or without coal. In the process of pellet making following parameters are important while considering pellet quality-pressure, temperature, moisture content and particle size. As presented by Carone et al. (2011), temperature is the most important parameter which affects mechanical properties of pellets. Initial moisture content is the second and particle size the third most important parameter. Good-quality pellets are favoured by low moisture, high process temperature, and smaller particle size.

As the demand for pellet and briquette is increasing, the stress is on (a) to search for new waste materials that can be used in pellet production, and (b) in developing the pellet stabilizing raw materials. In a study by Zawiślak (2020), it is shown that lignocellulosic materials can be used for pellet making and that the raw material composition affects the calorific value of the pellet; ash and heavy metal generation at the end; energy consumption during pellet production; physical properties affecting the durability while in transport and storage.

11.7 BIOMASS SUPPLY CHAIN MANAGEMENT

The expansion of the bioenergy industry from its heavy dependence on sugar and starch to cellulose needs novel infrastructure and management. Advanced logistics is already available for the cornstarch-based ethanol system. However, that is not the case for a large variety of the feedstock for the cellulosic system. In an age of completion and cost-cutting, the transportation of low-density, low-priced, and wide varieties of feedstock from diverse locations to the processing unit demands diligent planning and coordination. The cost is often balanced by the presence of co-products and multiple modes of shipping. Despite this, the distance between the place of origin of the biomass and the processing destination and the quality of transport infrastructure influence the supply chain of the biomass types.

Biomass trade is often coupled with sustainable land use. Sustainable land management integrating food, feed, fibre, and fuel crops can be the way forward to more dynamic agriculture that can contribute to food supply as well as biomass production. Subject to soil conditions, under-productive agricultural fields can be used for biomass production to yield better revenue for the farmer. Strategic landscape management can produce agricultural residues as well as energy crops. Above all, it can offer economic and environmental benefits to the farmers and strengthen the biomass supply chain (CAST, 2012).

Recently, based on the software simulations, several Integrated Biomass Supply Analysis and Logistics (IBSAL) models have been developed to propose more dynamic and cost-saving biomass supply chain prototypes (Sokhansanj et al., 2006, 2008). The IBSAL models have taken cognizance of energy feedstock such as forest resources (Mahmoudi et al., 2009; Mobini et al., 2011), perennial grasses (Kumar and Sokhansanj, 2007; Sokhansanj et al., 2009), and agricultural residues (Ebadian et al., 2014, 2011; Sokhansanj et al., 2010; Stephen et al., 2010).

IBSAL models are particularly efficient in wide coverage of operations along with the impact of a process. Other supply-chain models are not able to capture such aspects and they miss the information of effects on harvesting and downstream processing. Besides, making the use of popular file formats for data storage and input for the model helps in sharing among a wider audience and bigger group of researchers (Lautala et al., 2015).

The IBSAL model of Lautala et al. (2015) displays the movement of biomass items through a continuum of harvest, storage, preprocessing, and transportation. The model includes the mechanical process and its impacts on the biomass at various stages of their flow. The conditions and processes in the input chain are simulated using equations and algorithms to predict the quality of the output at different scenarios that include economic analysis and optimization of biomass supply chain designs, tracking biomass quality changes, estimation of the required resources, and operational performance (bottlenecks, delays, and inefficiencies).

11.8 SUMMARY

Based on the discussions made in the previous sections, Table 11.2 sums up the key strategies for the conservation and management of biomass. The principal factors involved are: (1) availability of biomass, (2) pre-conversion supply chain, (3) biomass conversion, (4) post-conversion supply chain, and the commercial viability of the bioenergy. To sum up, on the one hand, the quality and the market competitiveness of the bioenergy depends upon the quality and the supply of the biomass. On the

TABLE 11.2
Sustainable Practices: Production and Conversion Management

• **Availability of Biomass**	• Identifying the locations of forest, agricultural, industrial and urban waste, surplus, and residues
	• Developing a nomenclature of biomass fuel types, their scale of availability, and their level of suitability
	• Biomass diversification, i.e., up-scaling the range and variety of biomass types
	• Describing the legal and environmental issues of the accessibility and utilization of biomass for energy
	• Identifying areas for future expansion to cope with the increasing demand for biomass
• **Pre-conversion supply chain**	• Cost-effective collection of biomass from diverse sources
	• Transportation from the place of collection to the industry
• **Conversion**	• Cost-effective conversion technology
	• Product quality optimization and maximum asset utilization
	• Effective environmental life-cycle management
	• Scope for future technological enhancements and cost savings
• **Post-conversion supply chain**	• High-quality storage facility
	• Cost-effective transport from industry to user
• **Commercial viability**	• Finding markets and buyers
	• Attractive power sales agreements
	• Analysing the cost and volume and supply chain of competing fuels

other hand, the encashment of the biomass types depends on finding or developing a market/customer base. This inter-dependent trajectory is displayed in Table 11.2.

Education and public communication of key information and the public policy related to biomass are as important as technology and finance. This can be achieved in several ways. The ecosystem for the production, conservation, and management of biomass should be identified. The key stakeholders should be given the necessary training to manage the entire life cycle of biomass including the environmental management systems such as ISO 14001.

According to Miles and Miles (1992), Geographic Information Systems (laced with sophisticated simulation, mapping, and graphics potentials), their timely updates and the integration and correlation of the information from diverse sources can be hugely advantageous. Integrated systems can be useful to monitor biomass production and conversion by facilitating the flow of information regarding the eco-friendly and sustainable development and utilization of biomass.

11.9   CONCLUSIONS

Conservation and management of biomass is crucial for sustainable development. The production of biomass, whether in solid or liquid form, needs to be sustainable and renewable too. The transportation of the feedstock to the conversion plant by road, rail, or waterways should be cost-effective. This means the material should cast into a form that will be easy to store, handle, and utilize. In this context, the low-bulk and energy density of several biomass types like agricultural residues make their transportation a specific challenge.

The biomass feedstock must be available over the life of the bioenergy plants. The quality and moisture content of the feedstock needs on-time assessment to guarantee effective conversion and reasonable means of payment. The certification of the source of imported biomass and the adoption of low-cost delivery methods, in financial and energy terms, are important.

The project developer has to pay specific attention to the design and construction of the bioenergy conversion plant. The selection of the geographic location of the plant should be based on its proximity to power, gas, and water supplies. Compliance with the legal-environmental framework and obtaining the license and permission is important.

The size of the plant and the technology for energy conversion relies on the volume and richness of the biomass, the reliability and maturity of technology. Markets or customers for the bioenergy outputs such as heat, electricity, gaseous fuels, liquid biofuels, or solid fuels such as pellets need to be identified, negotiated, and monitored, and prior buy-sell agreements make this process smooth.

REFERENCES

Åmand, L.-E., Leckner, B., Eskilsson, D., and Tullin, C. (2006) 'Deposits on heat transfer tubes during co-combustion of biofuels and sewage sludge', *Fuel*, 85(10), pp. 1313–1322. doi: https://doi.org/10.1016/j.fuel.2006.01.001.

Balat, M. and Ayar, G. (2005) 'Biomass energy in the world, use of biomass and potential trends', *Energy Sources*, 27(10), pp. 931–940. doi: 10.1080/00908310490449045.

Biagini, E., Barontini, F. and Tognotti, L. (2014) 'Gasification of agricultural residues in a demonstrative plant: Corn cobs', *Bioresource Technology*, 173, pp. 110–116. doi: https://doi.org/10.1016/j.biortech.2014.09.086.

Carone, M. T., Pantaleo, A. and Pellerano, A. (2011) 'Influence of process parameters and biomass characteristics on the durability of pellets from the pruning residues of Olea europaea L.', *Biomass and Bioenergy*, 35(1), pp. 402–410. doi: https://doi.org/10.1016/j.biombioe.2010.08.052.

CAST (Council for Agricultural Science and Technology) (2012) Energy issues affecting corn/soybean systems: challenges for a sustainable production. Issue Paper 48, CAST, Ames, Iowa.

Corona, B., Shen, L., Sommersacher, P. and Junginger, M. (2020) 'Consequential Life Cycle Assessment of energy generation from waste wood and forest residues: The effect of resource-efficient additives', *Journal of Cleaner Production*, 259, p. 120948. doi: https://doi.org/10.1016/j.jclepro.2020.120948.

Dahunsi, S.O., Olayanju, A., Izebere, J.O. and Oluyori, A.P. (2018) 'Data on energy and economic evaluation and microbial assessment of anaerobic co-digestion of fruit rind of Telfairia occidentalis (Fluted pumpkin) and poultry manure', *Data in Brief*, 21, pp. 97–104. doi: 10.1016/j.dib.2018.09.065.

Dahunsi, S. O. (2019) 'Mechanical pretreatment of lignocelluloses for enhanced biogas production: Methane yield prediction from biomass structural components', *Bioresource Technology*, 280, pp. 18–26. doi: 10.1016/j.biortech.2019.02.006.

Dahunsi, S. O. (2020) 'Electricity generation from food wastes and spent animal beddings with nutrients recirculation in catalytic fuel cell', *Scientific Reports*, 10(1), p. 10735. doi: 10.1038/s41598-020-67356-0.

Dahunsi, S. O., Olayanju, T. M. A. and Adesulu-Dahunsi, A. T. (2019) 'Data on optimization of bioconversion of fruit rind of Telfairia occidentalis (Fluted pumpkin) and poultry manure for biogas generation', *Chemical Data Collections*, 20, p. 100192. doi: https://doi.org/10.1016/j.cdc.2019.100192.

Dobbs, R., Oppenheim, J., Thompson, F., Brinkman, M. and Zornes, M. (2011) *Resource revolution: Meeting the world's energy, materials, food, and water needs.* Available at: www.mckinsey.com/business-functions/sustainability/our-insights/resource-revolution#. (Last accessed: 15 February 2021).

Dwivedi, P., Khanna, M., Bailis, R. and Ghilardi, A. (2014) 'Potential greenhouse gas benefits of transatlantic wood pellet trade', *Environmental Research Letters*, 9(2), p. 24007. doi: 10.1088/1748-9326/9/2/024007.

Ebadian, M., Sowlati, T., Sokhansanj, S., Stumborg, M. and Townley-Smith, L. (2011) 'A new simulation model for multi-agricultural biomass logistics system in bioenergy production', *Biosystems Engineering*, 110(3), pp. 280–290. doi: https://doi.org/10.1016/j.biosystemseng.2011.08.008.

Ebadian, M., Sowlati, T., Sokhansanj, S., Smith, L.T. and Stumborg, M. (2014) 'Development of an integrated tactical and operational planning model for supply of feedstock to a commercial-scale bioethanol plant', *Biofuels, Bioproducts and Biorefining*, 8(2), pp. 171–188. doi: https://doi.org/10.1002/bbb.1446.

FAO (2013) Food and Agriculture Organization of the United Nations. *Summary Report of Food Wastage Footprint: Impacts on Natural Resources* (FAO, Rome, 2013). ISBN: 978-92-5-107752-8.

FAO (2017) FAOSTAT. Food and Agriculture Organization (FAO), Rome.

Gangil, S. (2015) 'Superiority of intrinsic biopolymeric constituents in briquettes of lignocellulosic crop residues over wood: A TG-diagnosis', *Renewable Energy*, 76, pp. 478–483. doi: https://doi.org/10.1016/j.renene.2014.11.071.

Garcia, D.P., Caraschi, J.C., Ventorim, G. and Vieira, F.H.A. (2016) 'Trends and challenges of Brazilian pellets industry originated from agroforestry', *CERNE*, 22, pp. 233–240. Available at: www.scielo.br/scielo.php?script=sci_arttextandpid=S0104-77602016000300233andnrm=iso.

Jenkins, B. M., Baxter, L.L., Miles Jr., T.R. and Miles, T.R. (1998) 'Combustion properties of biomass', *Fuel Processing Technology*, 54(1), pp. 17–46. doi: https://doi.org/10.1016/S0378-3820(97)00059-3.

Joshi, A. (2019) '160,000 jobs can be created by manufacturing petrochems using green methanol: CEEW', *ETEnergyWorld*, 17 October. Available at: https://energy.econom ictimes.indiatimes.com/news/oil-and-gas/160000-jobs-can-be-created-by-manufa cturing-petrochems-using-green-methanol-ceew/71423985#:~:text=New%20De lhi%3A%20India%20could%20produce,Environment%20and%20Water%20 (CEEW). (Last accessed: 15 February 2021).

Kim, M.-S., Na, J.-G., Lee, M.-K., Ryu, H., Chang, Y.-K., Triolo, J.M., Yun, Y.-M. and Kim, D.-H. (2016) 'More value from food waste: Lactic acid and biogas recovery', *Water Research*, 96, pp. 208–216. doi: https://doi.org/10.1016/j.wat res.2016.03.064.

Kumar, A. and Sokhansanj, S. (2007) 'Switchgrass (Panicum vigratum, L.) delivery to a biorefinery using integrated biomass supply analysis and logistics (IBSAL) model', *Bioresource Technology*, 98(5), pp. 1033–1044. doi: https://doi.org/10.1016/j.biort ech.2006.04.027.

Lautala, P. T., Hilliard, M. R., Webb, E., Busch, I., Hess, J. R., Roni, M. S., Hilbert, J., Handler, R. M., Bittencourt, R., Valente, A. and Laitinen, T. (2015) 'Opportunities and Challenges in the Design and Analysis of Biomass Supply Chains', *Environmental Management*, 56(6), pp. 1397–1415. doi: 10.1007/s00267-015-0565-2.

Lohri, C. R., Rajabu, H. M., Sweeney, D. J. and Zurbrugg, C. (2016) 'Char fuel production in developing countries – A review of urban biowaste carbonization', *Renewable and Sustainable Energy Reviews*, 59, pp. 1514–1530. doi: https://doi.org/10.1016/j.rser.2016.01.088.

Mahmoudi, M., Sowlati, T. and Sokhansanj, S. (2009) 'Logistics of supplying biomass from a mountain pine beetle-infested forest to a power plant in British Columbia', *Scandinavian Journal of Forest Research*, 24(1), pp. 76–86. doi: 10.1080/02827580802660397.

Marchenko, O., Solomin, S., Kozlov, A., Shamanskiy, V. and Donskoy, I. (2020) 'Economic efficiency assessment of using wood waste in cogeneration plants with multi-stage gasification', *Applied Sciences (Switzerland)*, 10(21), pp. 1–15. doi: 10.3390/app10217600.

Miles, T. R. (Sr.) and Miles, T. R. (Jr.) (1992) *Environmental Implications of Increased Biomass Energy Use*, Final Report. National Renewable Energy Laboratory, NREL/TP-230-4633 • UC Category: 247 • DE92001219, p. 14.

Mobini, M., Sowlati, T. and Sokhansanj, S. (2011) 'Forest biomass supply logistics for a power plant using the discrete-event simulation approach', *Applied Energy*, 88(4), pp. 1241–1250. doi: https://doi.org/10.1016/j.apenergy.2010.10.016.

Patuzzi, F., Prando, D., Vakalis, S., Rizzo, A. M., Chiaramonti, D., Tirler, W., Mimmo, T., Gasparella, A. and Baratieri, M. (2016) 'Small-scale biomass gasification CHP systems: Comparative performance assessment and monitoring experiences in South Tyrol (Italy)', *Energy*, 112, pp. 285–293. doi: https://doi.org/10.1016/j.ene rgy.2016.06.077.

Planning Commission, Govt. of India, (2003). *Integrated Energy Policy*. New Delhi.

Planning Commission, Govt. of India, (2006). *Integrated Energy Policy*. New Delhi.

Rödl and Partner (2020) *Market overview: Bioenergy in India | Rödl and Partner, Rödl and Partner*. Available at: www.roedl.com/insights/renewable-energy/2020-02/market-overview-bioenergy-india. (Last accessed: 15 February 2021).

Sawadogo, M., Tanoh, S. T., Sidibe, S., Kpai, N. and Tankoano, I. (2018) 'Cleaner production in Burkina Faso: Case study of fuel briquettes made from cashew industry waste', *Journal of Cleaner Production*, 195, pp. 1047–1056. doi: https://doi.org/10.1016/j.jclepro.2018.05.261.

Situmorang, Y. A., Zhao, Z., Yoshida, A., Abudula, A. and Guan, G. (2020) 'Small-scale biomass gasification systems for power generation (<200 kW class): A review', *Renewable and Sustainable Energy Reviews*, 117, p. 109486. doi: https://doi.org/10.1016/j.rser.2019.109486.

Sokhansanj, S., Mani, S., Turhollow, A., Kumar, A., Bransby, D., Lynd, L. and Laser, M. (2009) 'Large-scale production, harvest and logistics of switchgrass (*Panicum virgatum L.*) – current technology and envisioning a mature technology', *Biofuels, Bioproducts and Biorefining*, 3(2), pp. 124–141. doi: https://doi.org/10.1002/bbb.129.

Sokhansanj, S., Mani, S., Tagore, S. and Turhollow, A. F. (2010) 'Techno-economic analysis of using corn stover to supply heat and power to a corn ethanol plant – Part 1: Cost of feedstock supply logistics', *Biomass and Bioenergy*, 34(1), pp. 75–81. doi: https://doi.org/10.1016/j.biombioe.2009.10.001.

Sokhansanj, S., Kumar, A. and Turhollow, A. F. (2006) 'Development and implementation of integrated biomass supply analysis and logistics model (IBSAL)', *Biomass and Bioenergy*, 30(10), pp. 838–847. doi: https://doi.org/10.1016/j.biombioe.2006.04.004.

Sokhansanj, S., Turhollow, A. and Wilkerson, E. (2008) *Development of the integrated biomass supply analysis and logistics model (IBSAL), Oak Ridge National*. Available at: http://wiki.ornl.gov/sites/publications/Files/Pub10657.pdf. (Last accessed: 15 February 2021).

Spirchez, C., Lunguleasa, A. and Matei, M. (2018) 'Particularities of hollow-core briquettes obtained out of spruce and oak wooden waste', *Maderas-Cienc Tecnol*, 20(1), pp. 139–152. Available at: http://revistas.ubiobio.cl/index.php/MCT/article/view/3016.

Stephen, J. D., Sokhansanj, S., Bi, X., Sowlati, T., Kloeck, T., Townley-Smith, L. and Stumborg, M. A. (2010) 'The impact of agricultural residue yield range on the delivered cost to a biorefinery in the Peace River region of Alberta, Canada', *Biosystems Engineering*, 105(3), pp. 298–305. doi: https://doi.org/10.1016/j.biosys temseng.2009.11.008.

TERI (2010) 'Bioenergy in India'. A backgound paper prepared for the International Institute for Environment and Development (IIED) for an international ESPA workshop on biomass energy, 19–21 October 2010, Parliament House Hotel, Edinburgh by The Energy and Resource Institute (TERI). Available at: http://pubs.iied.org/pdfs/G02989.pdf. (Last accessed: 15 February 2021).

Thi, N. B. D., Lin, C.-Y. and Kumar, G. (2016) 'Electricity generation comparison of food waste-based bioenergy with wind and solar powers: A mini review', *Sustainable Environment Research*, 26(5), pp. 197–202. doi: https://doi.org/10.1016/j.serj.2016.06.001.

Veziroglu, T. N. (2001) 'Hydrogen energy system for sustainability', *Journal of Advanced Science*, 13(3), pp. 101–116. doi: https://doi.org/10.2978/jsas.13.101.

Vis, M., Mantau, U., Allen, B. (Eds.), 2016. Study on the Optimised Cascading Use of Wood, pp. 1–337. No 394/PP/ENT/RCH/14/7689. Final report 394.

Voelklein, M. A., Rusmanis, D. and Murphy, J. D. (2019) 'Biological methanation: Strategies for in-situ and ex-situ upgrading in anaerobic digestion', *Applied Energy*, 235, pp. 1061–1071. doi: https://doi.org/10.1016/j.apenergy.2018.11.006.

Wall, D. M., McDonagh, S. and Murphy, J. D. (2017) 'Cascading biomethane energy systems for sustainable green gas production in a circular economy', *Bioresource technology*, 243, pp. 1207–1215. doi: 10.1016/j.biortech.2017.07.115.

Wang, L., Hustad, J. E., Skreiberg, O., Skjevrak, G. and Gronli, M. (2012) 'A Critical Review on Additives to Reduce Ash Related Operation Problems in Biomass Combustion Applications', *Energy Procedia*, 20, pp. 20–29. doi: https://doi.org/10.1016/j.egy pro.2012.03.004.

WBA (2014) Pellets: a fast growing energy carrier. World Bioenergy Association (WBA), Stockholm.

'WEO (World Energy Outlook) 2007- Executive Summary - China and India Insights' (2007) in. IEA Publications, Paris, France. Available at: www.mofa.go.kr/eng/brd/m_5658/down.do?brd_id=11049andseq=308729anddata_tp=Aandfile_seq=1. (Last accessed: 15 February 2021).

Yank, A., Ngadi, M. and Kok, R. (2016) 'Physical properties of rice husk and bran briquettes under low pressure densification for rural applications', *Biomass and Bioenergy*, 84, pp. 22–30. doi: https://doi.org/10.1016/j.biombioe.2015.09.015.

Zawiślak, K., Sobczak, P., Kraszkiewicz, A., Niedziolka, I., Parafiniuk, S., Kuna-Broniowska, I., Tanas, W., Zukiewicz-Sobczak, W. and Obidzinski, S. (2020) 'The use of lignocellulosic waste in the production of pellets for energy purposes', *Renewable Energy*, 145, pp. 997–1003. doi: https://doi.org/10.1016/j.renene.2019.06.051.

Zhang, C., Su, H., Baeyens, J. and Tan, T. (2014) 'Reviewing the anaerobic digestion of food waste for biogas production', *Renewable and Sustainable Energy Reviews*, 38, pp. 383–392. doi: https://doi.org/10.1016/j.rser.2014.05.038.

12 Revalorization of Waste Biomass for Preparing Biodegradable Composite Materials

K. R. Srivastava[1], D. B. Pal[3], P. K. Mishra[1] and P. K. Srivastava[2]

[1]Department of Chemical Engineering and Technology, Indian Institute of Technology, (BHU), Varanasi – 221005, Uttar Pradesh, India

[2]School of Biochemical Engineering, Indian Institute of Technology, (BHU), Varanasi – 221005, Uttar Pradesh, India

[3]Department of Chemical Engineering and Technology, Birla Institute of Technology, Mesra, Ranchi – 835215, Jharkhand, India

CONTENTS

DOI: 10.1201/9781003196358-12

12.1   INTRODUCTION

Waste biomass is a potentially useful material which has not yet been utilized for the welfare of mankind. While advances in science and technology have promoted the discovery and application of new and versatile materials, focus was somewhere lost from sustainability and applicability gained more importance. As a result the suitability of natural materials in everyday usage declined and factory manufactured materials took their place owing to their better properties. However, the advantage of these improvements in properties is debatable today as the man-made materials available today have also caused one of the greatest problems in the history of mankind. Petroleum-derived plastics were supposed to be the wonder material but today the accumulation of plastic waste has posed serious and global environmental threat.

Current research has been investigating the synthesis of composite materials, using agricultural waste biomass and other biodegradable polymers, as an alternative to plastics thereby increasing the economical viability of such solutions. Lignocellulosic biomass has emerged as surprising filler, improving the overall mechanical and barrier aspects of resulting composites.

Over the years, polyvinyl alcohol (PVA) has been used with polymers such as starch, chitosan, etc. and lignocellulosic polymers from a number of agricultural biomass, which is treated as waste. The agricultural residue such as stalks, straw and husk from food crops, cobs of corn, hemp fibre, pineapple leaf fibre, coconut coir, kenaf fibre, etc. are a rich source of cellulose, hemicelluloses, and lignin. These natural polymers have impressive properties at both macro and micro scale. Cellulose, when extracted, shows remarkable strength in its nano form. Lignin and hemicelluloses also have interesting properties. The extraction and separation of cellulose, hemicelluloses, and lignin is not quite economical yet and therefore products based on these are not widespread among the general public as of yet.

A brief overview of the plastic waste problem, lignocellulosic biomass, its structure and properties follows. This is followed by a comprehensive review of composite packaging films prepared using polyvinyl alcohol and lignocellulosic residue.

12.1.1 PLASTIC WASTE

It is astonishing to know that more than half of all of the plastic produced in the world since the beginning of its mass production in 1950s still exists today (Geyer et al., 2017; Hoornweg et al., 2013). It is estimated that around 8 billion tonnes of plastic has been produced since industrial production began in 1950 and out of that around 6 billion tones has ended up as waste (Geyer et al., 2017) (Figure 12.1). This plastic has accumulated as waste in our landfills, clogging our water ways, choking up digestive tracts of animals that accidently consume them, inadvertently reaching into systems of aquatic life in micro plastic forms (Allen et al., 2020; Barnes et al., 2009; Dris et al., 2017; Wang et al., 2020), and ultimately collecting in the oceans in the form of a large artificial island called the great pacific garbage patch (Lebreton et al., 2018; NOAA, 2020). The great pacific garbage patch is one of the five such garbage patches in the oceans where all the debris in the world, mostly plastic (as its density is less than water), is accumulated by the ocean currents. It is roughly 1.6 million square kilometres in size and predicted to be having at least 80,000 tons of plastic waste floating in this area (Lebreton et al., 2018). Needless to say a vast amount of plastic has been generated and then discarded because of urbanization and modern living standards, which have somehow prioritized the disposable culture.

Around 40% of the plastic produced today is for the purpose of packaging and therefore is the quickest to be disposed off and converted to waste, typically within a period of six months (Geyer et al., 2017). Geyer et al. also estimate that in 2015, out of majority of plastics produced, 42% was for the purpose of packaging and composed of PP and PE mainly. Segregated by sector and polymer type also show a similar picture, with the major plastic waste being generated in the form of PP and PE polymers. The second largest user sector of plastics is the construction and infrastructure sector, which consumes around 20% of plastics produced globally but the plastic here has a life span of around 35 years and therefore the major waste production is due to plastics used in packaging.

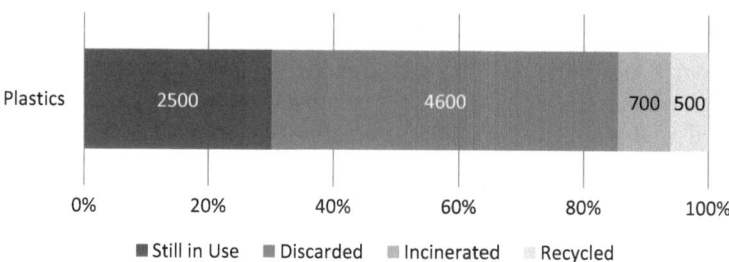

FIGURE 12.1 Fate of total plastic produced in the world since the 1950s (values in million metric tonnes).

Source: Geyer et al. (2017).

12.1.2 Lignocellulosic Fibres

Lignocellulosic fibres are an abundant renewable resource available throughout the world. These are natural fibres which can be extracted from various types of biomass and can be classified into following categories:

1. **Bast** (ramie, jute, flax, hemp, and kenaf),
2. **Leaf** (sisal, pineapple, and abaca),
3. **Seed** (coir, cotton, and kapok),
4. **Core** (jute, kenaf, and hemp),
5. **Grass and reed** (rice, wheat, and corn), and
6. Other types (wood and roots).

All the natural fibres majorly consist of lignin, cellulose, and hemicelluloses along with lower amounts of extractives, pectin, and pigments. Natural fibres have a complicated chemical composition and cell structure (Satyanarayana et al., 1990). The lignocellulosic fibres are in the form of a complex composite material with lignin and hemicelluloses forming the matrix in which cellulosic microfibrils are embedded (Rong et al., 2001).

Cellulosic fibres from different plants and from different parts of the same plant have different properties that depend upon a number of factors such as the fibre structure, cell dimension, and chemical composition and so on. Different cellulose types have different crystalline organizations impacting the mechanical properties hence the nature of cellulose present in the fibre determines its mechanical properties (Bledzki and Gassan, 1999).

Cellulosic fibres have interesting chemical bonding in which cellulose is linked to lignin and hemicelluloses by means of hydrogen bonds. Similarly chemical bonds exist between hemicelluloses and lignin also. Cellulose fibres consist of cellulose microfibrils bound within the lignin–hemicellulose matrix. The cellular structure of lignocellulosic fibre is composed of cells containing both amorphous (disordered) and crystalline (highly ordered) regions interconnected through hemicelluloses and lignin fragments (Klemm et al., 2005).

Structure-wise, a plain plant fibre is basically a single cell with length ranging from 1 to 50 mm and diameter around 10–50 μm. These plant fibres may be imagined as microscopic structures of tubular morphology consisting of multiple cell walls formed of cellulose microfibrils with diameter 10–30 nm, each formed by 30–100 cellulose molecules. Cellulosic fibre is made of concentric cylindrical structures with lumen and a central channel in the centre, surrounded by cell walls which are also continuous and cylindrical. The lumen is for transportation of water and nutrients. The cell wall of each fibre is made up of three main layers viz. middle lamella, primary wall, and secondary wall in order from outer to inner layer. All these layers are formed by crystalline microfibrils embedded in amorphous regions of lignin, hemicelluloses, and amorphous cellulose matrix. The secondary wall is further subdivided into external, middle, and internal secondary walls denoted by S1, S2, and S3, respectively. These layers are composed of crystalline microfibrils oriented at specific angles with respect to each layer, called the microfibrillar angle (Abdul Khalil et al., 2012; Pietak et al., 2007). The primary wall consists of disordered arrangement of cellulose fibrils (amorphous cellulose regions) embedded in a matrix of lignin, protein, pectin, and hemicelluloses. Crystalline cellulose microfibrils arranged helically make up the secondary walls. This layer is composed of cellulose microfibrils arranged in a helical formation. The microfibrils, of diameter around 10 to 30 nm, are arranged in an amorphous region formed of lignin and hemicelluloses. The outer most layer of cell, the middle lamella, is predominantly composed of pectin acting as the cement between fibres (Abdul Khalil et al., 2012; Pietak et al., 2007).

Cellulose is the main part of cellulosic fibres. Cellulose is a linear polysaccharide polymer formed by condensation of β-D glucose monomer building blocks that are bound together through the bonds between 1 and 4 carbons, called as β (1→ 4) glycosidic linkages (Thygesen et al., 2007).

Hemicelluloses are one complex group of carbohydrates surrounding the cellulose fibres of plant cells with other carbohydrates (e.g., pectin). Hemicelluloses is considered to be the compatible agent between cellulose and lignin (Rong et al., 2001). Mostly hemicelluloses consist of xylans (many five-carbon sugar molecules combined), uranium acid (i.e. sugar acid), and arabinose. Xylan is a polysaccharide of five carbon sugar pentose, D-xylose, linked by β (1→4). Hemicelluloses are highly hydrophilic and soluble in alkaline medium and hydrolyzed in presence of acids.

Lignin is the strongly interlinked, amorphous, and molecularly complex structure. It is the binding material for fibre cells and fibrils creating the cell wall. It is basically a three-dimensional and amorphous macromolecule composed of phenyl propane units. Lignin is mainly attributed for rigidity of the plant cell wall producing an impact and compression resistant structure. Lignin acts as a biological protection and a reinforcement agent to resist the gravity forces and the wind. By joining the fibres together lignin increases the compressive strength of natural fibres to form a rigid structure (Bjerre and Schmidt, 1997)

12.1.3 CELLULOSE, MICROCRYSTALLINE CELLULOSE, AND NANOCELLULOSE

Cellulose is the most abundant renewable polymer on earth. Majority of cellulose is synthesized by plants while some aquatic organisms, amoebae, bacteria, and fungi can also synthesize cellulose (Heux et al., 1999). Cellulose is a linear homopolymer of D-glucopyranose repeating units linked by α-1–4-glycosidic linkages. The D-glucopyranose monomer contains three hydroxyl groups which confer to it the ability to form hydrogen bonds. This hydrogen bonding directs the crystalline packing and thus governs the crystallinity of cellulose (John and Thomas, 2008). Cellulose microfibrils are the elementary structural constituent of cellulose created during the biosynthesis. These fibrils are long chains of poly-β-(1–4)-D-glucosyl residues aggregated to create long filament-like bundle of molecules. Intermolecular hydrogen bonds are responsible for lateral stabilization of these threads (Andresen et al., 2006; Dufresne et al., 1997; Stenstad et al.,

2008). Diameters of individual cellulose microfibrils range from 2 to 20 nm. Each microfibril is a filament of cellulose crystals joined as per the microfibril axis by amorphous domains (Azizi Samir et al., 2005). Siró and Plackett (2010) have reviewed the differences persistent in the scientific literature on the categorization applied to cellulosic fibres. The expression microfibril commonly refers to the 2–10 nm cellulose fibres of high aspect ratio, which are formed during biosynthesis of cellulose in higher plants. However, the microfibril diameters may vary depending on their origin. Nanofibrils and nanofibres are synonyms used for microfibril. Even though MFC nano elements of thickness 3–10 nm have been found, typically they are in the 20–40 nm range as bundles of cellulose microfibrils (Svagan et al., 2007). Acid hydrolysis causes transverse cleavage in the amorphous regions of cellulose microfibrils and sonication results in the formation of rod-like material with relatively low aspect ratio and is referred to as cellulose whiskers. Cellulose whiskers are also called nanowhiskers (Garcia de Rodriguez et al., 2006), nanorods (Dujardin et al., 2003), and rod-like cellulose crystals (Iwamoto et al., 2009). Strong hydrogen bonding among the cellulose crystals causes aggregation (Levis and Deasy, 2001) leading to a cellulose structure called microcrystalline cellulose (MCC). MCC is available for commercial use and is basically utilized as a viscosity enhancer and in the pharmaceutical industry as a binder (Janardhnan and Sain, 2006).

12.2 PRETREATMENTS/SURFACE MODIFICATIONS OF LIGNOCELLULOSIC FIBRES

There have been numerous attempts to improve upon the adhesion between matrix and fibre using chemical modifications of the fibre. Such chemical modifications can be brought upon by chemical treatments with alkali (sodium hydroxide, potassium hydroxide), acetyl, silane, benzyl, acryl, permanganate, peroxide, isocyanate, titanate, zirconate and acrylonitrile treatments and use of maleated anhydride grafted coupling agent (Faruk et al., 2014; Singh et al., 1996). With maleated anhydride grafted coupling agent and alkali, acetyl, silane treatments being conventionally used and due to their effect on environment, enzyme treatment is slowly being favoured upon due to its environmental friendliness (Faruk et al., 2014).

12.2.1 ALKALI TREATMENT

Alkali treatment removes hemicelluloses, lignin, and pectin, and exposes cellulose. It also increases surface roughness and surface area to offer better interfacial interaction. Alkali action can modify cellulose structure, mild treatments enhance cellulose crystallinity while harsher treatments convert crystalline cellulose to amorphous material (Beckermann and Pickering, 2008; Kabir et al., 2012a). Alkali treatment has been reported to improve fibre strength also (Beckermann and Pickering, 2008; Bera et al., 2010). Improvements in properties such as tensile strengths, Young's modulus and impact strength have been reported in many studies along with improvement in fracture toughness and flexural properties of composites as well as thermal stability and long-term moisture resistance. The moisture resistance is reported to improve due to the reduced moisture uptake observed with alkali treated natural fibres (Bera et al., 2010; Goda et al., 2006; Gomes et al., 2007; Islam et al., 2010; Kabir et al., 2012b; Nor Azowa Ibrahim et al., 2010).

12.2.2 ACETYLATION TREATMENT

Acetylation treatment causes esterification by reaction between acetyl groups and hydroxyl groups on the fibres and increases hydrophobicity (Hill et al., 1998). Acetylation has been reported to improve interfacial bonding, mechanical properties as well as heat stability and microbe resistance in natural fibre composites (Bledzki et al., 2008; Khalil et al., 2001; Tserki et al., 2005). Excess treatment can be unfavourable to mechanical properties. It has been reported that degradation of cellulose and cracking of fibres occurs due to the catalysts used as a result acetylation lowers the mechanical properties (Bledzki et al., 2008). Commonly acetylation treatment is preceded by the alkaline treatment.

12.2.3 SILANE TREATMENT

Silanes are versatile reagents that are used for the treatment of fibres. Silanes are advantageous when composites of natural fibre with hydrophobic polymers are to be made as they are unique molecules with different chemical groups at both ends and both ends can join covalently with hydrophilic groups of the fibre and hydrophobic groups of the matrix respectively. In silane modification of natural fibres, water hydrolyzes alkoxy groups of silane and forms silanol (Si–OH) groups which forms bonds with hydroxyl groups on the fibre surface (Rachini et al., 2012; Xie et al., 2010) to form hydrogen or covalent bonds. The silanes encountered most often are alkyl, amino, glycidoxy and methacryl silanes. Silanes increase the hydrophobicity of natural fibres and strength of NFCs. Significant improvements are seen when silane and the matrix form covalent bonds (Pickering et al., 2003; Rachini et al., 2012).

12.2.4 MALEIC ANHYDRIDE GRAFTING

Maleic anhydride (MA) grafted polymers have been widely used as linking agents to improve composite properties. Scientific literature about maleic anhydride grafting shows that MAPP, a maleic anhydride grafted polypropylene is the most commonly used graft polymer. MAPP improves the composite properties by reacting with hydroxyl group on fibre surface by covalent or hydrogen bonding while the polymer attached to the maleic anhydride provides compatibility with the matrix, usually PP. Franco–Marques found that maleic anhydride grafting to PP improved mechanical strength of PP matrix composites (Franco-Marquès et al.,

2011). Maleic anhydride grafting can be regarded as the best among other surface treatments to improve matrix-fibre interactions as shown by Bera et al., who found MAPP to give two times higher composite strength compared to silane treatment (Bera et al., 2010). Kazayawokoo et al., explained that MAPP could improve mechanical strength of composites due to its ability for better dispersion and wetting of the fibre (Kazayawoko et al., 1999). The MA grafted PLA has shown improvements in mechanical properties and thermal stability of the PLA matrix–natural fibre composite (Avella et al., 2008; Kang et al., 2014; Yu et al., 2014).

12.3 BIODEGRADABLE POLYMERS USED FOR COMPOSITE PREPARATION

Biopolymers refer to polymers that are either have a biological origin for e.g. Originating from plants or animals; the prefix "bio" derived from Greek "bios" refers to human life. Recently the prefix has been used quite liberally leading to confusions. The European Bioplastics Association has therefore classified the polymers into three broad types referring to three types of bioplastics (Figure 12.2):

1. Bio-derived and biodegradable/ compostable polymers (PLA, PHAs etc..); 2. Fossil fuel / petrochemical derived but biodegradable (PCL, PVA etc.) and 3. Bio derived and non-biodegradable (bio-PE, bio PET, etc.)

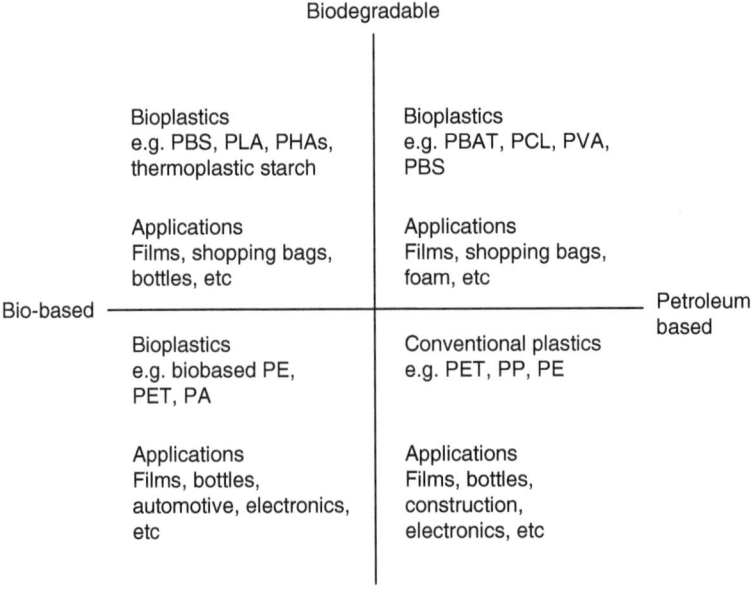

FIGURE 12.2 Classification of biopolymer types as per the European Bioplastics Association.

Biopolymers can also be classified according to their origin as follows (Petersen et al., 1999; van Tuil et al., 2000):

1. Polymers such as polysaccharides (e.g. cellulose, chitin, starch), proteins, lipids, and lignins directly derived from natural sources such as plants or animals
2. Polymers resulting from chemical synthesis of renewable bio-derived monomers (PLA obtained by polymerization of lactic acid produced from dextrose. Also, bio derived polyethylene produced by polymerization of ethylene derived from ethanol from sugar also falls under this category.
3. Polymers produced by microbial fermentation. Examples include Polyhydroxyalkanoates or PHAs, also consisting of polyhydroxybutyrate (PHB) and copolymers of hydroxybutyrate and hydroxyvalerate (PHBV).
4. Polymers derived by synthesis of petrochemicals; (Table 12.1)

Major focus is necessary to develop alternative to plastic. While we already have naturally occurring polymers, which are also renewable (poly saccharides such as cellulose and derivatives, starch, proteins such as casein, gluten, and carrageenan etc. hemicelluloses such as Xylan etc.), a number of polymers have been developed over the years which have properties similar to plastics and are derived from biomass for e.g. Polylactic acid, Polyhydroxy butyrates, and its derivatives etc. Such kinds of biopolymers (bioplastics) are produced as a result of bacterial fermentation. For example, PLA is produced by bacterial fermentation of starch to lactic acid which is then polymerized to produce PLA. While other bacteria produce PHBs when grown in special media. There are some other bioplastics which are although produced synthetically from petroleum derived chemicals but are still biodegradable for instance polyvinyl alcohols, poly caprolactones etc. Some of the popular biopolymers are briefly discussed here:

12.3.1 Polylactic Acid

LA is a polyester formed by the polymerization of lactic acid (Garlotta, 2001). It is made from lactic acid which in turn is derived from renewable polysaccharide polymers rich in glucose such as sugarcane, corn etc. (Farah et al., 2016; Madhavan Nampoothiri et al., 2010). USFDA has put PLA under the GRAS category (generally recognized as safe) which means PLA is considered safe for various uses. PLA has high modulus similar to PS, has high strength while also being biocompatible, and its films are transparent. PLA however is not efficient at moisture prevention and therefore needs either a barrier material coating (SiOx/AlOx93/ PP or PE) in order to be used as packaging for moisture sensitive items. PLA is not biodegradable as such and needs to be either recycled or treated in an industrial composting facility to completely degrade into plant fertilizer (Farah et al., 2016; Madhavan Nampoothiri et al., 2010).

TABLE 12.1
Classification of Biopolymer Types as per their Source of Origin

Environmentally degradable polymers

Natural origin		Synthetic origin	
Biomass/ derived from agri-resources	Derived from microorganisms	Biotechnology products	Petrochemical products
Polysaccharides	Polyhydroxyalkanoates (PHA)	Polylactides	Conventionally synthesized petrochemicals
Proteins	Polyhydroxybutyrate (PHB)	Polyglycolides	Polyvinyl alcohol (PVA)
	Poly-hydroxybutyrate – co-valerate (PHBV)	Poly lactide-co-glycolide	Polycaprolactone (PCL)
Polysaccharides — albumin Fibrinogen/ fibrin		Polylactic acid	Aromatic co-polyesters (e.g. PBAT)
Cellulose and derivatives — Soy protein		Poly glycolic acid	Polyesteramides (PEA)
Starches and derivatives — Zein			Polyurethane (PU)
Plant/ algal polysaccharides — Polynucleotides			Poly (ortho ester amide copolymers)
Alginate — Wheat gluten			Alipathic copolyesters (e.g. PBSA)
Pectin — Casein/ Collagen/ gelatin			
Insulin			
carrageenan			

Sources: Adapted from Averous and Boquillon (2004); Satyanarayana et al. (2009).

12.3.2 POLYHYDROXYALKANOATES

PHAs are produced by bacterial fermentation of renewable substrates. These consist of aliphatic polyesters and are stored inside cells as granules which serve as energy storage within bacterial cells. The major advantage with PHAs is their thermal and barrier properties match those of PET and PVC making them excellent alternative biopolymers but their high cost of production has inhibited their widespread use. (Albuquerque and Malafaia, 2018; Pascault et al., 2012; Raza et al., 2018).

PHB is the most widely studied PHA but has poor processibility due to its crystalline nature which makes it brittle. PHBV (3-hydroxybutyrate-3-hydroxyvalerate) is the polymer formed when 3-hydroxyvalerate (HV) monomers are incorporated into the PHB polymer. It has better properties than PHB but again is costlier (Anjum et al., 2016; Wang et al., 2013). In fact all the PHAs are costly due to the high cost of fermentation facilities and downstream processing systems (Snell and Peoples, 2009; Volova, 2004). PHA can be the choice of polymer for packaging in the future but certain improvements are required in its brittleness, high cost, and poor processability (Volova, 2004). Different approaches to improve PHA film properties have been pursued. PHA has been laminated/ blended with other polymers such as PLA, TPS, PBAT etc. or plasticization and reinforcement with inorganic or organic fillers has also been carried out with promising results. (Khosravi-Darani and Bucci, 2015).

12.3.3 POLYBUTYLENE SUCCINATE

PBS is a biodegradable polymer formed by condensation polymerization of 1–4 butanediol (BDO) and succinic acid. While succinic acid can be synthesized by fermentation of glucose and sucrose, BDO is currently petroleum-derived. But there is possibility of bio based BDO and thus PBS can be made completely bio based and biodegradable. PBS is promising biopolymer as it has mechanical properties similar to PP and PE, good thermal stability and is easily processable. Although highly crystalline nature of PBS causes it to become brittle and loose ductility over time and also makes it slower to natural biodegradation compared to other aliphatic polyesters such as PHA (Zeng et al., 2012).

12.3.4 STARCH-BASED THERMOPLASTICS

Starch is a wonderful carbohydrate polymer consisting long chains of glucose which are either branched (amylopectin) or straight (amylose). Usually there is 20–25% of amylose and 75–80% of amylopectin in starch (Vilpoux and Averous, 2004). Starch is abundant and can be sourced from various crops such as corn, sugarcane, maize, sweet potato etc., thereby reducing its cost. Starch however in itself cannot form packaging films due to its hygroscopic nature, high crystallinity, which makes it brittle and hard poorly processable. Also the tensile strength of starch is not suitable for packaging. Addition of plasticizers to starch combined with the thermo mechanical treatment in extruder improves the starch properties improve the mechanical and barrier properties (Vilpoux and Averous, 2004). Starch has been combined with several biopolymers to improve its properties. Starch has been combined with PVA, PLA, PBS, PHA, PCL, PBAT, PP, and PE also to improve flexibility and barrier properties. In order for thermoplastic starch based films to become mainstream packaging material, further research is needed to develop films with higher performance and stability at lower cost (Ortega-Toro et al., 2017; Versino et al., 2016).

12.3.5 Polyvinyl Alcohol

PVA is a biodegradable and nontoxic polymer derived from petrochemicals. It is not formed via polymerization of vinyl alcohol; instead, it is formed by the hydrolysis of polyvinyl acetate in presence of caustic to generate polyvinyl alcohol in varying degrees of hydrolysis. The molecular weight and degree of hydrolysis of PVA in general dictates its properties. For example, tensile strength, water resistance, and solvent resistance have been found to increase as molecular weight and degree of hydrolysis increases. Conversely, water solubility, and flexibility of films decrease with such an increase in the molecular weight and degree of hydrolysis (Mousa et al., 2016; Tang and Alavi, 2011).

PVA has high mechanical strength and good chemical resistance but is susceptible to moisture and therefore does not possess good barrier properties (Chiellini et al., 2003; Xiong et al., 2008). Barrier properties and thermal stability can be improved by addition of plasticizers which greatly improve the processibility also. PVA is a semi crystalline polymer comprising of predominantly amorphous phases and only a small amount of crystallinity (Guimarães et al., 2015a). Depending on the degree of hydrolysis of polyvinyl acetate, PVA may consist of 1, 3-diol units, or 1, 2-diol units. The abundant hydroxyl groups on PVA impart hydrophillicity to the polymer and its blends and composite materials are therefore widely used as packaging materials (Mousa et al., 2016). While PHAs and PHBs are excellent materials of choice, they are produced via bacterial fermentation and not economical yet. PLA is promising material and somewhat cheaper too but requires industrial composting facility to be able to degrade naturally otherwise its degradation is similar to other plastics. Polyvinyl alcohol like PLA is cheaper, economical, and is comparatively easier to biodegrade without composting. It is readily soluble in water and does not pose any harm to aquatic life at up to 5% v/v concentration. PVA can also be cast easily into films and processed just as plastic polymers are.

Having seen the suitability of this biodegradable polymer (PVA) in composite preparation, we further continue to review the various composites prepared using PVA and lignocellulosic materials, PVA, starch and lignocellulosic materials, and composites prepared using PVA and nanocellulose, with a brief overview of various methods used to extract nanocellulose.

12.4   PVA COMPOSITE FILMS REINFORCED WITH LIGNOCELLULOSIC FIBRES

Polyvinyl alcohol is an excellent film-forming polymer with high gas barrier properties and water solubility. It is also biodegradable which has made PVA a material of choice for biodegradable composite films. PVA has been successfully incorporated with polymers such as starch, chitosan, and xylan etc. to improve properties and/or to reduce costs. Table 12.2 lists the properties of different composite films prepared using PVA and lignocellulosic fibers; and PVA-starch and lignocellulosic fibers.

Asgher et al. (2017) prepared bacterial lignolytic enzyme mixture and treated wheat straw with the lignolytic enzymes which enhanced the cellulose and reduced the lignin content. The delignified wheat straw was further modified by bacteria which deposited bacterial cellulose fibres over the delignified wheat straw. Modified straw-PVA composites were prepared using compression moulding. The bacteria modified wheat straw showed best results of tensile strength. Glycerol addition improved the tensile strength further. However, the water uptake rate was very high in case of bacteria modified wheat straw–PVA composite. Mittal et al. (2016) prepared PVA-starch composite films with barley husk and palmitic acid graft copolymerized barley husk. The films were cross linked with urea-formaldehyde and plasticized with glycerol. Addition of starch to PVA films reduced the tensile strength and elongation at break-point while cross linking helped in increasing the tensile strength and reducing elongation at breaking point. They reported increase in %elongation and tensile strength with reinforcement of barley husk and graft copolymerized barley husk, the observed increase was greater in case of the latter. Water uptake and water vapour permeability increased with addition of natural barley husk while it decreased upon cross linking. Similarly, the graft copolymerized barley husk reinforced composite films showed comparatively lower values of water uptake and water vapour permeability. Meng et al. (2018) prepared PVA-carrageenan composite films plasticized with polyols derived from liquefied banana pseudo-stem. They studied the effect of novel plasticizer on thermal, mechanical and hydrophobic properties of carrageenan–PVA composite films. The polyol addition progressively reduced tensile strength and increased % elongation. However, incorporation of inorganic ions such as calcium and potassium ions improved the tensile strength as well as %elongation. Swelling in water also got reduced after plasticization with liquefied banana pseudo-stem derived polyol. Shi et al. (2017) used spirulina, PVA, and glycerol to cast composite films. Tensile strength and water resistance increased while %elongation decreased with increasing PVA concentration. Ali et al. (2015) studied the physico-mechanical properties of biocomposite films of polyvinyl alcohol (PVA) reinforced with kenaf lignocellulosic fibre. Kenaf fibres were chemically treated with either chromium sulphate alone or with chromium sulphate followed by sodium bicarbonate. The PVA composite films having treated fibres showed higher tensile strength compared to composite films having untreated fibre. The chromium sulphate-sodium carbonate treated kenaf fibre had higher mechanical properties and lower moisture uptake than chromium sulphate treated kenaf fibre.

Kalambettu et al. (2015) prepared PVA-pineapple leaf fibre composite films and evaluated their water uptake behaviour. The films with higher fibre content showed lower water uptake exhibited higher tensile strength. A soil burial test was performed to analyse the biodegradability by burying the composite films in soil in laboratory and taking out after a fixed period to measure biodegradability. The samples were thoroughly washed with water to remove all PVA and the degradation of leaf fibres in terms of weight loss was measured. This method might estimate the degradation of fibres but that of PVA could not be determined. Shen et al. (2015) prepared composite films of PVA and sugar-beet pulp. The water vapour permeability and tensile strength increased with increase in sugar-beet pulp content but %elongation at breakpoint decreased. Sorbitol was found to be a better plasticizer than glycerol for SBP-PVA composite films. Solikhin et al. (2018) studied the properties of PVA-chitosan films reinforced with lignocellulosic nanofibres (LCNF). The films made with 0.5% LCNFs had the highest tensile strength which decreased with increased loading. PVA/chitosan polymer blends with 0.5% LCNFs resulted in regular and smooth external surfaces and enhanced tensile strength of the films. Sutka et al. (2015) prepared electro-spun fibrous mats of PVA and lignocellulosic nanowhiskers (LCNWs) from hemp shives prepared via ball milling, ultrasonication, and steam explosion. They reported an increase in the tensile strength with addition of LCNWs up to 10 wt%.

12.5 PVA-STARCH COMPOSITE FILMS REINFORCED WITH LIGNOCELLULOSIC FIBRES

Priya et al. (2014) investigated the plasticizing effect of citric acid and cross linking effect of glutaraldehyde on the properties of *Grewia optiva* fibre reinforced composite films. Best values of TS and EB were displayed by films having 20 wt% citric acid, 0.2 wt% glutaraldehyde, and 20 wt% fibres. The films also exhibited antibacterial activity against gram-positive and gram-negative bacteria. Singha et al. (2015) prepared PVA-starch composite films reinforced with delignified and MMA grafted delignified *Grewia optiwa* fibres. The films were also plasticized with citric acid and cross linked with glutaraldehyde. The composite films exhibited maximum tensile strength and %elongation at 20 wt% for ungrafted fibre and at 15 wt% for MMA grafted fibre. The increase in tensile strength in case of ungrafted fibre was due to the better adhesion between fibre and polymers while in case of grafted polymer composite films it was due to the rough surface created by the grafting of MMA which led to better adhesion between matrix and fibre. Guimarães et al. (2015b) studied thermal, mechanical and structural properties of PVA-modified cassava starch (MCS) films. The films showed best mechanical properties at the PVA: starch ratio of 20: 80. Higher proportions of MCS made the composites opaque, rigid and brittle while also being thermally stable and water vapour resistant. Higher amount of PVA (60 and 80 %) resulted in

TABLE 12.2

Properties of PVA and PVA – Starch Composite Films Reinforced with Lingo-cellulosic Fibers

Polymer matrix	Reinforcement	Additives	Tensile strength MPa	Elongation at break %	Water Vapour permeability g.m/m².s.Pa	Remarks	Reference
PVA	Bacteria modified wheat straw	Glycerol	64.73	6.73	-	High water uptake rate due to impregnation with bacterial cellulose	(Asgher et al., 2017)
PVA – Starch	Barley Husk (Graft copolymerized)	Urea Formaldehyde	18.54	48.37	1.64E-10		(Mittal et al., 2016)
PVA	carrageenan	KCl, liquefied wood flour derived polyols as plasticizer	44.3	648.25	-	High water uptake rate.	(Meng et al., 2018)
PVA – Starch	Spirulina	Glycerol	32.5	11.6	-		(Shi et al., 2017)
PVA	Alkali treated Kenaf fiber		13.8	80	-		(Ali et al., 2015)
PVA	Pineapple leaf fiber		25.6	68.9	-		(Kalambettu et al., 2015)
PVA	Sugar beet pulp		53.84	12.45	1.61		(Shen et al., 2015)
PVA – Chitosan	Oil palm empty fruit bunch LCNF		32.15	-			(Solikhin et al., 2018)
PVA – Starch	MMA grafted Grewia optiva fibre	Citric acid, glycerol	45.62	264.64	-		(Singha et al., 2015b)
PVA – Starch	Grewia optiva fibre	Citric acid, Glutaraldehyde	38.53	182.1	-		(Priya et al., 2014)
PVA – Starch	Lignocellulosic nanowhiskers from hemp	Glycerol	26.4	183	1.39E-11		(Guimarães et al., 2015a)
PVA			8.81	9.35	-	PVA LCNW electro spun mat	(Sutka et al., 2015)
PVA – Starch	Bamboo nanofibrils	Glycerol	32.6	280	1.09E-11		(Guimarães et al., 2015b)
PVA	Bacterial cellulose nanocrystals	Glycerol, Boric acid	72.84	293.43	-		(Rouhi et al., 2017)
PVA	Nanocellulose		63.5	183.7	-		(Frone et al., 2011a)
PVA – Starch	Jute nanofibrils	Glycerol	46.8	-	-		(Das et al., 2011)

flexible, ductile, less thermally stable, transparent, and water-soluble composites. Guimarães et al. (2015a) prepared bamboo nanofibrils through mechanical defibrillation of refined bamboo pulp after treatment with alkali and bleaching. The best values for tensile strength, %elongation and water vapour permeability were exhibited by films having 6.5% bamboo nanofibrils reinforcement. The improved distribution of nanofibres led to compact structure and efficient packing. Das et al. (2011) prepared jute nanofibrils by acid hydrolysis of jute fibres. They reinforced starch-PVA composite films with jute nanofibrils and reported improvement in tensile strength at 15wt% loading. The water uptake rate was reported to be lower with 15 wt% jute nanofibrils reinforcement. The availability of lesser number of free hydrogen bonds due to interaction with nanofibrils to interact with water molecules was responsible for the lowering of the water uptake.

12.6 NANOCELLULOSE FROM AGRICULTURAL WASTE AND EFFECT OF ITS REINFORCEMENT IN PVA COMPOSITE FILMS

12.6.1 METHODS OF EXTRACTION OF NANOCELLULOSE FROM MICROCRYSTALLINE CELLULOSE

Broadly speaking nano-cellulose can be prepared by two methods: chemical and mechanical. Mechanical methods for preparing nanocellulose include microgrinding, homogenization under high-pressure, high intensity ultrasonication and PFI milling. The high-energy consumption is the major drawback of physical methods (Spence et al., 2011). Chemical methods to prepare nanocellulose are hydrolysis by acid and enzyme, and TEMPO [(2,2,6,6-tetramethylpiperidin-1-yl) oxidanyl] mediated oxidation. Chemical methods may be easy to use but they have certain drawbacks, including long duration, low-yield, and harmful effect on the environment as the chemicals used are toxic and not easily recyclable (Tang et al., 2011). To diminish energy utilization, usually mechanical and chemical methods are used in conjunction; fibres are treated with chemical reagents followed by mechanical processing to prepare nanocellulose (Li et al., 2011).

First report of nanocellulose preparation dates back to 1947, where acid hydrolysis of cotton and wood fibres was done to get nanocellulose (Nickerson and Habrle, 1947). Malainine et al. (2005) prepared cellulose microfibrils by means of homogenization under pressure during alkaline treatment. Abe et al. (2007) obtained nanocellulose fibres using grinding treatment while (Cheng et al., 2009) utilized high intensity ultrasonication to prepare nanocellulose. TEMPO-mediated oxidation followed by mechanical treatment of cotton linters was performed to obtain surface carboxylated cellulose nanocrystals by (Montanari et al., 2005). Man et al. (2011) used ionic liquid to prepare cellulose nanocrystals. Prepared nanocellulose fibres from cellulose pulps by treating with 17.5 wt% NaOH followed by mechanical fibrillation. Shankar and Rhim (2016) proposed a simplistic method to prepare nanocellulose from micro-crystalline cellulose using NaOH/urea dissolution of micro-crystalline cellulose followed by ultrasonication. Similar methods have been reported by Adsul et al. (2012) and Majdzadeh-Ardakani and Nazari (2010).

The most commonly used method to prepare nanocellulose is acid hydrolysis. But the requirement of highly concentrated acid is a major drawback of this method (Man et al., 2011). Using concentrated acid is corrosive to the reaction container and generates acidic liquid waste which is hard to recycle thereby causing grave pollution issues. Further, controlling the reaction and the degree of hydrolysis is difficult resulting in reduced yields usually less than 50%. While the TEMPO mediated oxidation treatment is toxic to environment, and enzymatic treatment is time-consuming and highly expensive.

A facile approach for preparing nanocellulose fibres is therefore needed and to this effect, the alkali–urea method gives a hope. The chilled alkali–urea solution is used to dissolve microcrystalline cellulose followed by regeneration with excess water and subsequent ultra-sonication treatment is done to further break the fibre bundles and destroy the amorphous regions. Cai and Zhang (2005) used chilled sodium hydroxide- urea aqueous solution with weight ratio of NaOH: urea: H_2O as 7:12:81, to dissolve cellulose. Adsul and others have also utilized similar alkali–urea dissolution–regeneration of cellulose followed by ultrasonication for preparation of nanocellulose (Adsul et al., 2012; Majdzadeh-Ardakani and Nazari, 2010). Compared with usual pretreatment reagents, like TEMPO, ionic liquid and concentrated sulphuric acid, NaOH/urea pretreatment is economical and relatively less harmful to environment. The probable mechanism of the breakdown of microcrystalline cellulose to nanocellulose by NaOH-urea aqueous solution is that the intermolecular and intra molecular hydrogen bonds of cellulose are disrupted by sodium hydroxide and subsequently the amino group of urea associates with hydroxyl groups cellulose to form hydrogen bonds. This combined effect of urea and alkali averts intra and intermolecular hydrogen bonding between cellulose molecules and thus weakens the macrostructure of cellulose molecules (Cai and Zhang, 2005; Yan and Gao, 2008). As compared to the alkali solution alone, urea is thought to "further enhance the effect of pretreatment due to its synergy with NaOH and lead to better swelling of the cellulose" (Wang et al., 2016).

12.6.2 Nanocellulose Reinforced PVA Composite Films

Various authors have investigated nanocellulose incorporation in PVA composite films to improve upon the composite's properties. Table 12.3 lists the properties of different composite films prepared using PVA or other polymer and nanocellulose. Asad et al. (2018) prepared TEMPO-oxidized nanocellulose (TONC) from empty-fruit-bunches of oil palm pulp using NaOCl and NaBr at 25°C and at 10 pH. Nanocomposite films of TONC and PVA were prepared by solution casting using 0.5 to 6% (w/w) TONC loading with highest tensile strength (122% increase) and modulus (291%) at 4% (w/w) TONC content. The elongation at breakpoint decreased by about 42.7%. Through FTIR and ^{13}C NMR the authors confirmed formation of hydrogen bonding between TONC and PVA.

Rouhi et al. (2017) investigated the effects of bacterial cellulose nanocrystals (BCNC) (as reinforcement), glycerol (plasticizer), and boric acid (cross-linker) on the mechanical properties of PVA films using response surface methodology (RSM). Their models showed that UTS as high as 72.84 MPa %elongation at break up to 293.43% could be obtained using glycerol concentration 13.89%, BCNC concentration 5.00% and a boric acid content of 1.96%. They also reported the formation of intra and intermolecular hydrogen and ether cross linkages by FTIR analysis in boric acid cross linked films. Frone et al. (2011) reported noteworthy improvement in tensile and thermal properties of PVA films when reinforced with cellulose nanofibres. The cellulose nanofibres were obtained from microcrystalline cellulose (MCC) by ultrasonication and used at 1 to 5 wt % loading in polyvinyl alcohol films. They attributed the enhancement of mechanical properties of PVA-cellulose nanofibres composites to the hydrogen bonding between the OH groups of cellulose fibres and the PVA matrix. Also, mechanical properties were reported to be better in case of smaller-size fibres compared to when large-size fibres were used, attributed to higher surface area and the larger possibility of hydrogen bonding. Li et al. (2013) prepared nanocellulose fibrils (NCFs) from bleached hard Kraft pulp (BHKP) using high-intensity ultrasonication and investigated effect of their reinforcement in PVA films. NCFs had higher thermal stability and crystallinity than BHKP due to removal of hemicelluloses. Composites of PVA with 4 wt% NCFs content showed best mechanical properties, with tensile strength and Young's modulus 1.86 and 1.63 times better than plain PVA. Shankar and Rhim (2016) prepared nanocellulose from micro crystalline cellulose using NaOH/urea treatment. The prepared nanocellulose was used as reinforcement in agar films. The addition of nanocellulose up to 3 wt% increased the tensile strength up to 52.8 MPa from 46.7 MPa. The % elongation of films also increased marginally at higher concentration of nanocellulose. The WVP of films reduced with increasing nanocellulose concentration and best results were obtained at 3 wt%.

Ching et al. (2015) prepared PVA Nanocomposite by solution casting PVA reinforced with nanosilica and nanocellulose derived from oil palm empty fruit bunches. Nanosilica reinforcement improved thermal stability of PVA/nanocellulose composites by reducing the mobility of matrix molecules. The PVA composites with 3 wt% nanocellulose and 0.5 wt% nanosilica showed best thermo mechanical properties while retaining good optical properties due to effective dispersion and polymer–filler interaction.

Li et al. (2014) studied the thermal stability and mechanical properties of PVA composites reinforced with nanocellulose fibrils (NCFs) obtained from chemical-thermo mechanical pulps obtained using ultrasonication. The NCFs (diameters from 50 to 120 nm) showed higher crystallinity (72.9%) compared to chemical-thermo mechanical pulp (61.5%). The best thermal stability, light transmittance, and mechanical properties were achieved at 6 wt% NCF content. Tensile strength and young's modulus increased by 2.8 and 2.4 times respectively compared to plain PVA. Zhou et al. (2012) isolated nanocellulose from microcrystalline cellulose using three methods viz., acid hydrolysis (AH), 2,2,6,6-tetramethylpiperidine-1-oxyl radical (TEMPO)-mediated oxidation (TMO) and ultrasonication (US).

TABLE 12.3
Behaviour of PVA Composite Films Reinforced with Nanocellulose

Matrix Polymer	Additives used (w/w % of polymers)	TS (MPa)	EB (%)	WVP g.m/ s.m².Pa	WST (%)	Preparation Method	Application	Reference
PVA 10 wt%	BHKP NCF 4 wt%	34.7	-	-	-	Solvent casting	Packaging	(Li et al., 2013)
PVA		29.3	121.5	-				(Cabuk et al., 2017)
Agar 2 wt%	Glycerol 30 wt% Nanocellulose 3 wt%	52.8	15.8	0.97 E-9	-	Solvent casting	Packaging	(Shankar and Rhim, 2016)
PVA 10 wt%	Nanosilica 0.5 wt% Nanocellulose 3 Wt%	37.5	198.5	-	-	Solvent casting	Packaging	(Ching et al., 2015)
Pectin 5 wt%	Glycerol 30 wt% Nanocellulose 5 wt%	13.7	29.4	9.06 E-11	28.76	Solvent casting	Packaging	(Chaichi et al., 2017)
Cassava starch 72.5 wt%	Glycerol 25 wt% Rice husk cellulose nanofibrils 2.5 wt%	18	15	1.8 E-10	-	Blow film extrusion	Packaging	(Nascimento et al., 2016)
PVA 10 wt%	NCF 6 wt%	58.7	-	-	-	Solvent casting	Packaging	(Li et al., 2014)
PVA 10 wt%	TEMPO oxidized OPEFB nanocellulose	136	4.25	-	-	Solvent casting	Packaging	(Asad et al., 2018)
PVA 10 wt%	TEMPO oxidized nanocellulose	120	35	-	-	Solvent casting	Packaging	(Zhou et al., 2012)
PVA 10 wt%	Acacia mangium nanocellulose 10 wt%	29.3	103.4	-	-	Solvent casting	Packaging	(Jasmani & Adnan, 2017)
PVA 10 wt%	Rice straw NFC 3 wt%	59.8	117.6	-	108.4	Solvent casting	Packaging	(Z. Wang et al., 2018)

Nanocellulose isolated by TMO was higher in yield (37%) and had high aspect ratio. Nanocellulose isolated by AH treatment had higher crystallinity index (88.1%) and better size dispersion. Both the AH-derived and TMO-derived nanocellulose was dispersed homogeneously in the PVA matrices. Elongation at break was higher for AH/PVA films (51.59% at 6 wt% nanocellulose loading) compared to TMO/ PVA. The TMO/PVA films showed increments of 21.5% and 10.2% in tensile modulus and strength at 6 wt% nanocellulose loading. Jasmani and Adnan (2017) incorporated NCC isolated from acacia wood pulp via acid hydrolysis in PVA films. 2wt % loading of NCC into the PVA film improved tensile strength by 30% while maximum value of the tensile strength was obtained at 10 wt% loading of NCC. The results indicate reinforcement potential of acacia derived NCC. Wang et al. (2018) obtained cellulose nano-fibrils from TEMPO mediated oxidation of microfibrils obtained after alkali treatment of steam exploded rice straw fibres. Rice straw CNF and CMF reinforced polyvinyl alcohol (PVA) composite films were prepared by solution casting method. The authors reported CNFs/PVA composite films to be having uniform fibre dispersion, better mechanical properties, and transparency but weaker water resistance compared to the rice straw CMFs/ PVA composite films. More polar groups in rice straw CNFs compared to CMFs were cited as the reason for this.

12.7 CONCLUSION

A summary of the available reported information on the utilization of agricultural biomass in preparation of biodegradable composite films has been reviewed in this chapter. The potential of such agricultural residue as a source of nanocellulose and various methods of extracting nanocellulose from agricultural resides have been presented. Finally, the effect of such derived nanocellulose on properties of prepared composite films has been reviewed.

The success of composite films produced using agricultural reside and polyvinyl alcohol and starch demonstrates that such composite materials can replace plastics thereby reducing the production and consumption of petroleum derived plastics.

However, the growing problem of plastic pollution requires a three-pronged approach: first some way to utilize and thus degrade the accumulated plastic waste in a way that is environmentally friendly. Presently we are only aware of thermal treatment by pyrolysis that can completely breakdown plastics but it will require a great amount of energy and the gas emissions will have to be taken care of. Second, to facilitate recycling of plastics by generating awareness about its types for proper segregation practices which would make recycling easier. Third, to reduce the consumption of plastics so that no new plastic waste is generated; this would require alternatives to be available which are sustainable and biodegradable. The solution also needs to be less focused on chemicals and processing and more directed towards ease of production requiring little or no chemical pretreatment at all.

REFERENCES

Abdul Khalil, H.P.S., Bhat, A.H., Ireana Yusra, A.F., 2012. Green composites from sustainable cellulose nanofibrils: A review. Carbohydr. Polym. https://doi.org/10.1016/j.carbpol.2011.08.078

Abe, K., Iwamoto, S., Yano, H., 2007. Obtaining cellulose nanofibers with a uniform width of 15 nm from wood. Biomacromolecules 8, 3276–3278. https://doi.org/10.1021/bm700624p

Adsul, M., Soni, S.K., Bhargava, S.K., Bansal, V., 2012. Facile approach for the dispersion of regenerated cellulose in aqueous system in the form of nanoparticles. Biomacromolecules 13, 2890–2895. https://doi.org/10.1021/bm3009022

Albuquerque, P.B.S., Malafaia, C.B., 2018. Perspectives on the production, structural characteristics and potential applications of bioplastics derived from polyhydroxyalkanoates. Int. J. Biol. Macromol. https://doi.org/10.1016/j.ijbiomac.2017.09.026

Ali, M.E., Yong, C.K., Ching, Y.C., Chuah, C.H., Liou, N.S., 2015. Effect of single and double stage chemically treated kenaf fibers on mechanical properties of polyvinyl alcohol film. BioResources 10, 822–838. https://doi.org/10.15376/biores.10.1.822-838

Allen, S., Allen, D., Moss, K., Le Roux, G., Phoenix, V.R., Sonke, J.E., 2020. Examination of the ocean as a source for atmospheric microplastics. PLoS One 15. https://doi.org/10.1371/journal.pone.0232746

Andresen, M., Johansson, L.S., Tanem, B.S., Stenius, P., 2006. Properties and characterization of hydrophobized microfibrillated cellulose. Cellulose 13, 665–677. https://doi.org/10.1007/s10570-006-9072-1

Anjum, A., Zuber, M., Zia, K.M., Noreen, A., Anjum, M.N., Tabasum, S., 2016. Microbial production of polyhydroxyalkanoates (PHAs) and its copolymers: A review of recent advancements. Int. J. Biol. Macromol. https://doi.org/10.1016/j.ijbiomac.2016.04.069

Asad, M., Saba, N., Asiri, A.M., Jawaid, M., Indarti, E., Wanrosli, W.D., 2018. Preparation and characterization of nanocomposite films from oil palm pulp nanocellulose/poly (Vinyl alcohol) by casting method. Carbohydr. Polym. 191, 103–111. https://doi.org/10.1016/j.carbpol.2018.03.015

Asgher, M., Ahmad, Z., Iqbal, H.M.N., 2017. Bacterial cellulose-assisted de-lignified wheat straw-PVA based bio-composites with novel characteristics. Carbohydr. Polym. 161, 244–252. https://doi.org/10.1016/j.carbpol.2017.01.032

Avella, M., Bogoeva-Gaceva, G., Bužarovska, A., Errico, M.E., Gentile, G., Grozdanov, A., 2008. Poly(lactic acid)-based biocomposites reinforced with kenaf fibers. J. Appl. Polym. Sci. 108, 3542–3551. https://doi.org/10.1002/app.28004

Averous, L., Boquillon, N., 2004. Biocomposites based on plasticized starch: thermal and mechanical behaviours 56, 111–122. https://doi.org/10.1016/j.carbpol.2003.11.015

Azizi Samir, M.A.S., Alloin, F., Dufresne, A., 2005. Review of recent research into cellulosic whiskers, their properties and their application in nanocomposite field. Biomacromolecules. https://doi.org/10.1021/bm0493685

Barnes, D.K.A., Galgani, F., Thompson, R.C., Barlaz, M., 2009. Accumulation and fragmentation of plastic debris in global environments. Philos. Trans. R. Soc. B Biol. Sci. 364, 1985–1998. https://doi.org/10.1098/rstb.2008.0205

Beckermann, G.W., Pickering, K.L., 2008. Engineering and evaluation of hemp fibre reinforced polypropylene composites: Fibre treatment and matrix modification. Compos. Part A Appl. Sci. Manuf. 39, 979–988. https://doi.org/10.1016/j.compositesa.2008.03.010

Bera, M., Alagirusamy, R., Das, A., 2010. A study on interfacial properties of jute-PP composites. J. Reinf. Plast. Compos. 29, 3155–3161. https://doi.org/10.1177/0731684410369723

Bjerre, A., Schmidt, A., 1997. Development of Chemical and Biological Processes for Production of Bioethanol: Optimisation of Pre-treatment Processes and Characterisation of Products, Riso-R-967. ed. Riso National Laboratory.

Bledzki, A.K., Gassan, J., 1999. Composites reinforced with cellulose based fibres. Prog. Polym. Sci. 24, 221–274.

Bledzki, A.K., Mamun, A.A., Lucka-Gabor, M., Gutowski, V.S., 2008. The effects of acetylation on properties of flax fibre and its polypropylene composites. Express Polym. Lett. 2, 413–422. https://doi.org/10.3144/expresspolymlett.2008.50

Cabuk, M., Alan, Y., Unal, H.I., 2017. Enhanced electrokinetic properties and antimicrobial activities of biodegradable chitosan/organo-bentonite composites. Carbohydr. Polym. 161, 71–81. https://doi.org/10.1016/j.carbpol.2016.12.067

Cai, J., Zhang, L., 2005. Rapid dissolution of cellulose in LiOH/urea and NaOH/urea aqueous solutions. Macromol. Biosci. 5, 539–548. https://doi.org/10.1002/mabi.200400222

Chaichi, M., Hashemi, M., Badii, F., Mohammadi, A., 2017. Preparation and characterization of a novel bionanocomposite edible film based on pectin and crystalline nanocellulose. Carbohydr. Polym. 157, 167–175. https://doi.org/10.1016/j.carbpol.2016.09.062

Cheng, Q., Wang, S., Rials, T.G., 2009. Poly(vinyl alcohol) nanocomposites reinforced with cellulose fibrils isolated by high intensity ultrasonication. Compos. Part A Appl. Sci. Manuf. 40, 218–224. https://doi.org/10.1016/j.compositesa.2008.11.009

Chiellini, E., Corti, A., D'Antone, S., Solaro, R., 2003. Biodegradation of poly (vinyl alcohol) based materials, Progress in Polymer Science (Oxford). https://doi.org/10.1016/S0079-6700(02)00149-1

Ching, Y.C., Rahman, A., Ching, K.Y., Sukiman, N.L., Chuah, C.H., 2015. Preparation and characterization of polyvinyl alcohol-based composite reinforced with nanocellulose and nanosilica. BioResources 10, 3364–3377. https://doi.org/10.15376/biores.10.2.3364-3377

Das, K., Ray, D., Bandyopadhyay, N.R., Sahoo, S., Mohanty, A.K., Misra, M., 2011. Physico-mechanical properties of the jute micro/nanofibril reinforced starch/polyvinyl alcohol biocomposite films. Compos. Part B Eng. 42, 376–381. https://doi.org/10.1016/j.compositesb.2010.12.017

Dris, R., Gasperi, J., Mirande, C., Mandin, C., Guerrouache, M., Langlois, V., Tassin, B., 2017. A first overview of textile fibers, including microplastics, in indoor and outdoor environments. Environ. Pollut. 221, 453–458. https://doi.org/10.1016/j.envpol.2016.12.013

Dufresne, A., Cavaille, J.Y., Vignon, M.R., 1997. Mechanical behavior of sheets prepared from sugar beet cellulose microfibrils. J. Appl. Polym. Sci. 64, 1185–1194. https://doi.org/10.1002/(SICI)1097-4628(19970509)64:6<1185::AID-APP19>3.0.CO;2-V

Dujardin, E., Blaseby, M., Mann, S., 2003. Synthesis of mesoporous silica by sol-gel mineralisation of cellulose nanorod nematic suspensions. J. Mater. Chem. 13, 696–699. https://doi.org/10.1039/b212689c

Farah, S., Anderson, D.G., Langer, R., 2016. Physical and mechanical properties of PLA, and their functions in widespread applications — A comprehensive review. Adv. Drug Deliv. Rev. https://doi.org/10.1016/j.addr.2016.06.012

Faruk, O., Bledzki, A.K., Fink, H.P., Sain, M., 2014. Progress report on natural fiber reinforced composites. Macromol. Mater. Eng. https://doi.org/10.1002/mame.201300008

Franco-Marquès, E., Méndez, J.A., Pèlach, M.A., Vilaseca, F., Bayer, J., Mutjé, P., 2011. Influence of coupling agents in the preparation of polypropylene composites reinforced with recycled fibers. Chem. Eng. J. 166, 1170–1178. https://doi.org/10.1016/j.cej.2010.12.031

Frone, A.N., Panaitescu, D.M., Spataru, D.D., Radovici, C., Trusca, R., Somoghi, R., 2011. Preparation and characterization of PVA composites with cellulose nanofibers obtained by ultrasonication. BioResources 6, 487–512. https://doi.org/10.15376/biores.6.1.487-512

Garcia de Rodriguez, N.L., Thielemans, W., Dufresne, A., 2006. Sisal cellulose whiskers reinforced polyvinyl acetate nanocomposites. Cellulose 13, 261–270. https://doi.org/10.1007/s10570-005-9039-7

Garlotta, D., 2001. A literature review of poly(lactic acid). J. Polym. Environ. 9, 63–84. https://doi.org/10.1023/A:1020200822435

Geyer, R., Jambeck, J.R., Law, K.L., 2017. Production, use, and fate of all plastics ever made. Sci. Adv. 3, e1700782. https://doi.org/10.1126/sciadv.1700782

Goda, K., Sreekala, M.S., Gomes, A., Kaji, T., Ohgi, J., 2006. Improvement of plant based natural fibers for toughening green composites-Effect of load application during mercerization of ramie fibers. Compos. Part A Appl. Sci. Manuf. 37, 2213–2220. https://doi.org/10.1016/j.compositesa.2005.12.014

Gomes, A., Matsuo, T., Goda, K., Ohgi, J., 2007. Development and effect of alkali treatment on tensile properties of curaua fiber green composites. Compos. Part A Appl. Sci. Manuf. 38, 1811–1820. https://doi.org/10.1016/j.compositesa.2007.04.010

Guimarães, M., Botaro, V.R., Novack, K.M., Teixeira, F.G., Tonoli, G.H.D., 2015a. Starch/PVA-based nanocomposites reinforced with bamboo nanofibrils. Ind. Crops Prod. 70, 72–83. https://doi.org/10.1016/j.indcrop.2015.03.014

Guimarães, M., Botaro, V.R., Novack, K.M., Teixeira, F.G., Tonoli, G.H.D., 2015b. High moisture strength of cassava starch/polyvinyl alcohol-compatible blends for the packaging and agricultural sectors. J. Polym. Res. 22. https://doi.org/10.1007/s10965-015-0834-z

Heux, L., Dinand, E., Vignon, M.R., 1999. Structural aspects in ultrathin cellulose microfibrils followed by 13C CP-MAS NMR. Carbohydr. Polym. 40, 115–124. https://doi.org/10.1016/S0144-8617(99)00051-X

Hill, C.A.S., Khalil, H.P.S.A., Hale, M.D., 1998. A study of the potential of acetylation to improve the properties of plant fibres. Ind. Crops Prod. 8, 53–63. https://doi.org/10.1016/S0926-6690(97)10012-7

Hoornweg, D., Bhada-Tata, P., Kennedy, C., 2013. Environment: Waste production must peak this century. Nature, 502, 615–617.

Ibrahim, N.Z., Hadithon, K.A., Abdan, K., 2010. Effect of Fiber Treatment on Mechanical Properties of Kenaf Fiber-Ecoflex Composites. J. Reinf. Plast. Compos. 29, 2192–2198. https://doi.org/10.1177/0731684409347592

Islam, M.S., Pickering, K.L., Foreman, N.J., 2010. Influence of alkali treatment on the interfacial and physico-mechanical properties of industrial hemp fibre reinforced polylactic acid composites. Compos. Part A Appl. Sci. Manuf. 41, 596–603. https://doi.org/10.1016/j.compositesa.2010.01.006

Iwamoto, S., Kai, W., Isogai, A., Iwata, T., 2009. Elastic modulus of single cellulose microfibrils from tunicate measured by atomic force microscopy. Biomacromolecules 10, 2571–2576. https://doi.org/10.1021/bm900520n

Janardhnan, S., Sain, M.M., 2006. Cellulose Microfibril Isolation enzymatic. BioResources 1, 176–188.

Jasmani, L., Adnan, S., 2017. Preparation and characterization of nanocrystalline cellulose from Acacia mangium and its reinforcement potential. Carbohydr. Polym. 161, 166–171. https://doi.org/10.1016/j.carbpol.2016.12.061

John, M.J., Thomas, S., 2008. Biofibres and biocomposites. Carbohydr. Polym. 71, 343–364. https://doi.org/10.1016/j.carbpol.2007.05.040

Kabir, M.M., Wang, H., Lau, K.T., Cardona, F., 2012a. Chemical treatments on plant-based natural fibre reinforced polymer composites: An overview. Compos. Part B Eng. 43, 2883–2892. https://doi.org/10.1016/j.compositesb.2012.04.053

Kabir, M.M., Wang, H., Lau, K.T., Cardona, F., Aravinthan, T., 2012b. Mechanical properties of chemically-treated hemp fibre reinforced sandwich composites. Compos. Part B Eng. 43, 159–169. https://doi.org/10.1016/j.compositesb.2011.06.003

Kalambettu, A., Damodaran, A., Dharmalingam, S., Vallam, M.T., 2015. Evaluation of Biodegradation of Pineapple Leaf Fiber Reinforced PVA Composites. J. Nat. Fibers 12, 39–51. https://doi.org/10.1080/15440478.2014.880104

Kang, J.T., Park, S.H., Kim, S.H., 2014. Improvement in the adhesion of bamboo fiber reinforced polylactide composites. J. Compos. Mater. 48, 2567–2577. https://doi.org/10.1177/0021998313501013

Kazayawoko, M., Balatinecz, J.J., Matuana, L.M., 1999. Surface modification and adhesion mechanisms in woodfiber-polypropylene composites. J. Mater. Sci. 34, 6189–6199. https://doi.org/10.1023/A:1004790409158

Khalil, H.P.S.A., Ismail, H., Rozman, H.D., Ahmad, M.N., 2001. Effect of acetylation on interfacial shear strength between plant fibres and various matrices. Eur. Polym. J. 37, 1037–1045. https://doi.org/10.1016/S0014-3057(00)00199-3

Khosravi-Darani, K., Bucci, D.Z., 2015. Application of poly(hydroxyalkanoate) in food packaging: Improvements by nanotechnology. Chem. Biochem. Eng. Q. https://doi.org/10.15255/CABEQ.2014.2260

Klemm, D., Heublein, B., Fink, H.P., Bohn, A., 2005. Cellulose: Fascinating biopolymer and sustainable raw material. Angew. Chemie - Int. Ed. https://doi.org/10.1002/anie.200460587

Lebreton, L., Slat, B., Ferrari, F., Sainte-Rose, B., Aitken, J., Marthouse, R., Hajbane, S., Cunsolo, S., Schwarz, A., Levivier, A., Noble, K., Debeljak, P., Maral, H., Schoeneich-Argent, R., Brambini, R., Reisser, J., 2018. Evidence that the Great Pacific Garbage Patch is rapidly accumulating plastic. Sci. Rep. 8, 1–15. https://doi.org/10.1038/s41598-018-22939-w

Levis, S.R., Deasy, P.B., 2001. Production and evaluation of size reduced grades of microcrystalline cellulose. Int. J. Pharm. 213, 13–24. https://doi.org/10.1016/S0378-5173(00)00652-9

Li, W., Wang, R., Liu, S., 2011. Nanocrystalline cellulose prepared from softwood kraft pulp via ultrasonic-assisted acid hydrolysis. BioResources 6, 4271–4281. https://doi.org/10.15376/biores.6.4.4271-4281

Li, W., Wu, Q., Zhao, X., Huang, Z., Cao, J., Li, J., Liu, S., 2014. Enhanced thermal and mechanical properties of PVA composites formed with filamentous nanocellulose fibrils. Carbohydr. Polym. 113, 403–410. https://doi.org/10.1016/j.carbpol.2014.07.031

Li, W., Zhao, X., Huang, Z., Liu, S., 2013. Nanocellulose fibrils isolated from BHKP using ultrasonication and their reinforcing properties in transparent poly (vinyl alcohol) films. J. Polym. Res. 20, 1–7. https://doi.org/10.1007/s10965-013-0210-9

Nascimento, P., Marim, R., Carvalho, G., Mali, S., 2016. Nanocellulose Produced from Rice Hulls and its Effect on the Properties of Biodegradable Starch Films. Mater. Res. 19, 167–174. https://doi.org/10.1590/1980-5373-MR-2015-0423

Madhavan Nampoothiri, K., Nair, N.R., John, R.P., 2010. An overview of the recent developments in polylactide (PLA) research. Bioresour. Technol. https://doi.org/10.1016/j.biortech.2010.05.092

Majdzadeh-Ardakani, K., Nazari, B., 2010. Improving the mechanical properties of thermoplastic starch/poly(vinyl alcohol)/clay nanocomposites. Compos. Sci. Technol. 70, 1557–1563. https://doi.org/10.1016/j.compscitech.2010.05.022

Malainine, M.E., Mahrouz, M., Dufresne, A., 2005. Thermoplastic nanocomposites based on cellulose microfibrils from Opuntia ficus-indica parenchyma cell. Compos. Sci. Technol. 65, 1520–1526. https://doi.org/10.1016/j.compscitech.2005.01.003

Man, Z., Muhammad, N., Sarwono, A., Bustam, M.A., Kumar, M.V., Rafiq, S., 2011. Preparation of Cellulose Nanocrystals Using an Ionic Liquid. J. Polym. Environ. 19, 726–731. https://doi.org/10.1007/s10924-011-0323-3

Meng, F., Zhang, Y., Xiong, Z., Wang, G., Li, F., Zhang, L., 2018. Mechanical, hydrophobic and thermal properties of an organic-inorganic hybrid carrageenan-polyvinyl alcohol composite film. Compos. Part B Eng. 143, 1–8. https://doi.org/10.1016/j.compositesb.2017.12.009

Mittal, A., Garg, S., Kohli, D., Maiti, M., Jana, A.K., Bajpai, S., 2016. Effect of cross linking of PVA/starch and reinforcement of modified barley husk on the properties of composite films. Carbohydr. Polym. 151, 926–938. https://doi.org/10.1016/j.carbpol.2016.06.037

Montanari, S., Roumani, M., Heux, L., Vignon, M.R., 2005. Topochemistry of carboxylated cellulose nanocrystals resulting from TEMPO-mediated oxidation. Macromolecules 38, 1665–1671. https://doi.org/10.1021/ma048396c

Mousa, M.H., Dong, Y., Davies, I.J., 2016. Recent advances in bionanocomposites: Preparation, properties, and applications. Int. J. Polym. Mater. Polym. Biomater. https://doi.org/10.1080/00914037.2015.1103240

Nickerson, R.F., Habrle, J.A., 1947. Cellulose Intercrystalline Structure. Ind. Eng. Chem. 39, 1507–1512. https://doi.org/10.1021/ie50455a024

NOAA, U., 2020. Garbage Patches | OR&R's Marine Debris Program [WWW Document]| OR&R's Marine Debris Program [WWW Document]. URL https://marinedebris.noaa.gov/info/patch.html (accessed 6.3.20).

Ortega-Toro, R., Bonilla, J., Talens, P., Chiralt, A., 2017. Future of Starch-Based Materials in Food Packaging, in: Starch-Based Materials in Food Packaging: Processing, Characterization and Applications. Elsevier Inc., pp. 257–312. https://doi.org/10.1016/B978-0-12-809439-6.00009-1

Pascault, J.P., Höfer, R., Fuertes, P., 2012. Mono-, Di-, and Oligosaccharides as Precursors for Polymer Synthesis, in: Matyjaszewski, K., Möller, M. (Eds.), Polymer Science: A Comprehensive Reference, 10 Volume Set. Elsevier, pp. 59–82. https://doi.org/10.1016/B978-0-444-53349-4.00254-5

Petersen, K., Nielsen, P., Bertelsen, G., Lawther, M., Olsen, M.B., Nilsson, N.H., Mortensen, G., 1999. Potential of biobased materials for food packaging. Trends Food Sci. Technol. 10, 52–68. https://doi.org/10.1016/S0924-2244(99)00019-9

Pickering, K.L., Abdalla, A., Ji, C., McDonald, A.G., Franich, R.A., 2003. The effect of silane coupling agents on radiata pine fibre for use in thermoplastic matrix composites. Compos. Part A Appl. Sci. Manuf. 34, 915–926. https://doi.org/10.1016/S1359-835X(03)00234-3

Pietak, A., Korte, S., Tan, E., Downard, A., Staiger, M.P., 2007. Atomic force microscopy characterization of the surface wettability of natural fibres. Appl. Surf. Sci. 253, 3627–3635. https://doi.org/10.1016/j.apsusc.2006.07.082

Priya, B., Gupta, V.K., Pathania, D., Singha, A.S., 2014. Synthesis, characterization and antibacterial activity of biodegradable starch/PVA composite films reinforced with cellulosic fibre. Carbohydr. Polym. 109, 171–179. https://doi.org/10.1016/j.carbpol.2014.03.044

Rachini, A., Le Troedec, M., Peyratout, C., Smith, A., 2012. Chemical modification of hemp fibers by silane coupling agents. J. Appl. Polym. Sci. 123, 601–610. https://doi.org/10.1002/app.34530

Raza, Z.A., Abid, S., Banat, I.M., 2018. Polyhydroxyalkanoates: Characteristics, production, recent developments and applications. Int. Biodeterior. Biodegrad. https://doi.org/10.1016/j.ibiod.2017.10.001

Rong, M.Z., Zhang, M.Q., Liu, Y., Yang, G.C., Zeng, H.M., 2001. The effect of fiber treatment on the mechanical properties of unidirectional sisal-reinforced epoxy composites. Compos. Sci. Technol. 61, 1437–1447. https://doi.org/10.1016/S0266-3538(01)00046-X

Rouhi, M., Razavi, S.H., Mousavi, S.M., 2017. Optimization of crosslinked poly(vinyl alcohol) nanocomposite films for mechanical properties. Mater. Sci. Eng. C 71, 1052–1063. https://doi.org/10.1016/j.msec.2016.11.135

Satyanarayana, K.G., Arizaga, G.G.C., Wypych, F., 2009. Biodegradable composites based on lignocellulosic fibers-An overview. Prog. Polym. Sci. 34, 982–1021. https://doi.org/10.1016/j.progpolymsci.2008.12.002

Satyanarayana, K.G., Sukumaran, K., Mukherjee, R.S., Pavithran, C., Piuai, S.G.K., 1990. Natural Fibre-Polymer Composites 12, 117–136.

Shankar, S., Rhim, J.-W., 2016. Preparation of nanocellulose from micro-crystalline cellulose: The effect on the performance and properties of agar-based composite films. Carbohydr. Polym. 135, 18–26. https://doi.org/10.1016/j.carbpol.2015.08.082

Shen, Z., Ghasemlou, M., Kamdem, D.P., 2015. Development and compatibility assessment of new composite film based on sugar beet pulp and polyvinyl alcohol intended for packaging applications. J. Appl. Polym. Sci. 132, 1–8. https://doi.org/10.1002/app.41354

Shi, B., Liang, L., Yang, H., Zhang, L., He, F., 2017. Glycerol-plasticized spirulina–poly(vinyl alcohol) films with improved mechanical performance. J. Appl. Polym. Sci. 134, 1–10. https://doi.org/10.1002/app.44842

Singh, B., Gupta, M., Verma, A., 1996. Influence of fiber surface treatment on the properties of sisal-polyester composites. Polym. Compos. 17, 910–918. https://doi.org/10.1002/pc.10684

Singha, A.S., Priya, B., Pathania, D., 2015. Cornstarch/Poly(vinyl alcohol) Biocomposite Blend Films: Mechanical Properties, Thermal Behavior, Fire Retardancy, and Antibacterial Activity. Int. J. Polym. Anal. Charact. 20, 357–366. https://doi.org/10.1080/1023666X.2015.1018491

Siró, I., Plackett, D., 2010. Microfibrillated cellulose and new nanocomposite materials: A review. Cellulose 17, 459–494. https://doi.org/10.1007/s10570-010-9405-y

Snell, K.D., Peoples, O.P., 2009. PHA bioplastic: A value-added coproduct for biomass biorefineries. Biofuels, Bioprod. Biorefining 3, 456–467. https://doi.org/10.1002/bbb.161

Solikhin, A., Hadi, Y.S., Massijaya, M.Y., Nikmatin, S., Suzuki, S., Kojima, Y., Kobori, H., 2018. Properties of Poly(Vinyl Alcohol)/Chitosan Nanocomposite Films Reinforced with Oil Palm Empty Fruit Bunch Amorphous Lignocellulose Nanofibers. J. Polym. Environ. 26, 3316–3333. https://doi.org/10.1007/s10924-018-1215-6

Spence, K.L., Venditti, R.A., Rojas, O.J., Habibi, Y., Pawlak, J.J., 2011. A comparative study of energy consumption and physical properties of microfibrillated cellulose produced by different processing methods. Cellulose 18, 1097–1111. https://doi.org/10.1007/s10570-011-9533-z

Stenstad, P., Andresen, M., Tanem, B.S., Stenius, P., 2008. Chemical surface modifications of microfibrillated cellulose. Cellulose 15, 35–45. https://doi.org/10.1007/s10570-007-9143-y

Sutka, Anna, Gravitis, J., Kukle, S., Sutka, Andris, Timusk, M., 2015. Electrospinning of poly(vinyl alcohol) nanofiber mats reinforced by lignocellulose nanowhiskers. Soft Mater. 13, 18–23. https://doi.org/10.1080/1539445X.2014.995309

Svagan, A.J., Azizi Samir, M.A.S., Berglund, L.A., 2007. Biomimetic polysaccharide nanocomposites of high cellulose content and high toughness. Biomacromolecules 8, 2556–2563. https://doi.org/10.1021/bm0703160

Tang, L. rong, Huang, B., Ou, W., Chen, X. rong, Chen, Y. dan, 2011. Manufacture of cellulose nanocrystals by cation exchange resin-catalyzed hydrolysis of cellulose. Bioresour. Technol. 102, 10973–10977. https://doi.org/10.1016/j.biortech.2011.09.070

Tang, X., Alavi, S., 2011. Recent advances in starch, polyvinyl alcohol based polymer blends, nanocomposites and their biodegradability. Carbohydr. Polym. 85, 7–16. https://doi.org/10.1016/j.carbpol.2011.01.030

Thygesen, A., Thomsen, A.B., Daniel, G., Lilholt, H., 2007. Comparison of composites made from fungal defibrated hemp with composites of traditional hemp yarn. Ind. Crops Prod. 25, 147–159. https://doi.org/10.1016/j.indcrop.2006.08.002

Tserki, V., Zafeiropoulos, N.E., Simon, F., Panayiotou, C., 2005. A study of the effect of acetylation and propionylation surface treatments on natural fibres, in: Composites Part A: Applied Science and Manufacturing. Elsevier, pp. 1110–1118. https://doi.org/10.1016/j.compositesa.2005.01.004

van Tuil, R., Fowler, P., Lawther, M., Weber, C.J., 2000. Properties of biobased packaging materials, in: Weber, C.J. (Ed.), Biobased Packaging Materials for the Food Industry: Status and Perspectives.

Versino, F., Lopez, O. V., Garcia, M.A., Zaritzky, N.E., 2016. Starch-based films and food coatings: An overview. Starch - Stärke 68, 1026–1037. https://doi.org/10.1002/star.201600095

Vilpoux, O., Averous, L., 2004. Chapter 18 Starch-based plastics, in: Technology, Use and Potentialities of Latin American Starchy Tubers. pp. 521–553.

Volova, T., 2004. Polyhydroxyalkanoates--plastic materials of the 21st century: production, properties, applications. Nova Science Publishers Inc, New York.

Wang, J., Li, Yao, Wang, Z., Li, Yujie, Liu, N., 2016. Influence of pretreatment on properties of cotton fiber in aqueous NaOH/urea solution. Cellulose 23, 2173–2183. https://doi.org/10.1007/s10570-016-0938-6

Wang, Y., Chen, R., Cai, J.Y., Liu, Z., Zheng, Y., Wang, H., Li, Q., He, N., 2013. Biosynthesis and Thermal Properties of PHBV Produced from Levulinic Acid by Ralstonia eutropha. PLoS One 8 . https://doi.org/10.1371/journal.pone.0060318

Wang, Y.L., Lee, Y.H., Chiu, I.J., Lin, Y.F., Chiu, H.W., 2020. Potent impact of plastic nanomaterials and micromaterials on the food chain and human health. Int. J. Mol. Sci. https://doi.org/10.3390/ijms21051727

Wang, Z., Qiao, X., Sun, K., 2018. Rice straw cellulose nanofibrils reinforced poly(vinyl alcohol) composite films. Carbohydr. Polym. 197, 442–450. https://doi.org/10.1016/j.carbpol.2018.06.025

Xie, Y., Hill, C.A.S., Xiao, Z., Militz, H., Mai, C., 2010. Silane coupling agents used for natural fiber/polymer composites: A review. Compos. Part A Appl. Sci. Manuf. 41, 806–819. https://doi.org/10.1016/j.compositesa.2010.03.005

Xiong, H.G., Tang, S.W., Tang, H.L., Zou, P., 2008. The structure and properties of a starch-based biodegradable film. Carbohydr. Polym. 71, 263–268. https://doi.org/10.1016/j.carbpol.2007.05.035

Yan, L., Gao, Z., 2008. Dissolving of cellulose in PEG/NaOH aqueous solution. Cellulose 15, 789–796. https://doi.org/10.1007/s10570-008-9233-5

Yu, T., Jiang, N., Li, Y., 2014. Study on short ramie fiber/poly(lactic acid) composites compatibilized by maleic anhydride. Compos. Part A Appl. Sci. Manuf. 64, 139–146. https://doi.org/10.1016/j.compositesa.2014.05.008

Zeng, J.B., Huang, C.L., Jiao, L., Lu, X., Wang, Y.Z., Wang, X.L., 2012. Synthesis and properties of biodegradable poly(butylene succinate-co-diethylene glycol succinate) copolymers. Ind. Eng. Chem. Res. 51, 12258–12265. https://doi.org/10.1021/ie300133a

Zhou, Y.M., Fu, S.Y., Zheng, L.M., Zhan, H.Y., 2012. Effect of nanocellulose isolation techniques on the formation of reinforced poly(vinyl alcohol) nanocomposite films. Express Polym. Lett. 6, 794–804. https://doi.org/10.3144/expresspolymlett.2012.85

13 Biomass of Microalgae as Potential Biodiesel Source for Future Energy Needs

Deen Dayal Giri[1], Juhi Khan[2], Ajay Giri[3], Dan Bahadur Pal[4] and Amit Kumar Tiwari[4]

[1]Department of Botany, Maharaj Singh College, Saharanpur (UP), India 247001

[2]Department of Botany, IFTM University, Moradabad (UP), India

[3]Department of Basic Education, Uttar Pradesh, India

[4]Department of Chemical Engineering, BIT, Mesra, Ranchi (Jharkhand), India

CONTENTS

13.1   INTRODUCTION

Reserves of petroleum are constantly depleting across the world. Fuel demand is increasing and crude oil price is rising rapidly day by day. Excessive use of petroleum fuel is adversely affecting environmental health. A suitable, clean, and renewable fuel will be required as an option of petroleum in need of near future. The production of fuel from the biological material is a possible alternate of petro-fuel. So, in past few decade production of biofuel has attracted attention of scientific community people around the world (Song et al., 2016; He et al., 2016). Obviously, biofuel is likely to fulfil demand the world energy after the depletion of fossil oil reserves (Abomohra et al., 2016). Using biofuels at present levels could reduce our dependence on fossil fuels and help protect environment by about 20% less atmospheric CO_2 contributed by transportation the worldwide (Hoppe et al., 2016; Bharathiraja et al., 2015). Researchers and governments are producing biofuels in the form of bioethanol, bio-butanol, biogas, and biodiesel because of its safe, renewable, sustainable, and environment-friendly nature. As per US Department of Energy, the major biofuels (bioethanol and biodiesel) are representative of the first generation of biofuel technology (Marwa et al., 2019).

Photosynthetic organisms have capability to harvest light energy and use it in transforming atmospheric carbon dioxide to the biomass. The inbuilt ability of light harvesting is found in prokaryotic cyanobacteria to the higher plants (Machado and Atsumi, 2012). The biomass of these organism release the entrapped energy as heat when burnt, and can be considered as natural battery storing light energy. It is possible to fulfil future energy needs sustainably by producing plant biomass for energy utilizing residues of the plants used in other purposes for energy production.

Biofuels are fuels derived from biomass, and include solid biomass, liquid fuels as well as biogases. The liquid biofuels are bioethanol and other oils like biodiesel whereas landfill gases and synthetic gases are gaseous biofuels. Bioethanol is a fermentation product of sugar and starch components of plant materials of sugar and starch crops. In near future, technology up-gradation can convert cellulosic biomass of trees and grasses for bioethanol production and their use in use in the vehicle instead of present-day petroleum products. At present, alcohol is added as an additive in the fuel to enhance octane level and to boost emissions in the vehicle. The vehicular use of such fuel in a vehicle is common in some countries, notably Brazil and the United States, for reducing the emission of particulate matter, hydrocarbons, and carbon monoxide compared to diesel-based vehicles; in the year 2010, its contribution in the world's transport fuel was around 2.7%. The biodiesels are basically group of esters (Rawat et al., 2013). Esters are produced by reaction between fatty acids and alcohol; the reaction is called transesterification. So, the biodiesel production requires fatty acids that are derived from animal fat or vegetable oil (Robles-Medina et al., 2009). The availability of the feed-stock is a major challenge in biodiesel production. The United States, South-East Asia, Europe, and South America generally use vegetable oil of corn, canola, soybeans, and palm for biodiesel production. These crops are not able to fulfil the future demand of diesel because they will increase cost of the crops as well as possibly food shortage. The top ten countries for potential biodiesel production (ML) are Malaysia (14,540), Indonesia (7595), Argentina (5255), United States (3212), Brazil (2567), Netherlands(2496), Germany (2024), Philippines (1234), Belgium (1213), and Spain (1073) (Atadashi et al., 2011; Balat and Balat, 2010; Johnston and Holloway, 2009; Sharma et al., 2009). About 70 billion gallons per year of diesel required in the United States cannot be met with other sources of biodiesel.

13.2 BENEFITS OF BIODIESEL

Biodiesel is highly combustion fuel due to rich oxygen content about 10–11% and it burn without soots (Balat, 2011; Demirbas, 2007; Jena et al., 2010). It emits about 78% less net CO_2 compared to petro-diesel (Van et al., 2005). This easily available, non-toxic, eco-friendly renewable fuel is sulphur-free and emit less particulate matter in the ambient air, and therefore decreases air pollution minimizes chance of many fatal diseases (Kafuku and Mbarawa, 2010; Shahid and Jamal, 2011).

Petroleum-based fuel production requiring drilling, transportation, or product refining is a time-consuming process whereas production of microalgae biomass is relatively easy and less time-consuming. A country producing fuel locally will naturally solve the issue of taxes and tariff posed by the fuel-producing country (Jain and Sharma, 2010; Jena et al., 2010; Silitongaet al., 2011). The better lubrication ability of biodiesel to improve pumps and injector units is likely to increases efficacy of the engine (Moser et al., 2010; Porte et al., 2010).

13.3 MICROALGAE

Microalgae are a promising source of plant-derived biofuels (Harman-Ware 2013, Shukla, et al 2017). These rapid-growing photosynthetic aquatic-microorganisms store energy in form of biomass (proteins, carbohydrates, cellulose, hemicellulose, lipids). The lipids are brilliant raw material for different type of liquid fuels. Triglycerides can be reacted directly from esters and selectively utilized as a biodiesel liquid fuel. Lipid depleted biomass can be used for feedstock for plastic additives, animal nutrition, and in production of other fuel (syngas, methane, etc.)

13.4   PRODUCTIVITY OF MICROALGAE

Biomass of crop plants as well as microalgae can be used for energy production. The algae grow 20–30 times faster compared to crop and their harvesting cycle is only of 1–10 days (Chisti, 2007).Per acre microalgal oil production is about 300 times higher compared to conventional crops, such as soybeans, rapeseed, palms, jatropha etc. The comparison of biomass production (metric tons/year/Acre) of algae in terms of lipid, carbohydrate and protein is 12.5, 8 and 7, and better the values for soybean crop (0.1, 0.4, and 0.4) and palm plant (1.7, 1.5, and 0.3). Clearly, microalgae are more efficient in producing biomass and oil content. As compared to the crop plant, a microalga has the potential to produce per year 5,000 to 15,000 gallons of biodiesel in one acre area of the open pond.

13.5   ALGAE CULTIVATION

In culture the microalgae growth follow the traditional growth pattern, having a lag phase, log phase, and stationary phase. Traditional microalgae growth in open ponds, suffer from problems of uncontrolled environmental condition, and open air contamination with bacteria or virus or other microbes in the culture. In open ponds there is no adequate light exposure to all cells and nutrient distribution is heterogeneous that refrain normal culture growth.

The preparation of photo-bioreactors (PBR), not only significantly solve the problem of culture contamination but also offer better control of culture condition such as proper light penetration, homogeneous medium, and easy carbon dioxide input in the PBRs, resulting in better biomass yield in small space, however PBRs require excessive manufacturing and maintenance.

Microalgae constitute a better alternative to the crop plants for producing renewable biofuels due to their high oil content and productivity. Algae cultivation for fuel production is in demand for generating renewable energy all over the world (Hu and Sommerfeld, 2008). Microalgae are unique in sequestering carbon dioxide and converting it into a environmental friendly source of biofuels worldwide. The other positive points with microalgae are their rapid growth, ability to grow in very inhospitable conditions, no issue of human food consumption, and growing non-competitively on land and water sources not used for conventional food production. Algae cultivation can be done on arid land, excessively saline soil, and drought-disturbed area without affecting land area for food crop production. Thus, for the production of liquid fuels and carbon sequestering, the microalgal process is a revolutionary and suitable renewable platform for biofuels. For biofuel production it is very crucial to select an optimal strain of microalgae. In addition, many other parameters should also be kept in mind during evaluation of algal strains for their suitability as a biofuel feedstock. Green algae such as *Botryococcus braunni* have a great potential for biodiesel. Some of its strain) growing at ambient temperature (23^0C), light intensity (30–60W/m^2), photoperiod of 12 hours and salinity (8.8%) get doubled their cells in about two days. They have high amount of hydrocarbons around 86% (dry weight basis), whereas oil content may vary between 25 and 75%.

13.5.1 ALGAL CULTIVATION IN THE OPEN POND

Algal cultivation in the open ponds is been broadly studied in the past few years (Boussiba et al., 1988; Tredici and Materassi, 1992; Hase et al., 2000). These open ponds may be of natural water resources like lakes, lagoons, naturally occurred ponds, man-made ponds, and water containers, etc. Circular ponds, shallow big ponds, tanks, and channel ponds etc. are the widely used water systems for microalgal cultivation. The main advantage of using an open pond is its easy fabrication and functioning compared to closed-water systems. The disadvantages of an open pond include poor distribution of light, losses due to evaporation, dissemination of CO_2 into the atmosphere, and the requirement of a huge area. Apart from these limitations, water contamination, predators and many other heterotrophs are responsible for restricting the growth of algae in open ponds, only a few organisms can grow well under these extreme situations; therefore, the commercial production of required microalgae can be inhibited or lowered by this type of contamination. The unskilful stirring mechanism and bad mass transfer rate will also result in low productivity of biomass. To sum up, the cost-effectiveness, easy cleaning (after cultivation), and good production of algal mass are the main advantages; on the other hand, problems in the control of cultural conditions, difficulties in cultivation for a long duration, less productivity, the requirement of large area, availability of fewer strains of algae, and chances of contamination are major drawbacks of open ponds cultivation.

13.5.1.1 Photo-bioreactors (PBRs)

PBRs are closed transparent vessels used for cultivating photosynthetic microalgae. The PBRs provide suitable environment for the growth of cultured algae as well as protects them from undesirable microbes. In PBRs, algal cultures are illuminated either by artificial light, solar light or both ways. The natural illumination of algal culture occurs in the large surface areas which include open-type ponds (Hase et al., 2000), horizontal tubular airlift (Camacho Rubio et al., 1999), flat-plate (Hu et al., 1996), inclined tubular PBRs (Ugwu et al., 2002), etc. In the laboratory, PBRs are illuminated artificially by using fluorescent lamps. The examples of such bioreactors are bubble type columns (Degen et al., 2001; ChiniZittelli et al., 2003), airlift column (Kaewpintong et al., 2007), stirred-tank (Ogbonna et al., 1999), helical tubular column (Hall et al., 2003), torus type columns (Pruvost et al., 2006), and seaweed-type columns (Chetsumon et al., 1998). A few PBRs are tempered by putting them in a constant temperature room. However, the compaction facility is limited in PBRs. The large-scale outdoor systems cannot be easily tempered without high technical skill. The tempering facility is introduced in some commercial PBRs (Biostat PBRs of Sartorius BBI Systems Inc.). Researchers and scientists in this field are trying to develop and design a temperature-controlled PBRs, including double-walled internally illuminated PBRs facilitated with a heating and cooling water circuit (Pohl et al., 1988).

13.5.1.2 Vertical-column PBRs

The vertical-column PBRs of various designs have been tested for algae cultivation (Choi et al., 2003; Garcı´a-MaleaLópez et al., 2006; Kaewpintong et al., 2007). These PBRs have some useful feature like compactness, cost effectivity, and ease of operation makes them suitable for large-scale algae cultivation(Sánchez Mirón et al., 2002). The bubble-column and airlift PBRs of up to 0.19 m in diameter could attain specific growth rate and algal biomass production comparable to narrow tubular PBRs (Sánchez Mirón et al., 2002).

13.5.1.2.1 Flat-plate PBRs

These PBRs having large illuminating surface area are appropriate for outdoor algal cultures. The biomass production in flat plate PBRs is cheaper, easy to clean, readily tempered and has very low O_2 build-up issue. Scaling up of these PBRs requires handling the issues of culture temperature control, algal growth on the walls, and hydrodynamic stress to the algae. These PBRs received much attention for cultivation of algae after its presentation by Milner (1953) was further equipped by fluorescent lamp, use of flat-panel thick transparent PVC materials and other modification in the designs take place for cultivation of algae in mass (Tredici and Materassi, 1992; Zhang et al., 2002; Hoekema et al., 2002). The use of transparent materials in PBRs maximizes utilization of light. Such PBR show relatively less accumulation of dissolved O_2 and a better photosynthetic efficiency (Hu et al., 1996; Richmond, 2000). The PBRs are suitable for mass cultivation.

13.5.1.2.2 Tubular PBRs

Tubular PBR constructed using plastic tubes or glasses are most suitable for outdoor mass cultures. Their large illuminated surface and well-arranged algal culture re-circulation through pump or an airlift system cost effectively yield good biomass. In such PBRs, dissolved O_2, gradients of pH, and CO_2 may develop along the tube. PBRs require large amounts of land for construction and it may be associated with the problem of fouling and wall growth to some extent. The tubular bioreactors could vary in forms such as horizontal, vertical, near horizontal, conical, or inclined (Molina et al., 2001, Tredici and ChiniZittelli, 1998; Pirt et al., 1983; Watanabe and Saiki, 1997; Ugwu et al., 2002).These PBRs suffer from poor mass transfer during scale up, easily build up a very high dissolved oxygen(Molina et al., 2001) and face frequent photoinhibition. To overcome such issues tube diameter is increased that results in decreased illumination surface to volume ratio. In scaled up of these PBRs, the tube length is reduced to most possible level whereas tube diameter is extended sufficiently to minimize the shadow effect and increasing cell growth. The proper culture mixing and equal light exposure to all the cells can result in good growth (Ugwu et al., 2003; Ugwu et al., 2005). In most tubular PBRs, control of culture temperature to desired range is difficult task even after equipping with thermostats. The temperature control system also expands the making cost. Frequent adherence of algae on the wall, development of gradients of oxygen and CO_2 transfer in long tubes are some common problems (Camacho Rubio et al., 1999; Ugwu et al., 2003). The increase in pH of the algal cultures requires frequent re-carbonation that increase the production cost.

13.5.1.2.3 Internally Illuminated PBRs

Some PBRs are internal illuminated using fluorescent lamp. These PBRs can be modified to use solar and artificial light depending upon availability of light (Ogbonna et al., 1999). In these PBR, the artificial light is switched off when solar radiation is high enough. There are reports regarding use of optic fibre for equal distribution of solar light in cylindrical PBRs. The internally illuminated PBRs naturally helps in heat-sterilization under pressure and reduces contamination chance. In these PBRs, supply of light is continuously maintained irrespective of day and night by integrating devices for artificial and solar light. Nevertheless, the outdoor mass cultivation of algae in such PBRs would require more technical knowledge.

13.5.2　HYDRODYNAMICS AND MASS TRANSFER IN PBRs

Hydrodynamics and mass transfer of phototrophic cultures in bioreactors have limited informations. The important parameters in this regards are overall mass transfer coefficient (kLa), velocity of liquid, mixing time and gas hold up. In performance assessing overall mass transfer coefficient (kLa) vary with temperature, agitation rate, sparger type and use of antifoaming agent or surfactants.

Mixing time and culture homogeneity are determined by injecting and monitoring trace of dye, signal-response, pH electrode or by advanced computerized fluid dynamics (Ugwu et al., 2003; Pruvost et al., 2006, Sato et al., 2006).Appropriate culture mixing helps in maintaining high cell density by proper nutrient distribution, low shading and minimized photo-inhibition and gas exchange without issue of thermal stratification (Janvanmardian and Palsson, 1991). That is why, under other optimum environmental condition turbulent flow results in higher algal biomass production (Hu et al., 1996).The culture mixing strategy differ depending on the PBRs. Paddle wheels are used in open ponds for inducing turbulent flow (Boussiba et al., 1988; Hase et al., 2000). Impellers are used in stirred-tank PBRs (MazzucaSobczuk et al., 2006) whereas air bubbling helps in culture mixing directly or indirectly in airlift systems. The static mixers sand baffles are installed inside the tubular and bubble-column PBRs (Tredici and ChiniZittelli, 1998; Merchuk et al., 2000; Degen et al., 2001, Ugwu et al., 2002).The increased aeration improves culture mixing and mass transfer between gas and liquid, however, it pose shear stress to algal cells (MazzucaSobczuk et al., 2006; Kaewpintong et al., 2007).

In tubular PBRs, velocity of gas bubble indicates flow of algal cultures. The spargers are used to decrease bubble size and increases gas dispersion. The large bubbles produced from small bubbles in tubular flow reduce contact area between the gas and liquid resulting in poor mass transfer. Velocity and size of the bubbles depends on flow rate of liquid. In fast flow rate, the size of bubble may increases to a level to interrupt baffles or static mixers used for breaking the large bubbles down into fine ones. The superficial gas velocity, bubble velocity, and the overall mass transfer coefficient are interrelated (Lu et al., 1995; Couvert et al., 2004).

Phaeodactylum tricornutum culture growing in concentric tube airlift PBRs at superficial gas velocity of 0.055 ms^{-1} resulted in the kLa value of ~ 0.02 s^{-1} (Contreras et al. (1998). As per Ogbonna et al. (1998) reported superficial gas velocity of 0.009 m s^{-1}for internally illuminated PBRs (3-L)having *Chlorella pyrenoidosa* cultures and Ugwu et al. (2002) reported kLa ~ 0.003 s^{-1} at gas velocity of 0.02 m s^{-1}for inclined tubular PBRs (6L) cultivating *Chlorella sorokiniana*. The superficial gas velocity between 5.4 to 82 × 10^{-4}in bubble-column PBRs (13-L) growing *Porphyridium* kLa ranged between 1.7–4.7 × 10^{-3} s^{-1}(Merchuk et al., 2000). For *Haematococcus pluvialis* culture in split-cylinder internal-loop airlift PBRs(2-L) with superficial gas velocity of 0.024 m s^{-1} showed the kLa about 0.009 s^{-1}(Vega-Estrada et al., 2005). An airlift tubular horizontal PBRs (200-L) running at gas velocity of 0.16 m s^{-1}, the kLa was around 0.014 s−1 (Camacho Rubio et al., 1999). Thus, it is not fair to compare kLa value only on the basis of superficial gas velocity. The other parameters like volume of PBRs, design and algal strains significantly affecting kLa should also be considered.

In the PBR liquid velocity states about the flow and turbulence ensuring light exposure to all the cells as well as uniform algal cells transportation along the tube length (Carlozzi, 2003 and Pruvost et al., 2006). Solid velocity also determines hydrodynamics and mass transfer. The shape, size, porosity and quantity of solids affect mass transfer (Couvert et al., 2004). In luxuriant algal growth, cells get aggregated to form clump that decrease tube internal diameter and clumps settling restricts uniform cell biomass circulation along the tubes in the PBR.

Gas holdup is an important hydrodynamics parameter. Gas holdup is the fraction of the reactor volume occupied by the gas. Gas holdup is estimated by the volume of the liquid displaced by the gas due to aeration. It affect overall mass transfer by determining the circulation rate and the gas residence time. Some studies states about relationship between bubble size, gas holdup, the overall mass transfer coefficient and gas–liquid interfacial surface area (Vandu et al., 2005).

13.5.3 Productivity of Algae in Outdoor PBRs

Most of the PBRs for mass algae production operate well in laboratory level, and a limited number are used in large-scale production by modifying length, height, diameter and other components of PBRs. PBRs scaled-up with optimum light, temperature and proper culture mixing and efficient mass transfer is challenging task. Some PBRs producing good biomass in grams per litre per day (g L^{-1} d^{-1}) have been constructed. An outdoor airlift tubular PBRs of 200-L capacity culti- vating *Phaeodactylum tricornutum* produced 1.20–1.50(Camacho Rubio et al., 1999) and 1.90 (Molina et al., 2001).An outdoor flat-plate PBR (440-L) produced ~ 0.27 *Nannochlorospis*(Cheng-Wu et al., 2001) whereas large-scale outdoor PBR of 25,000-L commercially produced astaxanthin using *Haematococcus pluvialis* 0.05–0.06 g (Olaizola, 2000; Garcı́a-MaleaLópez et al., 2006). In the inclined tubular PBRs of 6L *Chlorella sorokiniana* productivity was 1.47 (Ugwu et al., 2002) and an undular row tubular PBRs of 11L *Arthrospira platensis* productivity about 2.7 (Carlozzi, 2003).The productivity of the PBRs can be assessed in terms of per unit of time, per unit of reactor volume, per unit land area per unit time, bio- mass yield as well as photosynthetic efficiency.

13.5.4   PBRs with Mixotrophic Mode of Microalgae Cultivation

Microalgae cultivation in natural or man-made open ponds is simple but require high processing charges compelled researchers to develop closed PBRs having higher biomass yield in the controlled condition. Closed PBRs are not much better for open-pond cultures. Microalgae cultivation in heterotrophic mode in steriliz- able fermenters are better for commercial production, due to better specific growth rate of photosynthetic microalgae compared to heterotrophs as microalgae are able to utilize organic substrates as main source of carbon and energy and light remains supplementary source for photosynthesis. The unicellular alga *Chlorella* utilize glucose at low light intensity but both process operate concurrently. It could be the actual case, many economically important microalgae. The PBR design for mixotrophic algae culture presuming organic substrate is the main carbon and energy source and light/carbon dioxide as supplementary, create problem in optimising varying solar irradiance in outdoor cultures.

Outdoor productivity of chlorella ~ 127 g m^{-2}d^{-1}was achieved during the day through mixotrophic in a horizontal tubular PBR (Lee et al., 1996). The culture grow heterotrophically at night (68.7 g m^{-2} d^{-1}) using the organic substrate as source of carbon and energy source. Light supplemented energy only during the daytime. The maximum specific growth rate, *Chlorella vulgaris* growing mixotrophically, is the sum of the photosynthetic and heterotrophic specific growth rates. Chlorella can be acclimated for mixotrophic growth and light induction of chlorophyll syn- thesis by use of indoor stirred tank fermentor and pumping it outdoor through transparent tubing and tank outdoors (Lee et al., 1996).

13.6   MICROALGAE HARVESTING

The methods used to harvest microalgae are coagulation and flocculation, flota- tion, centrifugation, and filtration. These methods either alone or in combination are employed for harvesting micro algae due to the lack of unique harvesting method for microalgae. The preferred method for harvesting algal biomass varies depending upon the purpose of the biomass production. For biofuel production the most preferred method is coagulation and flocculation followed by filtration>>centrifugation > floatation. The electrolysis process is least preferred in all the process biofuel, human and animal food production, high value product production and water quality restoration (Singh and Patidar, 2018, Chew et al 2017). How easily an alga could be harvested by settling and filtering is dependent on its size. The fast-growing microalgae of very small sizes motile single cells pose difficulty in their harvesting. So, selection of algal strain and its effective harvesting technologies is important. It is necessary for explorer to have a clear understanding of available harvesting options and how to overcome these challenges.

13.7   OIL YIELD OF MICROALGAE

The amount of oil in an algae differ highly depending on the algae. Some algae with high oil content are *Botryococcus braunii (29–75)*,*Schiochytrium* (50–77), *Nannochloropsis* 31–68,*Hantzschia sp* (66),*Nannochloris sp* (31–63),*Stichococcus* (39–59), *Chlorella sp.* (15–55), *Neochloris oleoabundans* (35–54), *Dunaliella sp* (36–42), *Cyclotella sp* (42), *Ankistrodesmus* sp. (28–40), *Nannitzschia* (28–50) *Scenedesmus* (45) *Tetraselmis suecica* (15–32)*Phaeodactylum tricornutum* (31). The algae of smaller dimensions are generally preferred as a source for biomass production as compared to the macroalgae like seaweeds. The microalgae having less complex structure, high surface area for nutrient uptake for fast growth rate with high oil content are best for mass production. The macroalgae could be option for biofuel purpose in conditions like easy availability of biomass in case of seaweeds. Some important algal species which are supposed to be suitable for mass production for extracting oils across various locations the world are *Botryococcus* sp, *Chlorella* sp, *Dunaliella* sp, *Pleurochrysis* sp, *Sargassum* sp and *Gracilaria* sp.

13.8   BIODIESEL PRODUCTION

Biodiesel is composed of alkyl esters of fatty acids. The alkyl esters are product of transesterification of triglycerides. The microalgae used to accumulate lipid under stress condition and this lipid content may exceed 50% of its dry weight. The lipid of cell biomass is required to be extracted for the biofuel production. The most commonly used methods for lipid extraction are physical, solvent and supercritical fluids, ultrasound and microwave. There are two steps in biodiesel production

13.8.1   EXTRACTION OF LIPID

The technique used for lipid extraction depends upon the species, its growth stage, lipid content and biomass harvesting method.

13.8.2 TRANSESTERIFICATION OF THE LIPID

The reaction may take place in acidic or alkaline condition using homogeneous or heterogeneous catalyst.

Triglycerides + 3methenol = Methyl Esters + Glycerol

Plant and animal based fats and oils are generally triglycerides. In the transesterification process, the used alcohol (methanol or ethanol) is deprotonated with a base to make it a stronger nucleophile. Thus the reaction has only inputs of the triglyceride and the alcohol. The reaction proceeds very slowly in normal condition, and its rate can be accelerated by use of heat as well as use of an acid or base as catalyst that are not consumed in transesterification reaction. Biodiesel production from vegetable oils is generally base-catalysis as the process proceed at temperatures and pressures, produce around 98% conversion yield, when triglycerides have no free fatty acids and low moisture. The biodiesel productions requiring acid catalysis are much slower (Dubé et al., 2007).

13.9 PRODUCTION METHODS

13.9.1 BATCH PROCESS

It is necessary to monitor quantity of water and free fatty acids in the lipid (Knothe, 2005). High free fatty acid may lead saponification instead of esterification. The lipid is added to vessel containing agitated catalyst alcohol mix and vessel is closed to prevent the alcohol loss. For ensuring complete conversion the fat to esters, the temperature of the system is maintained around 70 °C for a period of one to eight hours with excess of alcohol. The biodiesel is lighter compared to the glycerine, so it can be separated under gravity or alternatively they can be separated by centrifugation. Flash evaporation or distillation is generally used to remove excessive alcohol present in each phase.

13.9.2 SUPERCRITICAL PROCESS

It is a catalyst-free method for transesterification. In this process supercritical methanol at high temperatures and pressures get mixed homogeneously with the oil leads to rapid spontaneous reaction (Bunkyakiat et al., 2006). In the process free fatty acid gets converted into methyl ester instead of soap; also, the process is tolerant to water in the feedstock. Though the process occurs at high temperatures and pressures, it is still cheaper compared to the catalytic process.

13.9.3 ULTRA- AND HIGH-SHEAR IN-LINE AND BATCH REACTORS

Such reactors facilitate biodiesel production in continuous, semi- continuous, and in batch-mode, and significantly decrease time of production with enhanced biodiesel production volume. The small sized droplets with large surface area facilitate fast catalysis. The fine droplets of oil fat or methanol are produced in high shear mixers.

13.9.4 ULTRASONIC-REACTOR METHOD

In the ultrasonic reactors, continuous generation and collapse of bubbles takes place. Such cavitations fulfil both heating and mixing of reactants required in the transesterification.

13.9.5 MICROWAVE METHOD

The microwave method is likely to be a potential application for efficient and cost-effective commercial application in biodiesel production process (Leadbetter, Nicholas E, et al 2006). The need of heat in esterification process can be fulfilled by use of commercial microwave ovens providing intense localized heating (Leadbetter, Nicholas E, et al 2007). There was a report regarding one fourth of energy use in a continuous flow (6L/min) and about 99% conversion rate compared to energy used in batch process.

13.9.6 LIPASE-CATALYZED METHOD

Many researchers internationally are interested in the enzymatic transesterification very good yields have been obtained from crude and used oils using lipases because this enzyme is very less sensitive to high free fatty acid that is problematic in standard biodiesel production process. The only issue related with the use of enzyme lipase is its inactivation after one batch with methanol; however this problem can be solved by use of methyl acetate for maintaining activity for the several batches. Scale-up of industrial biodiesel production about 40 m^3 have been achieved by use of liquid lipase (Price et al 2016)

13.10 MAJOR CHALLENGES IN ALGAL FUEL PRODUCTION

The technological advancement is continuously upgrading the commercial production strategies of algal biofuel; however such fuel production is still very costly. The major factors responsible for higher production cost are due to lack of proper temperature and growth control. Algal growth gets reduced in the low temperature in the winter. Invasion of other microbes and culture fouling results in lower biomass production. Dewatering of diluted cultures is energy dependent. From the harvested algal biomass, carbohydrate is processed to produce bioethanol whereas lipid fractions are used for production of gasoline, biodiesel and jet fuel and protein fraction is used in power generation. The harvested product fractionation, extraction, and product purification steps need further improvement for low energy input and large scale production for reducing cost. Further, the fuel produced is not much compatible with the present day engines and so its demand in the market (Graboski et al 1998). The issue of space management for producing, handling and storing biomass is challenge for the future up-scaling.

13.11 CONCLUSION

Microalgae have great possibility to be fuel of future need. The biodiesel produced from microalgae are not carbon neutral and its production at commercial scale has still many challenges. The better fundamental understanding algal strain having high natural oil content, its mass cultivation, harvesting and oil yield. Strain improvements using biotechnological tools and integrated chemical engineering for reactor design will be helpful in commercial production of biodiesel. Further, optimization of algal biomass production and the content of fuel molecules within the algal cell is required.

REFERENCES

Abomohra, A.; Jin, W.; Tu, R.; Han, S.; Eid, M.; and Eladel, H. 2016. Microalgal biomass production as a sustainable feedstock for biodiesel: current status and perspectives, Renew. Sustain. Energy Rev., 64: 596–606.

Atadashi, I.M.; Aroua, M.K.; and Abdul Aziz, A. 2011. Biodiesel separation and purification: a review. Renew Energy, 36(2): 437–443.

Babich, I.V.; Van der Hulst, M.; Lefferts, L.; Moulijn, J.A.; O'Connor, P.; and K. Seshan. 2011. Catalytic pyrolysis of microalgae to high-quality liquid bio-fuels. Biomass Bioenergy, 35, 3199–3207.

Balat, M. 2011. Potential alternatives to edible oils for biodiesel production – a review of current work. Energy Convers Manage, 52(2): 1479–1492.

Balat, M. and H. Balat. 2010. Progress in biodiesel processing. Appl Energy, 87(6): 1815–1835.

Bharathiraja, B.; Chakravarthy, M.; Ranjith Kumar, R.; Yogendran, D.; Yuvaraj, D.; Jayamuthunagai, J.; Praveen Kumar, R.; and S. Palani. 2015. Aquatic biomass (algae) as a future feedstock for bio-refineries: a review on cultivation, processing and products. Renew. Sustain. Energy Rev., 47, 634–653.

Boussiba, S., Sandbank, E., Shelef, G.; Cohen, Z.; Vonshak, A.; Ben- Amotz, A.; Arad, S.; and A. Richmond. 1988. Outdoor cultivation of the marine microalga Isochrysisgalbana in open reactors. Aquaculture 72, 247–253.

Bunkyakiat, Kunchana et al. 2006. Continuous Production of Biodiesel via Transesterification from Vegetable Oils in Supercritical Methanol. Energy and Fuels (American Chemical Society), 20 (2): 812–817.

Camacho Rubio, F.; Acién Fernández, F.G.; SánchezPérez, J.A.; Garcı́a Camacho, F.; and E. Molina Grima. 1999. Prediction of dissolved oxygen and carbon dioxide concentration profiles in tubular photobioreactors for microalgal culture. Biotechnol. Bioeng, 62, 71–86.

Carlozzi, P. 2003. Dilution of solar radiation through culture lamination in photobioreactor rows facing south–north: a way to improve the efficiency of light utilization by cyanobacteria (Arthrospira platensis). Biotechnol. Bioeng. 81, 305–315.

Cheng-Wu, Z.; Zmora, O.; Kopel, R.; and A. Richmond. 2001. An industrialsize flat glass reactor for mass production of Nannochloropsis sp. (Eustigmatophyceae). Aquaculture 195, 35–49.

Chetsumon, A.; Umeda, F.; Maeda, I.; Yagi, K.; Mizoguchi, T.; and Y.Miura. 1998. Broad spectrum and mode of action of an antibiotic produced by Scytonema sp. TISTR 8208 in a seaweed-type bioreactor. Appl. Biochem. Biotechnol. 70–72, 249–256.

Chew, K.W.; Yap, J.Y.; Show, P.L.; Suan, N.H.; Juan, J.C.; Ling, T.C.; and J. S. Chang. 2017. Microalgae biorefinery: High value products perspectives. Bioresour. Technol., 229, 53–62.

Chini Zittelli, G.; Rodolfi, L.; and M.R. Tredici. 2003. Mass cultivation of Nannochloropsis sp. in annular reactors. J. Appl. Phycol. 15, 107–114.

Chisti, Y., 1998. Pneumatically agitated bioreactors in industrial and environmental bioprocessing: hydrodynamics, hydraulic, and transport phenomena. Appl. Mech. Rev. 51, 33–112.

Choi, S.L.; Suh, I.S.; and C.G.Lee. 2003. Lumostatic operation of bubble column photobioreactors for Haematococcuspluvialis cultures using a\ specific light uptake rate as a control parameter. Enzyme Microb. echnol. 33, 403–409.

Contreras, A.; Garcıa, F.; Molina Grima, E; and J. C. Merchuk. 1998. Interaction between CO2-mass transfer, light availability and hydrodynamic stress in the growth of Phaeodactylumtricornutum in a concentric tube airlift photobioreactor. Biotechnol. Bioeng. 60, 318–325.

Couvert, A.; Bastoul, D.; Roustan, M.; and P. Chatellier. 2004. Hydrodynamic and mass transfer study in a rectangular three-phase airlift loop reactor. Chem. Eng. Proc. 43, 1381–1387.

Degen, J.; Uebele, A.; Retze, A.; Schmidt-Staigar, U.; Trosch, W.A. 2001. A novel airlift photobioreactor with baffles from improved light utilization through flashing light effect. J. Biotechnol. 92, 89–94.

Demirbas A. Importance of biodiesel as transportation fuel. 2007. Energy Policy; 35(9): 4661–70.

Du, Wei et al. 2004. Comparative study on lipase-catalyzed transformation of soybean oil for biodiesel production with different acyl acceptors. Journal of Molecular Catalysis B: Enzymatic 30 (3–4): 125–129.

Dubé, Marc A., et al. 2007. Acid-Catalyzed Transesterification of Canola Oil to Biodiesel under Single- and Two-Phase Reaction Conditions. Energy & Fuels 21: 2450–2459. American Chemical Society. Retrieved on 2007-11-01.

Garcıa-Malea López, M.C.; Del Rìo Sánchez, E.; Casas López, J.L.; Acién Fernández, F.G.; Fernández Sevilla, J.M.; Rivas, J; Guerrero, M.G.; and Molina Grima. 2006. Comparative analysis of the outdoor culture of *Haematococcuspluvialis* in tubular and bubble column photobioreactors. J. Biotechnol. 123, 329–342.

Graboski, M. S. and McCormick, R. L. 1998. Combustion of fat and vegetable oil derived fuels in diesel engines. Prog. Energy Combust. Sci. 24: 125–164.

Hall, D.O.; Fernandez, F.G.A.; Guerrero, E.C.; Rao, K.K.; and E.M. Grima. 2003. Outdoor helical tubular photobioreactors for microalgal production: modeling of fluid-dynamics and mass transfer and assessment of biomass productivity. Biotechnol. Bioeng. 82, 62–73.

Harman-Ware, A.E.; Morgan, T.; Wilson, M.; Crocker, M.; Zhang, J.; Liu, K.; and Debolt, S. 2013. Microalgae as a renewable fuel source: Fast pyrolysis of Scenedesmus sp. Renew. Energy, 60, 625–632.

Hase, R.; Oikawa, H.; Sasao, C.; Morita, M.; and Watanabe, Y. 2000. Photosynthetic production of microalgal biomass in a raceway system under greenhouse conditions in Sendai City. J. Biosci. Bioeng. 89, 157–163.

He Q, Yang H, and Hu C. 2016. Culture modes and financial evaluation of two oleaginous microalgae for biodiesel production in desert area with open raceway pond. Bioresource Technology; 218: 571–579.

Hoekema, S.; Bijmans, M.; Janssen, M.; Tramper, J.; and R. H. Wijffels. 2002. A pneumatically agitated flat-panel photobioreactor with gas recirculation: anaerobic photoheterotrophic cultivation of a purple nonsulfur bacterium. Int. J. Hydro. Energy. 27, 1331–1338.

Hoppe, W.; Bringezu, S.; and N. Thonemann. 2016. Comparison of global warming potential between conventionally produced and CO2-based natural gas used in transport versus chemical production, J. Clean. Prod.; 121, 231–237.

Hu, Q.; Guterman, H.; and A. Richmond. 1996. A flat inclined modular photobioreactor for outdoor mass cultivation of phototrophs. Biotechnol. Bioeng. 51, 51–60.

Jain, S. and M.P. Sharma. 2010. Biodiesel production from Jatropha curcas oil. Renew Sustain Energy Rev; 14(9): 3140–3147.

Janvanmardian, M.; and B. O. Palsson. 1991. High density photoautotrophic algal cultures: design, construction and operation of a novel photobioreactor system. Biotechnol. Bioeng. 38, 1182–1189.

Jena PC, Raheman H, Kumar GVP, Machavaram R. 2010. Biodiesel production from mixture of mahua and simarouba oils with high free fatty acids. Biomass Bioenergy, 34(8): 1108–16.

Johnston M, Holloway T. 2007. A global comparison of national biodiesel production potentials. Environ SciTechnol, 41(23): 7967–73.

Kaewpintong, K.; Shotipruk, A.; Powtongsook, S.; and P. Pavasant. 2007. Photoautotrophic high-density cultivation of vegetative cells of Haematococcuspluvialis in airlift bioreactor. Bioresource Technol. 98, 288–295.

Kafuku, G. and M. Mbarawa. 2010. Biodiesel production from Croton megalocarpus oil and its process optimization. Fuel; 89: 2556–60.

Knothe, G. 2005. Dependence of biodiesel fuel properties on the structure of fatty acid alkyl esters. Fuel Proc. Technol. 86: 1059–1070.

Leadbetter, Nicholas E, et al. 2006. Fast, Easy Preparation of Biodiesel Using Microwave Heating. Energy & Fuels 20: 2281–2283. American Chemical Society. Retrieved on 2007-11-01.

Leadbetter, Nicholas E, et al. 2007. Continuous-Flow Preparation of Biodiesel Using Microwave Heating. Energy & Fuels 21: 1777–1781. American Chemical Society. Retrieved on 2007-11-01.

Lee, Y.K. and C. S. Low. 1991. Effect of photobioreactor inclination on the biomass productivity of an outdoor algal culture. Biotechnol. Bioeng. 38, 995–1000.

Lu, W.J.; Hwang, S.J.; and C. M. Chang. 1995. Liquid velocity and gas hold up in three-phase internal loop airlift reactors with low density particles. Chem. Eng. Sci. 50, 1301–1310.

Machado, I.M. and S. Atsumi. 2012. Cyanobacterial biofuel production. J. Biotechnol., 162, 50–56.

Marwa, G.; Saad, Noura S.; Dosoky, Mohamed S. Zoromba, M.S.; and M. Hesham M. Shafik, 2019. Algal Biofuels: Current Status and Key Challenges. Energies., 12, doi:10.3390/en12101920

MazzucaSobczuk, T.; Garcìa Camacho, F.; Molina Grima, E.; Chisti, Y., 2006. Effects of agitation on the microalgae Phaeodactylumtricornutum and Porphyridiumcruentum. Bioproc. Biosyt. Eng. 28, 243–250.

Merchuk, J.C.; Gluz, M.; Mukmenev, I. 2000. Comparison of photobioreactors for cultivation of the microalga Porphyridium sp. J. Chem. Technol. Biotechnol. 75, 1119–1126.

Milner, H.W.; 1953. Rocking tray. In: Burlew, J.S. (Ed.), Algal Culture from Laboratory to Pilot Plant. Carnegie Institution, Washington, DC, p. 108, No. 600.

Molina, E.; Fernández, J.; Acién, F.G.; and Y. Chisti., 2001. Tubular photobioreactor design for algal cultures. J. Biotechnol. 92, 113–131.

Moser B.R and S. F. Vaughn. 2010. Evaluation of alkyl esters from Camelinasativa oil as biodiesel and as blend components in ultra low-sulfur diesel fuel. BioresourTechnol, 101(2): 646–53.

Ogbonna, J.C.; Ichige, E.; and H. Tanaka, 2002. Interactions between photoautotrophic and heterotrophic metabolism in photoheterotrophic cultures of Euglena gracilis. Appl. Microbiol. Biotechnol. 58, 532–538.

Olaizola, M., 2000. Commercial production of astaxanthin from Haematococcuspluvialis using 25,000-liter outdoor photobioreactors. J. Appl. Phycol. 12, 499–506.

Pirt, S.J.; Lee, Y.K.; Walach, M.R.; Pirt, M.W.; Balyuzi, H.H.M.; and M. J. Bazin. 1983. A tubular photobioreactor for photosynthetic production of biomass from carbon dioxide: design and performance. J. Chem. Tech. Biotechnol. 33B, 35–38.

Pohl, P.; Kohlhase, M.; Martin, M.; 1988. Photobioreactors for the axenic mass cultivation of microalgae. In: Stadler, T.; Mollion, J.; Verdus, M.C.; Karamanos, Y.; Morvan, H.; Christiaen, D. (Eds.), Algal Biotechnology. Elsevier Applied Science, London and New York, pp. 209–217.

Porte, A.F.; Schneider, R.; Kaercher, J.A.; Klamt, R.A.; Schmatz, W.L.; da Silva, W.L.T. and W.A.S. Filho. 2010. Sunflower biodiesel production and application in family farms in Brazil. Fuel; 89(12): 3718–24.

Price, J.; Nordblad, M.; Martel H.H. et al. 2016. Scale-up of industrial biodiesel production to 40 m3 using a liquid lipase formulation. *Biotechnol Bioeng*. 113, 1719–1728.

Pruvost, J.; Pottier, L.; and J. LegrandJ. 2006. Numerical investigation of hydrodynamic and mixing conditions in a torus photobioreactor. Chem. Eng. Sci 61, 4476–4489.

Rawat, I; Ranjith Kumar, R.; Mutanda, T.; and F. BuxF. 2013. Biodiesel from microalgae: a critical evaluation from laboratory to large scale production. Applied Energy ; 103: 444–467.

Richmond, A., 2000. Microalgal biotechnology at the turn of the millennium: a personal view. J. Appl. Phycol. 12, 441–451.

Robles-Medina, A.; González-Moreno, P.A.; Esteban Cerdán, L.; and E. Molina-Grima. 2009. Biocatalysis: towards ever greener biodiesel production. Biotechnology Advances ; 27(4): 398–408.

Sánchez Miron, A.; Cerón García, M.C.; García Camacho, F.; Molina Grima, E.; and Y. Chisti. 2002. Growth and characterization of microalgal biomass produced in bubble column and airlift photobioreactors: studies in fed-batch culture. Enzyme Microb. Technol. 31, 1015–1023.

Sato, T.; Usui, S.; Tsuchiya, Y.; and Y.Kondo. 2006. Invention of outdoor closed type photobioreactor for microalgae. Energy Convers. Manage. 47, 791–799.

Shahid, E.M. and J. Jamal. 2011. Production of biodiesel: a technical review. Renew Sustain Energy Rev; 15(9): 4732–45.

Sharma, Y.C. and B. Singh. 2009. Development of biodiesel: current scenario.. Renew Sustain Energy Rev; 13(6–7): 1646–51

Shukla S.P.; Gita, S.; Bharti, V.S.; Bhuvaneswari, G.R. and Wikramasinghe, W.A.A.D.L. 2017. Atmospheric Carbon Sequestration through Microalgae: Status, Prospects, and Challenges, Agro-Environmental Sustainability, 10.1007/978-3-319-49724-2, 219–235.

Shukla, S.P.; Singh J.S.; Kashyap, S.; Giri, D.D. and Kashyap A.K. 2008. Antarctic cyano-bacteria as a source of phycocyanin: an assessment. *Ind. J. Marine Sci.* 37, 446–449.

Silitonga, A.S.; Atabani, A.E.; Mahlia, T.M.I.; Masjuki, H.H.; Badruddin, I.A,. and S. Mekhilef. 2011. A review on prospect of Jatropha curcas for biodiesel in Indonesia. Renew Sustain Energy Rev; 15: 3733–3756.

Singh, G. and Patidar, S.K. 2018. Microalgae harvesting techniques: A review Journal of Environmental Management 217: 499–508.

Song, C.; Liu, Q; Ji, N; Deng, S.; Zhao, J.; Li, S. and Y. Kitamura. 2016. Evaluation of hydrolysis–esterification biodiesel production from wet microalgae. Bioresource Technology; 214: 747–754.

Tredici M.R.; Zittelli G.C. 1998. Efficiency of sunlight utilization: Tubular versus flat photobioreactors. *Biotechnol Bioeng*. 57, 187–197.

Tredici, M.R.; and R. Materassi. 1992. From open ponds to vertical alveolar panels: the Italian experience in the development of reactors for the mass cultivation of photoautotrophic microorganisms. J. Appl. Phycol. 4, 221–231.

Ugwu, C.U.; Ogbonna, J.C.; and H. Tanaka. 2002. Improvement of mass transfer characteristics and productivities of inclined tubular photobioreactors by installation of internal static mixers. Appl. Microbiol. Biotechnol. 58, 600–607.

Ugwu, C.U.; Ogbonna, J.C.; and H. Tanaka. 2003. Design of static mixers for inclined tubular photobioreactors. J. Appl. Phycol. 15, 217–223.

Ugwu, C.U.; Ogbonna, J.C.; and H. Tanaka. 2005. Light/dark cyclic movement of algal cells in inclined tubular photobioreactors with internal static mixers for efficient production of biomass. Biotechnol. Lett. 27, 75–78.

Van Gerpen, J.H.; Hammond, E.G.; Johnson, L.A.; Marley, S.J.; Yu, L.; Lee, I. and A. Monyem. 2005. Biodiesel processing and production. Fuel Process Technol; 86(10): 1097–107.

Vandu, C.O.; Liu, H.; and R. Krishna. 2005. Mass transfer from Taylor bubbles rising in single capillaries. Chem. Eng. Sci. 60, 6430–6437.

Vega-Estrada, J.; Montes-Horcasitas, M.C.; Domiñgues-Bocanegra, A.R.; and R. O. Cañizares-Villanueva. 2005. Haematococcuspluvialis cultivation in split-cylinder internal-loop airlift photobioreactor under aeration conditions avoiding cell damage. Appl. Microbiol. Biotechnol. 68, 31–35.

Watanabe, Y. and H. Saiki. 1997. Development of photobioreactor incorporating Chlorella sp. for removal of CO2 in stack gas. Energy Convers. Manage. 38, 499–503.

Zhang, K.; Kurano, N.; and S. Miyachi. 2002. Optimized aeration by carbon dioxide gas for microalgal production and mass transfer characterization in a vertical flat-plate photobioreactor. Bioproc. Biosys. Bioeng. 25, 97–101.

14 Waste Biomass Pretreatment Using Novel Materials

Vivek Kumar[1], Jay Mant Jha[2], Amarendra Kumar Dash[3], Balgovind Tiwari[4] and Rahul[5]

[1]Department of Chemical Engineering, Rajiv Gandhi University of Knowledge Technologies, RK Valley, AP – 516330, India

[2]Department of Chemical Engineering, Maulana Azad National Institute of Technology, Bhopal, MP – 462003, India

[3]Department of English, Rajiv Gandhi University of Knowledge Technologies, Nuzvid, AP – 521202, India

[4]Department of Physics, Rajiv Gandhi University of Knowledge Technologies, RK Valley, AP – 516330, India

[5]Department of Chemical Engineering, Jaipur National University, Jaipur, RJ – 302017, India

CONTENTS

DOI: 10.1201/9781003196358-14

14.1 BACKGROUND

In the 21st century, biomass has received wide acceptance as an alternative, renewable, and sustainable source of energy because of the advent of efficient technologies to extract fuel from them. Biomass can be found in abundance (of the order of 10^{10} MTPA) and can serve as the primary source of energy that has the potential to provide alternative fuels like bio-ethanol as well as biodiesel (Sun and Cheng, 2002). The alarming rise in pollution levels prompts us to look for alternate solutions for the energy demand problem. Catering to the ever-increasing demand for growth, biomass-derived fuels can provide high impact solutions for green economic development. With the continuous worldwide research and development, there is an availability of processes for the conversion of lignocellulosic biomass (LCB) to biofuels, other forms of bioenergy, and other value-added products (Adsul et al., 2011).

The main cause of slow growth worldwide in the production of energy from LCB feedstock is the lack of low-cost technology for dealing with the recalcitrance of such feedstock (Lynd et al., 2005). Within around last 20 years, there have been several studies on LCB feedstock, e.g., wheat and rice straw, fibrous residue of sugarcane, barley, and wood materials (wood chips, sawdust, naturally dried parts of trees etc.). These LCBs were examined for the production of bio-ethanol (Naik et al., 2010).

Biomass includes the remains of forestry, crops, and mills; scrap lumbers, animal excreta, and municipal solid waste; food processing wastes, and many similar residues. These materials are available in huge quantities and a large share of these resources is yet to be inducted into our energy chain. Similarly, a renewable, non-toxic, and biodegradable fuel, biodiesel is marked for its low-sulphur and high-lubrication properties. Production and use of biodiesel can provide many benefits to our society including the reduction in the scale of global warming (Kiss et al., 2008) and the Opportunities To Create New Jobs, Especially Rural Jobs.

14.2 POTENTIAL OF WASTE BIOMASS

Biodiesel has been positioned as the key product alternative for conventional fossil fuels since decades. Primary production of biodiesel is done from the biological sources, e.g., vegetable oils and fats via esterification or trans-esterification process (Jain and Sharma, 2010; Aransiola et al., 2012). The major challenges in the production of bio-ethanol and other bio-based products are the accessibility, assortment, and circulation of favourable raw materials followed by the specificity and optimization of pretreatment technologies and fermentation processes (Akhtar et al., 2016).

Biomass, in general, includes energy crops, forest and agricultural residues, as well as industrial residue, and even municipal solid wastes (Knauf et al., 2004). However, in many parts of the world, a large share of this LCB goes unutilized as it is often burnt at the site. In recent past, researchers worldwide have focused on LCB due to its renewable nature. It has been expected that a big fraction of worldwide LCB can be changed into various high-end products, e.g., bio-based fuels, fine chemicals, as well as low-cost energy sources for special cases, e.g., fermentation of microbes and production of enzymes.

Agricultural wastes of fast-growing plants that can be frequently harvested play a big role in the production of bio-ethanol (Knauf et al., 2004; Kim and Dale, 2004). Besides, *Miscanthus giganteus* (Brosse et al., 2009), switchgrass (Xu et al., 2010), and *Poplar sp.* (Wang et al., 2012) have been proven as important sources of LCB for bio-ethanol production. Research supports that municipal waste, by-products from the food-processing industry and products as well as by-products from the paper industry can be used as raw materials for bio-based fuel production (Ciciora, 2011).

14.3 STRUCTURE OF BIOMASS

Due to their complex cell structure, plants pose a challenge to separate their polysaccharides in a straightforward method. This property is known as biomass recalcitrance. Therefore, the bioconversion of lignocellulosic biomass needs improved technology and efficient methods of pretreatment. The saccharification step for a given lignocellulosic biomass is in the development phase as the quest for better techniques for the digestion of cellulose continues. The prevailing techniques have severe limitations due to several physicochemical, structural, and compositional factors (Mosier et al., 2005).

Pectins, glycosylated proteins, and several other inorganic materials are present in lignocelluloses: however, the major components are cellulose $(C_6H_{10}O_5)_n$, hemicellulose $(C_5H_8O_4)_m$, and lignin $[C_9H_{10}O_3(OCH_3)_{0.9-1.7}]_x$ which are present in the range of 30–50%, 15–35%, and 10–20%, respectively (Pettersen, 1984; Knauf et al., 2004).

The cellulose and hemicellulose part of LCB is joined firmly with the lignin part with the help of covalent as well as hydrogenic bonds. Therefore, the structure of LCB becomes very much resilient, and hence, does not show any degradation or reaction to any treatment (Knauf et al., 2004). This recalcitrant structure and structural complexity have been the main reason behind technical challenges towards the production of LCB-based biofuel (Zhang et al., 1995). Because of this, the enzymatic hydrolysis step is able to provide not more than 20% of its theoretical glucose yield without a proper pretreatment (Kim and Lee, 2005).

14.3.1 CELLULOSE

LCB material is composed of two types of cellulose – crystalline and amorphous. The crystalline form has bundles of cellulosic chains that are arranged in parallel to each other. Due to their inter-chain hydrogen bonding, these bundles become closely packed. On the other hand, the amorphous form has unordered layers. The amorphous part of cellulose with more accessibility is quickly hydrolysed by cellulase (Taherzadeh and Karimi, 2008).

14.3.2 HEMICELLULOSE

The structural constituents (pentoses, methyl pentoses, hexoses, carboxylic acids, etc.) of the natural polymer hemicellulose can be used for producing ethanol and related value-added products through bioconversion. Singh et al. (2008) have been able to characterize the LCBs and found that among the family of hemicellulose, xylan forms a big portion of hard-wood and softwood has glucomannan as a major constituent. In an LCB, hemicelluloses are connected to cellulose through hydrogen bonds. Hemicelluloses show physical linking with lignin via covalent bonds. Hemicelluloses, being amorphous and branched polymer, are easily hydrolysed by acids.

14.3.3 LIGNIN

After cellulose and hemicellulose, lignin is the most widely accessible biopolymer on earth. In lignin, the phenyl propane unit along with hemicellulose undergoes a complex formation that encloses the cellulose and due to this fact, biomass becomes non-reacting to enzymatic hydrolysis as well as chemical hydrolysis. This makes lignin a very complex heteropolymer that helps plants in making a cell wall strong enough to withstand the pathogens' attack (Brown and Chang, 2014). Due to this lignin cover around cellulose, the cellulolytic enzymes find it hard to reach to cellulose molecules and not able to hydrolyse it. According to Itoh et al. (2003), the lignin network must be broken before any attempt of enzymatic hydrolysis. Lignin is found in a greater amount in softwood than in hardwood. It is observed that during the depolymerization process, lignin may lead to the creation of furan compounds (furfural and hydroxymethyl-furfural) that might pose inhibition on fermentation at a later stage of processing (Zaldivar et al., 1999).

To produce ethanol from lignocellulosic biomass (LCB) it is required to perform enzymatic hydrolysis (saccharification) of cellulose and hemicellulose followed by the fermentation of monomeric sugars obtained in the previous step, the separation of lignin, and, at last, the recovery of ethanol. Ethanol is further distilled to match desired specifications (Alvira et al., 2010). Before any enzymatic hydrolysis, it is advisable to pretreat the biomass to make most of it accessible for the hydrolysis step.

14.4 BIOMASS PRETREATMENT

For each pretreatment method, it is important that it has effective access to cellulose and it can remove the cellulose out for further processing. Thus, pretreating the biomass helps to enhance the conversion of the enzymatic hydrolysis step which is carried out after the pretreatment step.

Pretreatment processes have to be selected depending on the characteristics of the biomass. The physical and chemical properties of biomass make a basis to choose the primary physical steps involved in biomass processing which are followed by other chemical-physical, thermochemical, chemical, and biotechnological processes. The knowledge of characterizing properties of biomass, i.e., physical, structural, and shearing, and targeted particle size helps to choose the suitable size reduction machinery and the operating conditions while maximizing the performance at minimum energy requirement per unit mass of biomass (Liu et al., 2016).

Pretreatment technologies are significant due to the fact that they are able to increase the effective surface area of cellulose sites so that it becomes accessible to hydrolytic enzymes even in small dosage and it helps further to complete the conversion process within less time and minimize the formation of unwanted products which might slow down the process or decrease the yield. In the past, several pretreatment methods and techniques had been tested and documented for hydrolysis of LCB (Alvira et al., 2010; Taherzadeh and Karimi, 2008).

Research on feedstock characterization reveals distinct physicochemical properties (Naik et al., 2010) needed to be specified for a given feedstock. This demands the selection of a suitable pretreatment technique that would be beneficial for saccharification and fermentation steps. Furthermore, rapid de-crystallization of cellulose and de-polymerization of hemicellulose, adoption of energy-efficient processes, minimum inhibitors formation, value-added products formation, and recovery are key characteristics of a scientific pretreatment technique. Major potential pretreatment techniques with recent updates as well as critical review are presented below:

14.4.1 WATER TREATMENTS

Aqueous systems have emerged as one of the eco-friendly and robust processes for the pretreatment of biomass. In these hydrothermal systems, the use of heat at different levels plays a major role. Cantero et al. (2019) observed three temperature zones to be important: from 150°C to 225°C, from 225°C to 350°C, and from 350°C to 400°C.

14.4.1.1 Temperature between 150°C and 225°C

As is known that water, when used at mild temperatures, shows favourable results in the extraction step and the hydrolysis step for hemicellulose and other less recalcitrant biomass components. Hemicelluloses can be separated from biomass using – a) fractionation technique to separate chemicals from lignocellulosic matter or b) and pre-extraction technique to separate hemicelluloses from lignocelluloses.

14.4.1.1.1 Fractionation

After the fractionation step, cellulose becomes readily available for the saccharification step. These pretreatments remove the hemicellulose as well as lignin part as is presented by Sun and Cheng (2002) and Mosier et al. (2005). Among them are steam explosion (with and without catalyst), liquid hot water, and dilute-acid hydrolysis. Research on alkaline hydrolysis is quite limited.

i. Steam explosion with further extraction steps makes it successful to fractionate lignocellulosic materials. Hemicelluloses are separated by water extraction. On the other hand, lignin can be removed by the use of alkali, aqueous acetic acid, or aqueous ethanol (Hongzhang and Liying, 2007).

ii. It has been reported that steam explosion coupled with acids (e.g., H_2SO_4), and gases (e.g., SO_2, CO_2) increases enzymatic hydrolysis, reduces the formation of inhibitory compounds, and results in better hemicellulose reduction (Wayman et al., 1986; Morjanoff and Gray, 1987). Morjanoff and Gray (1987) have also shown that the steam explosion pretreatment of sugarcane bagasse gives best results at 220°C, residence time of 30s, and water taken in double quantity to that of solid, and 1% H_2SO_4. Wayman et al. (1986) also observed that up to 89% wood hemicelluloses can be removed using up to 2.6% SO_2. Puri and Mamers (1983) used steam and high-pressure CO_2 for the pretreatment of LCB. They observed that the pretreatment process had effectively helped solubilize the hemicellulose fraction of the *Eucalyptus regnans*. At last, a xylose-rich liquid with 70% monomeric sugars was obtained.

iii. Hot water, at very high pressure, can be used for pretreatment; however, a high pressure needs to be maintained to keep the water in the liquid phase at high temperatures. This pretreatment can be called by either of the names – hydro-thermolysis, aqueous or steam/aqueous fractionation, uncatalyzed solvolysis, and aquasolv (Mosier et al., 2005). At the temperature between

200°C and 230°C, the 15-minute exposure to hot water helps dissolve up to 60% of biomass. This process can remove the hemicellulose up to 100%, cellulose in the range of 4–22%, and the lignin in the range of 35–60% (Mosier et al., 2005). In presence of liquid hot water, the hemicellulosic acetic acid as well as other organic acids are generated. Due to this two opposite kinds of effects are observed, which speed the pretreatment process as well as slow it down. The release of these acids catalyzes the generation and removal of oligosaccharides. On the other hand, there is a chance that the hemicellulose along with polysaccharides, may also be hydrolyzed to monosaccharides which are then degraded to aldehydes, to some extent due to the presence of the acid used in the process. Furfural thus produced from pentoses and 5-hydroxymethyl furfural produced from hexoses, are the main inhibitory compounds for the subsequent fermentation process (Palmqvist et al., 2000).

14.4.1.1.2 Pre-extraction

Heiningen (2006) reported the prehydrolyzing of wood chips just before the normal pulping step; the prehydrolyzing also covered low-hemicellulose-content dissolving pulps. The water prehydrolysis in the temperature range of 100°C to 170°C for the softwood showed a maximum extraction yield of 16% and in the temperature range of 170°C to 180°C for hardwood it was 18% (duration 45 minutes). Further increase in the severity of the reaction conditions helps in lignin degradation and dissolution.

The studies influenced by the work of Abatzoglou et al. (1992) on hydrothermal treatments for hemicellulose (xylan) hydrolysis covering several LCB tested to evaluate the severity factor. The activation energy for xylan degradation was determined to be 130 kJ/mol. Studies on auto-hydrolysis of E. globulus wood chips have shown that xylan up to 90.4% of initial content was removed by hydrothermal treatments, along with up to 13.8% removal of lignin. The amount of cellulose removed in this treatment was negligible and found intact in the solid phase which can be utilized for the production of pulp and paper (Garrote, Domínguez and Parajó, 2002; Garrote and Parajó, 2002).

Hot-water extraction is also known as pre-hydrolysis. It has also been in use at a commercial scale in the paper industry within the kraft process. The wood chips are exposed for 30–120 minutes to a high-temperature environment in the range of 160°C–180°C by using steam. Lignin precipitates are formed in this process step at high temperature and acidic environment. These precipitates pose an operational problem (Gütsch et al., 2012). It is also observed that hydrolysate is displaced by black as well as white liquor. An aqueous prehydrolysis process based on formic acid has been devised for the reduction of lignin precipitates (Heiningen et al., 2017). In this new process, at 20 g/L formic acid prehydrolysis, it is observed that the molecular weight and polydispersity of lignin thus precipitated comes down to approximately 1,100 g/mol (from 2,600 g/mol) and 2.0 (from 3.0) at autohydrolysis conditions.

However, there is still very less information about the techniques for the subtraction of lignin and other components by hot water pretreatment. There have been some studies which have tried to explain the process (Conner, 1984). In another study, Song et al. (2012) examined the finely ground spruce wood under a hot-water extraction system to confirm that initially the process is diffusion-limited (in the fibre wall) and the hemicelluloses-diffusion was relatively slower than that of monosaccharides and acetic acid.

14.4.1.2 Temperature between 225°C and 350°C Range

Broadly, hydrothermal processes can be classified into four groups of processes: aqueous phase reforming, hydrothermal liquefaction (HTL), hydrothermal carbonization (HTC), and hydrothermal gasification. In all of these processes, water acts as a reactant, a catalyst, and a solvent. Unlike conventional methods, hydrothermal processes are suitable to handle wet biomass and no prior dewatering is needed. The operating conditions of these hydrothermal processes produce a range of solid, liquid, or gaseous chemicals as well as fuels.

HTC is a thermochemical process that makes it possible to convert an organic feedstock into high carbon chemicals. HTC can be operated at conditions of temperature between 180°C to 250°C, and pressure between 2 MPa to 10 MPa in the presence of liquid water (Libra et al., 2011). In this process, the carbohydrates become hydrolyzed and polymerized again to form a solid known as hydro-char.

The hydro-char is easily breakable and highly hydrophobic and, therefore, can be separated easily from the liquid. When compared with the raw biomass, the hydro-char has better mass and energy density as well as dewaterability. It also has better combustibility as a solid fuel (Wang et al., 2018).

Hydro-char is obtained in the HTC process and its properties depend on feedstock, residence time, and temperature. Wiedner et al. (2013) presented an HTC study on the olive residues, poplar wood, and wheat straw to show that chemical properties of hydro-chars were related to temperature. At low-temperature processing, the properties were also dependent on the type of feedstock. In this study, it was shown that an increasing trend in temperature decreased polarity but the aromaticity of hydro-chars was increased. At elevated temperature, biomass loses its lignin content and gains in lignin oxidation as well as highly condensed black carbon sites. In comparison with biochar from pyrolysis, HTC produces the hydro-char characterized by less aromaticity and very low density. The property of the hydro-char is dependent on feedstocks at low temperatures. At carbonization temperatures above 180°C, it showed that properties of hydro-char were almost independent of feedstock. Gao et al. (2016) studied HTC of eucalyptus bark and found that process yield and properties of hydro-char were mainly dependent on temperature; however, residence time showed little effect. Increased carbonization temperature showed an increment in the hydrothermal conversion of feedstock, fixed carbon content, fuel ratio, heating value, and thermal stability of hydro-char respectively. But it also showed a decrement in yield and oxygenated functional groups respectively. Studies have been presented to show that a range of feed-stock including sewage sludge, municipal solid waste, animal manures, food waste, agricultural residue, aquaculture, and algal residues can also be used for HTC (Wang et al., 2018).

A HTL process is operated in the temperature range from 250°C to 370°C and suitable high pressure (Zhu et al., 2014). In HTL, it is targeted to prepare a liquid that can later be used as a fuel. In this process, a liquid phase, called bio-oil, is obtained which is rich in water-insoluble chemicals. Besides, an aqueous solution along with a gaseous mixture is also obtained. This aqueous phase contains few valuable products (Brunner, 2014). In the HTL process, obtained bio-oil shows high heating value and high viscosity. Comparatively, HTL is better than pyrolysis because at optimized conditions there is no solid formation, and due to comparatively lower process temperature, the oil does not degrade (Kruse and Dahmen, 2015).

Bio-oil has a comparatively high heating value (30–36 MJ/kg) than that of oil produced from pyrolysis (20–25 MJ/kg) due to the presence of polar compounds and water in the latter. These are produced during the pyrolysis process and are not separated even after cooling. On the other hand, in HTL, such polar compounds are carried away in the aqueous phase, and hence, bio-oil contains only low oxygen content (e.g., phenols) compounds (Kruse and Dahmen, 2015). Several studies have been conducted to perform HTL on a range of biomass, (Cao et al., 2017; Xu et al., 2018) and showed success in optimizing the operating conditions with better yield. Also, from their results it has been clear that there is no certain relationship between the hemicellulose, cellulose, and lignin contents and bio-oil/bio-char yield. It is found that for a biomass hydrothermal conversion, in addition to the effect of the three components, i.e., hemicellulose, cellulose, and lignin contents and feedstock species, the presence of extracts, e.g., fats, pigments, pectins, resins, proteins, terpenoids, and alkaloids are also responsible for bio-oil/hydro-char yield (Cao et al., 2017).

A specialized HTL is focused on lignin decomposition with an emphasis on producing phenols, which are further useful in the production of resins. Due to the lower reactivity of lignin in comparison to LCB, catalyzed HTL at higher temperatures is often required (Kang et al., 2013).

14.4.1.3 Temperature between the 350°C to 400°C Range

Hydrothermal processes operating in the near-critical range and supercritical (SC) (i.e., temperature range of 350°C to 400°C) make use of biomass to produce several bio-based products and bioenergy. Biomass pretreatment can be performed for hydrolysis of polymeric saccharides or phenols into oligomers and monomers (Saka and Ueno, 1999). These saccharide monomers can then be converted into fuels, chemicals, and materials production. In some cases, biomass pretreatment can help in separating the constituent polymers which may be utilized for the production of a range of bio-based products.

As is known, the critical point of water is characterized by 374°C, 22.1 MPa and the supercritical water (SCW) is the water at conditions above these values.

A temperature range from 350°C to 374°C and a pressure more than the liquid-vapour equilibrium can be termed as near critical. Though the exploration of the pretreatment of biomass using SCW has been an old topic (Bobleter, 1994), but very few studies were successful in reaching further development stage or commercial scale. Some studies show that SCW can be used to pretreat the biomass for cellulose or lignin (Saka and Ueno, 1999). However, some polymers are easily hydrolysed or extracted, e.g., starch, pectin, proteins, and hemicellulose they show over-reactivity in presence of SCW. This behaviour is due to the fact that the rate of reaction at such a temperature becomes too high to produce any viable yield of monomeric saccharides. To get a reasonable yield for hemicellulose and starch hydrolysis/extraction in SCW, it demands to modify the design to have extremely low reaction times (Cantero et al., 2015). Researchers have observed that at SC conditions, the rate of hydrolysis for cellulose is much more than as given by the Arrhenius equation at the subcritical temperature conditions (Sasaki et al., 2004). This motivated the researchers working on LCB feedstock for the development of reactor design to achieve desired (extremely less) reaction time as well as improve the selectivity of the process (Cantero et al., 2013). Studies reported designs to produce oligosaccharides as well as monosaccharides with reaction kinetic models along with the reacting mechanisms (Cantero et al., 2013) altogether improving the equipment for cellulose hydrolysis (Cocero, 2018). Besides, Cantero et al. (2015) developed the process of cellulose hydrolysis for LCB feedstock.

There have been some examples of lignin hydrolysis which utilized ionic mediums (subcritical conditions, targeting ion product) to maximize the yield and some others which proposed process designs for the rebonding of monomers and oligomers to the external alcohols providing stabilized linkage (Okuda et al., 2004). Abad-Fernandez et al. (2019) developed a process that has effectively minimized the rebonding while the reaction was operated at SCW conditions and it did not need to add a secondary reagent alcohol. To effectively control the reaction time at SCW conditions, this process makes use of the sudden expansion micro-reactor (Cantero et al., 2013), with a dose of sodium hydroxide for improving lignin solubility (Abad-Fernandez et al., 2019). However, the exact mechanism of the process is not clear. It is proposed to be either ionic based (Torry et al., 1992) or free-radical based (SCW nature).

14.4.2   Chemical Pretreatments

14.4.2.1   Acid Pretreatment

Acid pretreatment is the most used technique among various pretreatment methods available for delignifying LCBs (Anwar et al., 2014). This technique makes use of a concentrated (conc.) or a dilute acid which dissolves the hemicellulose–lignin network and improves the accessibility of enzymes to the cellulose for hydrolysis.

Pretreatment with dilute acid releases fewer inhibitors, minimizes the corrosion of equipments, and supports efficient separation of chemicals. There is wider adoption of the technique in the fermentation process. Also, for short periods, the dilute acid pretreatment suits with high operating temperature (>160°C) and low concentration of substrate (5–10 wt%). Besides, at some lower temperatures (<160°C), it can handle a high concentration of substrate (up to 40 wt%) for a longer retention period from 30 minutes to 90 minutes. However, conc. H_2SO_4 and conc. HCl are in use at large in the pretreatment of LCBs (Sun and Cheng, 2002).

Normally, in hydrolysis with dilute acid (1–3% acid), a temperature from 200°C to 240°C is required to degrade the cellulose crystals. After that, hexose and pentose dissociation starts along with the formation of some toxic compounds which are detrimental to an effective saccharification (Sun and Cheng, 2002). Dilute H_2SO_4 has been the most common acid used at commercial scale to pretreat various biomass materials – switch-grass (Li et al., 2010), corn stover (Du et al., 2010), spruce (Shuai et al., 2010), and poplar (Kumar and Wyman, 2009a). There have been studies on using HCl and HNO_3 (Himmel et al., 1997) and H_3PO_4 (Marzialetti et al., 2008) for acid pretreatment. Nguyen et al. (2000) used soft woods as feedstock and optimized sugar yield with the help of dilute H_2SO_4 pretreatment. Further, it is also shown that after acid pretreatment (1.5% H_2SO_4) followed by a 48-hour incubation period, the enzymatic hydrolysis of bermudagrass and rye straw showed 20% and 23% yield, respectively (Sun and Cheng, 2005). In a similar study it is shown that 56% yield is possible by enzymatic saccharification of acid (0.75% H_2SO_4) pretreated wheat straw (Saha et al., 2005). On the other hand, Marzialetti et al. (2008) placed loblolly pine in several different acids (Trifluloror-acetic acid, HCl, H_2SO_4, HNO_3, and H_3PO_4) and obtained up to 70% yield soluble monosaccharides based on the available hemicellulose.

Under the dilute acid category pretreatment, alternatively, organic acids such as maleic acid and fumaric acid have also been utilized to pretreat wheat straw (Kootstra et al., 2009). Furthermore, it is reported that the amount of furfural formed in these two organic acid pretreatments is comparatively less than that with sulphuric acid (Kootstra et al., 2009). Wang et al. (2015) have reported the ethanol yield of 77.3% via simultaneous saccharification and fermentation (SSF) for their study over acid (0.5% H_2SO_4, 170°C, 10 minutes) pretreated rice straw with substrate at 15% (w/w).

14.4.2.2 Alkaline Hydrolysis

Other than acids, an alkali can also be used as a pretreatment agent. In such case, NH_4OH, KOH, NaOH, $Ca(OH)_2$, NH_3, and $(NH_4)_2SO_3$ can be used for pretreating agricultural LCB. In presence of an alkali, LCB undergoes lignin deformation, partial de-crystallization of cellulose (Cheng et al., 2010; McIntosh and Vancov, 2011) and partial dissolution of hemicellulose (McIntosh and Vancov, 2011). Such alkaline pretreatment has been useful for agri-based LCB with a better yield of sugars (Kumar and Wyman, 2009b). For hardwood, NaOH has been useful in the digesting process and has been able to reduce lignin fraction from 24–50% to 20% (Kumar and Wyman, 2009b). Lime treatment presents similar benefits for enzymatic hydrolysis. Switchgrass, wheat straw, hardwoods, and softwoods having maximum lignin content up to 26% were studied (Zhao et al., 2008) for the efficacy of alkali pretreatment.

Saha and Cotta (2006) achieved a sugar yield of 97% for wheat straw pretreated in alkaline peroxide. Also, lime has been used in the pretreatment of wheat straw (85°C, 3-hour) (Chang, Nagwani and Holtzapple, 1998), poplar wood (150°C, 6-hour, 14 atm) (Chang et al., 2001), switchgrass (100°C, 2 hours) (Chang et al., 1997) and corn stover (100°C, 13 hours) (Kaar and Holtzapple, 2000). Alkali pretreatment of cotton stalks (2% NaOH, 121°C, 90 minutes) showed that delignification was 65% and cellulose conversion to be 61% (Silverstein et al., 2007), while alkali treatment of spruce (3% NaOH) in presence of urea (12%) at low temperature (15°C) gives 60% cellulose conversion (Zhao et al., 2008). It has been shown that, for wheat straw, sugar yield can be improved by five times at optimized conditions (2% NaOH, 60°C, 90 minutes) (McIntosh and Vancov, 2011). Zhu et al. (2010) applied different loadings from 1% w/w to 7.5% w/w of NaOH to corn stover, which was later put to biogas production through anaerobic digestion (solid-state), and found that the highest yield of biogas (372 L/kg) was possible with 5% NaOH with an explanation that increased depolymerization resulted in enhanced biogas production; however, at higher alkali concentration the methanogenesis was slowed down because of very fast rates of hydrolysis and acidogenesis.

14.4.2.3   Solvent Extraction/Organosolv

In the solvent extraction pretreatment technique, organic solvent or mixture of water-organic solvents with water is used for the separation of hemicellulose and lignin. Examples of such liquids are ethylene glycol and water with butanol/benzene/ethanol. Such systems can separate pure lignin. It has been reported that several solvents have been tested for pretreatment – low BP (methyl and ethyl alcohol), high BP (glycerol, ethylene glycol, and tetrahydrofurfuryl alcohol), and dimethyl sulfoxide, ethers, ketone, and phenols. Also, these solvents are later separated post-treatment due to the possibility of inhibition with enzymatic hydrolysis (Sun and Cheng, 2002). Beetle-killed Lodgepole pine showed a sugar yield of 97% after the ethanol-organosolv pretreatment (Pan et al., 2007). In a research by Sun and Chen (2007) on the wheat straw, 95% cellulose recovery along with 70% lignin removal was achieved using glycerol-organosolv pretreatment. Araque et al. (2008) have conducted a study over *Pinus radiata* chips and showed that pretreatment with acetone-organosolv (195°C, 5 minutes, pH 2.0) resulted in an ethanol yield of 99%.

14.4.2.4   Oxidation

Oxidizing agents such as ozone, hydrogen peroxide, oxygen, or air can be utilized for the delignification of LCB. Pretreatment with ozone is known as "ozonolysis", and is useful in the breaking of aromatic ring structures of lignin, while it offers no effect on hemicellulose and cellulose. It has been tested on number of LCB, for example, wheat straw, bagasse, pine, peanut, cotton straw, and poplar sawdust

(Sun and Cheng, 2002). Straw, reed, and residues of some cereal crop which have a thick wax coat of silica and protein can be treated by wet oxidation (Schmidt et al., 2002). Also, Na_2CO_3 treatment of wheat straw has shown 96% cellulose recovery along with 65% glucose yield (Klinke et al., 2002). Similarly, wet oxidation, in comparison to steam explosion, of sugarcane bagasse has resulted in 57% cellulose conversion and more useful by-products (Martín and Thomsen, 2007). Ozonolysis of wheat straw and rye straw is shown to have helped in enzymatic hydrolysis yield (García-Cubero et al., 2009). Lignins also present themselves as an important free radical scavenger and a good source of antioxidants. Studies have shown that lignins from different sources can serve as a source of antioxidants. García et al. (2010) have found that the ultrafiltration process separates lignins that have useful antioxidant properties.

14.4.2.5 Ionic Liquids

In the recent past, ionic liquids (IL) have been found as efficient and cost-effective pretreatment technology. Ionic liquids are liquid at room temperature with very low vapour pressure and have a small anion and a large organic cation. Examples of IL are 1-ethyl-3- methylimidazolium diethyl phosphate ([Emim]Dep), N-methyl morpholine N-oxide (NMMO), 1-allyl-3 methylimidazoliumchloride ([Amim]Cl), 1-n-butyl-3-methylimidazolium chloride ([Bmim]Cl), 1-buthyl-3-methylimidazolium acetate [BMIM][OAc], and 1-ethyl-3- methylimidazolium acetate ([Emim]Ac). A range of ionic liquids have been tested to treat various LCBs for example, corn stover (Cao et al., 2010), cotton (Zhao et al., 2009), bagasse (Wang et al., 2009), switchgrass (Samayam and Schall, 2010), wheat straw (Li et al., 2009), and woods (eucalyptus, pine, poplar, and oak) of varying hardness (Fort et al., 2007).

NMMO has been useful in the commercial production of Tencel fibre which can dissolve several LCBs (Kuo and Lee, 2009) and also offers more than 99% solvent recovery rate (Perepelkin, 2007). NMMO treatment offered faster rates of cellulose hydrolysis and it had been able to break through the crystalline zone of cellulose (Kuo and Lee, 2009). It has been shown that bio-ethanol yield from hard and softwood increased from 19% to 85% and 7% to 89% respectively after NMMO treatment (130°C, 3hour, 85 wt%) assisted nonisothermal SSF. It is found that enzymatic hydrolysis is improved by two times after NMMO pretreatment over sugarcane bagasse at 130°C for one hour (Kuo and Lee, 2009). Similarly, it is observed that hydrolysis yield increased from 16.5% to 72.5% for switchgrass after it was pretreated with 1-ethyl-3- methylimidazolium acetate (Singh et al., 2009). Poornejad et al. (2013) reported the enhancement in saccharification and bio-ethanol yield after the rice straw was pretreated by NMMO and [BMI-M][OAc]) at 120°C. Sugarcane bagasse was studied for the effect of pretreatment with an IL ([EMIM][OAc]) over the enzymatic hydrolysis (Qiu et al., 2012). In another study, Shafiei et al. (2014) reported that NMMO pretreatment was able to improve the bi-ethanol yield up to 68.1–86.1% (g/g) of theoretical yield for pinewood powder and up to 12.6–51.2% (g/g) for wood chips.

14.4.3 PHYSICOCHEMICAL PRETREATMENTS

14.4.3.1 Explosion/Autohydrolysis

Both, steam explosion (uncatalyzed or catalyzed) and steaming without explosion (autohydrolysis) have been successful in pretreating the LCB for bio-ethanol production. Explosion works through injecting the high-pressure saturated steam to the LCB reactor. It saturates the dry LCB with the help of steam at high temperature and pressure and then pressure is released all of a sudden to initiate flash evapouration. LCB gets ruptured because of the thermo-mechanical force exerted by water and opened up structure becomes available for enzymatic-hydrolysis. Saccharification of cellulose, henceforth increased, is, however, not correlated very well with rupturing of LCB (Brownell et al., 1986). There have been studies to produce ethanol, butanol, and methane from corn stover (Tucker et al., 2003), municipal solid waste (Liu et al., 2002), and wood (Claassen et al., 1999) after steam-explosion pretreatment. A study over sunflower stalks after steam explosion (5 minutes, 220°C) pretreatment before enzymatic hydrolysis shows maximum hemicellulose recovery (Ruiz et al., 2006).

14.4.3.2 Ammonia Pretreatment

Pretreatment of LCB with aqueous ammonia at high temperature destroys the structure of lignin, decrystallizes the cellulose, removing some of the hemicellulose as well. In a broad sense, this pretreatment involves soaking in aqueous ammonia (SAA), ammonia recycle percolation (ARP), and the ammonia fibre explosion-method (AFEX). Evaluation of SAA (solid: liquid ratio 1:12, 75°C, aqueous Ammonia 15%, no agitation, 48 h) pretreatment method for LCB showed that 66% lignin was solubilized and hydrolysis yield was 63% (xylan) and 83% (glucan) (Kim et al., 2008). During AFEX treatment, LCB comes in contact with aq. ammonia at higher temperature in the range of 90°C–100°C for 30-minute results in a sudden drop in pressure. The main operational parameters are temperature, ammonia concentration, blow-down pressure, water loading, cycles of treatments, and time (Holtzapple et al., 1991). A study on AFEX for switchgrass determined the optimum conditions – temperature of 100°C, time of retention five minutes, and ammonia to dry biomass ratio 1:1 (Alizadeh et al., 2005). Another study on over-optimizing AFEX parameters in converting sweet sorghum bagasse and forage into bio-ethanol was conducted (Li et al., 2010). A yield of 90% is reported in bio-ethanol production with AFEX pretreatment assisted saccharification of fibre from palm fruit in the 72 h duration of fermentation (Lau et al., 2010).

14.4.3.3 Supercritical Fluid Pretreatment

A supercritical fluid (SCF) is a material existing in the supercritical region (above its critical temperature and pressure). In such case, liquid becomes a novel solvent with a diffusivity and viscosity of a gas and it can pass through tiny pores of LCB and it bypasses the limitations of mass transfer (McHardy and Sawan, 1998). Supercritical CO_2 (Tc of 31°C and Pc= 73 atm) has been very effective in pretreatment of cellulose resulting in 100% yield for glucose in enzymatic hydrolysis step (Park et al., 2001) and 32% for rice straw (Gao et al., 2010). A study reported a yield of 15% for sugar from wheat straw which was pretreated with supercritical CO_2 (Alinia et al., 2010). However, CO_2 explosions have not been able to give better results when compared to ammonia or steam explosion (Sun and Cheng, 2002).

14.4.4 Biological Pretreatments

In a biological pretreatment soft-rot, brown, and white fungi are utilized which help in separating the cellulose from lignin and hemicellulose. This pretreatment technique degrades hemicellulose and lignin; however, the cellulose is not affected. This technique needs normal operating parameters and low energy input. However, it demands a longer retention time because of its low rate of reaction (Sun and Cheng, 2002).

14.4.4.1 White-Rot Fungi

Among saprophytic fungi group, *Lignolytic basidiomycetes* is responsible for a white-rot in woods and hence it is known as white-rot fungi. There have been studies on the application of white-rot fungi with high specificity for degradation of lignin, for example, *Pycnoporus sanguineus* (Lu et al., 2010), *Irpex lacteus* (Yu et al., 2010), *Echinodontium taxodii* (Yu et al., 2009) and *Phlebia spp.* (Arora and Sharma, 2009). There has been simultaneous carbohydrate degradation with *Phlebia radiata, P. floridensis, P. brevispora* (Sharma and Arora, 2010), *Pleurotus sajorcaju* (Kannan et al., 1990), *Trametes versicolor* (Yu et al., 2009), *Oxysporus sp.* (Haddadin et al., 2002), *Trichoderma reesei* (Singh et al., 2008), *Echinodontium taxodii* (Yu et al., 2009), *Ceriporiopsis subvermispora* (Wan and Li, 2010) and *Gonoderma sp.* (Haddadin et al., 2002; Tripathi et al., 2008) which also showed a maximum of 40–60% lignin reduction. The yield of bio-ethanol production has been related to biodelignification as it helps in the recovery of sugar. In some cases, biodelignification if used before any chemical pretreatments, has shown 80% delignification (Yu et al., 2010; Yu et al., 2009).

14.4.4.2 Brown-Rot Fungi

Brown-rot fungi can degrade the hemicellulose and cellulose very efficiently but it is comparatively less efficient in degrading lignin. Though lignin is degraded, it is, however, incomplete and leaves a wood brown-rotted and this is the reason that fungi has been given such name. It has been found that it preferably acts on the hemicellulose and cellulose part, leaving the lignin part unoxidized (Hastrup et al., 2012). There have been biodegradation studies on various species, for example, *Gleophyllum trabeum, Coniophora puteana, Laetoporeus sulphureus, Meruliporia incrassata,* and *Serpula lacrymans* (Monrroy et al., 2011).

14.4.4.3   Soft-Rot Fungi

There are two kinds of soft-rot fungi available – type I and II. Type I has cavities of biconical/cylindrical shape which are inside secondary walls and type II is a degraded erosion form (Blanchette, 2000). Among type II, *Daldinia concentric* is very effective in working on hardwood (Narayanaswamy et al., 2013) and it has been reported that, in 2 months, 53% conversion is possible for birchwood (Nilsson et al., 1989). *Cadophora spp.* (Chandel et al., 2015) has been found useful in the very fast biodelignification of LCB.

14.4.5   COMBINED PRETREATMENTS

14.4.5.1   Oxidative Lime Pretreatment

It has been reported that normal thermal-lime pretreatment does not succeed in the delignification of high lignin biomass and hence, an oxidant, for example, oxygen, needs to be used in the pretreatment step (Chang et al., 2001). Chang and Holtzapple (2000) found that the lime pretreatment can make the substrate sites approachable for hydrolysis. In another case, when tested for hybrid poplar, corn stover, and switchgrass in presence of $Ca(OH)_2$ (10% w/w of substrate), the pretreatment showed 13-times better enzymatic hydrolysis along with less sugar degradation, and lime recovered up to 21% using CO_2 (Chang et al., 2001). Optimization studies showed that at 240-minute duration at 6.89 bar pressure, 110°C temperature along with a milling step for switchgrass, *Panicum virgatum* resulted in 68% xylan yield and 93% glucan yield (Falls and Holtzapple, 2011).

14.4.5.2   Supercritical CO_2 with Steam Explosion

There have been studies on combining supercritical CO_2 with steam to degrade lignin and prepare the cellulose for saccharification. For example, dry and wet wheat straw tested for pretreatment over a range of operating conditions in the reactor (Alinia et al., 2010). In another study, over various LCBs (corn stover, switchgrass, and wood) under pretreatment with supercritical CO_2 with steam at 200 bar with 40% loadings (high) has shown glucose yields of 85%, 81%, and 73% respectively (Luterbacher et al., 2010).

14.4.5.3   Dilute Acid Pre-soaking before Organosolv

In a study, 67% glucose yield has been reported when the sugarcane bagasse was placed for presoaking (24 h) in dilute acid before undergoing an organosolv (30% v/v ethanol) with NaOH (catalyst) at 195°C for 60 minutes (Mesa et al., 2011). In another study over dilute acid pre-soaking before aq. ethanol organosolv for *Miscanthus sp.* showed 20% xylans recovery and 98% cellulose conversion to glucose (Brosse et al., 2009).

14.4.5.4   Alkaline Peroxide Treatment Coupled with Steam Explosion

It has been reported that reducing sugar with a maximum amount of 568 mg/g of initial dry biomass basis have been recovered from bamboo via enzyme saccharification using H_2O_2 (1% v/v) and NaOH (1% w/w) (Yamashita et al., 2010). In a study by Chen and Qiu (2010) with steam explosion (20 atm), a sugar yield of 460 mg/g (based on initial dry biomass) from bamboo was obtained in presence of 10% (w/w) NaOH. Steam explosion (1.5 MPa, 10 minutes, 198°C) along with alkali-H_2O_2 pretreatment for wheat straw, resulted in 81% bio-ethanol yield using SSF (Chen et al., 2008). Similarly, 92–99% of lignin was removed in a two-stage process where steam explosion (15–22 bar, 200°C–220°C) was employed coupled with alkaline peroxide (50°C, 2% H_2O_2, pH 11.5, 5 h) (Sun et al., 2005).

14.4.5.5   Additional Combined Pretreatment Techniques

The cumulative effect of [Emim]Ac and ammonia was examined on rice straw by Nguyen et al. (2010) and observed 82% of the cellulose extraction and 97% conversion glucose at a later step. Also, in the case of the rice straw, a treatment with [BMIM]Cl and steam explosion, it led to complete enzymatic hydrolysis (Liu and Chen, 2006). Ethanol (40%) recovery coupled with steam explosion (20 minutes. at 180°C) pretreatment of wheat straw demonstrated 82% delignification with cellulose extraction of 94%. A pretreatment by combining HCW with ball milling increased the effectiveness of the enzymatic hydrolysis of Eucalyptus wood chips and yielded 70% sugar (Inoue et al., 2008).

14.5   CONCLUSIONS

For a long time, researchers around the globe have focused on the development of new technologies for the conversion of LCBs to biofuels and other bio-based materials. In case of bio-ethanol, reduction in the cost of ethanol production depends on the efficient use of the raw material for high yields, high productivity, greater concentration of ethanol in the distillation feed, and the integration of processes to lower the demand for energy. However, the failure to find robust and appropriate technologies on a broad industrial scale continues to hinder the research.

Biomass pretreatment has been in practice in conventional ways since a long time. However, to meet the worldwide increasing demand the new technologies have to be optimized for faster conversion, better yield, and useful by-products. Among all available options for pretreatment of LCBs, chemical methods are the first choice for industrial applications as they offer high reactivity even at normal reaction conditions. However, the recovery of unused chemicals is a major challenge. Advantages of acid pretreatment techniques include their economic feasibility and robustness to be applied on more than one type of LCB to offer high scalability for large scale production of biofuels. Similarly, in a steam explosion process, the inhibitory products which are formed during downstream processing, high energy input, and the least energy efficiency are major challenges.

There have been innovative pretreatment technologies where some of the existing pretreatment techniques are combined for better results in biomass digestion, good recovery along with minimizing undesired product formation. Some of the thermo-chemical and chemical pretreatment techniques were coupled with bioconversion processes; however, it has not been very successful to reach commercial scale. There has been a constant search to find a suitable pretreatment technique coupled with efficient process integration to achieve such conversion levels that it could become economically viable for a given biomass.

Diversity in LCB presents various types of cells and molecules for biofuel processing. Due to this, it is a challenge to develop an efficient pretreatment technique to cover wide range of LCB. Technology for the production of biofuels will guide the future of business models in the area of feedstock cultivation and will serve as a limit up to which the societal and economic benefits could be reaped. The evidences from the research work at global level have conveyed to our policymakers that there is a need to record and combine climate change and its effects on biofuels, bioenergy production, consumption, and distribution as well as changes in land use in connection with feedstock cultivation. Thus, there is a scope of developing a coordinated research plan to establish optimized and integrated bio-refining, elaborated pretreatment mechanisms to achieve better yield for fuels as well as co-products. It also calls for detailed feasibility analysis covering energy balances, enzyme loadings, improved enzymatic hydrolysis, and good solvent recovery before implementing these technologies at a commercial-scale.

CONTRIBUTION

This chapter offers a review of the conventional as well as latest developments in pretreatment processes for the hydrolysis of biomass. It focuses on the physical, chemical, physicochemical, and biological processes that are largely used for pre-treatment. At the same time, it also discusses some of the unique and integrated pretreatment strategies used to boost the digestibility of lignocellulosic biomass (LCB). This chapter brings to the fore some of the latest methods, techniques, and materials available for the pretreatment of waste biomass.

REFERENCES

Abad-Fernández, N., Pérez, E. and Cocero, M. J. (2019) 'Aromatics from lignin through ultrafast reactions in water', *Green Chemistry*, 21(6), pp. 1351–1360.

Abatzoglou, N. et al. (1992) 'Phenomenological kinetics of complex systems: The development of a generalized severity parameter and its application to lignocellulosics fractionation', *Chemical Engineering Science*, 47(5), pp. 1109–1122.

Adsul, M. G. et al. (2011) 'Development of biocatalysts for production of commodity chemicals from lignocellulosic biomass', *Bioresource Technology*, 102, pp. 4304–4312.

Akhtar, N. et al. (2016) 'Recent advances in pretreatment technologies for efficient hydrolysis of lignocellulosic biomass', *Environmental Progress and Sustainable Energy*, 35(2), pp. 489–511.

Alinia, R. et al. (2010) 'Pretreatment of wheat straw by supercritical CO2 and its enzymatic hydrolysis for sugar production', *Biosystems Engineering*, 107(1), pp. 61–66.

Alizadeh, H. et al. (2005) 'Pretreatment of switchgrass by ammonia fiber explosion (AFEX)', *Applied Biochemistry and Biotechnology*, 124(1), pp. 1133–1141.

Alvira, P. et al. (2010) 'Pretreatment technologies for an efficient bioethanol production process based on enzymatic hydrolysis: A review', *Bioresource Technology*, 101(13), pp. 4851–4861.

Anwar, Z., Gulfraz, M. and Irshad, M. (2014) 'Agro-industrial lignocellulosic biomass a key to unlock the future bio-energy: A brief review', *Journal of Radiation Research and Applied Sciences*, 7(2), pp. 163–173.

Aransiola EF (2012) 'Production of biodiesel from crude neem oil feedstock and its emissions from internal combustion engines', *African Journal of Biotechnology*, 11(22), pp. 6178–6186.

Araque, E. et al. (2008) 'Evaluation of organosolv pretreatment for the conversion of Pinus radiata D. Don to ethanol', *Enzyme and Microbial Technology*, 43(2), pp. 214–219.

Arora, D. S. and Sharma, R. K. (2009) 'Comparative ligninolytic potential of *Phlebia* species and their role in improvement of *in vitro* digestibility of wheat straw', *Journal of Animal and Feed Sciences*, 18(1), pp. 151–161.

Blanchette, R. A. (2000) 'A review of microbial deterioration found in archaeological wood from different environments', *International Biodeterioration and Biodegradation*, pp. 189–204.

Bobleter, O. (1994) 'Hydrothermal degradation of polymers derived from plants', *Progress in Polymer Science*. Pergamon, pp. 797–841.

Brosse, N., Sannigrahi, P. and Ragauskas, A. (2009) 'Pretreatment of Miscanthus x giganteus using the ethanol organosolv process for ethanol production', *Industrial and Engineering Chemistry Research*, 48(18), pp. 8328–8334.

Brown, M. E. and Chang, M. C. Y. (2014) 'Exploring bacterial lignin degradation', *Current Opinion in Chemical Biology*. Elsevier Current Trends, pp. 1–7.

Brownell, H. H., Yu, E. K. and Saddler, J. N. (1986) 'Steam-explosion pretreatment of wood: Effect of chip size, acid, moisture content and pressure drop', *Biotechnology and Bioengineering*, 28(6), pp. 792–801.

Brunner, G. (2014) 'Processing of biomass with hydrothermal and supercritical water', in Brunner, G. (ed.) *Supercritical Fluid Science and Technology*, vol 5. Amsterdam: Elsevier B.V., pp. 395–509.

Cantero, D. et al. (2019) 'Pretreatment processes of biomass for biorefineries: Current status and prospects', *Annual Review of Chemical and Biomolecular Engineering*, 10, pp. 289–310.

Cantero, D. A., Martínez, C., et al. (2015) 'Simultaneous and selective recovery of cellulose and hemicellulose fractions from wheat bran by supercritical water hydrolysis', *Green Chemistry*, 17(1), pp. 610–618.

Cantero, D. A., Dolores Bermejo, M. and José Cocero, M. (2013) 'High glucose selectivity in pressurized water hydrolysis of cellulose using ultra-fast reactors', *Bioresource Technology*, 135, pp. 697–703.

Cao, L. et al. (2017) 'Hydrothermal liquefaction of agricultural and forestry wastes: State-of-the-art review and future prospects', *Bioresource Technology*, 245 (part A), pp. 1184–1193.

Cao, Y. et al. (2010) 'Structure and properties of novel regenerated cellulose films prepared from cornhusk cellulose in room temperature ionic liquids', *Journal of Applied Polymer Science*, 116(1), pp. 547–554.

Chandel, A. K. et al. (2015) 'Biodelignification of lignocellulose substrates: An intrinsic and sustainable pretreatment strategy for clean energy production', *Critical Reviews in Biotechnology*, pp. 281–293.

Chang, V. S., Nagwani, M., et al. (2001) 'Oxidative lime pretreatment of high-lignin biomass: Poplar wood and newspaper', *Applied Biochemistry and Biotechnology - Part A Enzyme Engineering and Biotechnology*, 94(1), pp. 1–28.

Chang, V. S., Burr, B. and Holtzapple, M. T. (1997) 'Lime pretreatment of switchgrass', *Applied Biochemistry and Biotechnology*, 63(1), p. 3.

Chang, V. S. and Holtzapple, M. T. (2000) 'Fundamental factors affecting biomass enzymatic reactivity', *Applied Biochemistry and Biotechnology - Part A Enzyme Engineering and Biotechnology*, 84, pp. 5–37.

Chang, V. S., Nagwani, M. and Holtzapple, M. T. (1998) 'Lime pretreatment of crop residues bagasse and wheat straw', *Applied Biochemistry and Biotechnology*, 74(3), pp. 135–159.

Chen, H., Han, Y. and Xu, J. (2008) 'Simultaneous saccharification and fermentation of steam exploded wheat straw pretreated with alkaline peroxide', *Process Biochemistry*, 43(12), pp. 1462–1466.

Chen, H. and Qiu, W. (2010) 'Key technologies for bioethanol production from lignocellulose', *Biotechnology Advances*, 28(5), pp. 556–562.

Cheng, Y.-S. et al. (2010) 'Evaluation of high solids alkaline pretreatment of rice straw', *Applied Biochemistry and Biotechnology*, 162(6), pp. 1768–1784.

Ciciora, P. (2011) *A billion tons of biomass a viable goal, but at high price, new research shows*, 16 Feb, Illinois News Bureau, University of Illinois at Urbana-Champaign, viewed 10 December 2020, Available at: https://news.illinois.edu/view/6367/205414.

Claassen, P. A. M. et al. (1999) 'Utilisation of biomass for the supply of energy carriers', *Applied and Environmental Microbiology*, 52(6), pp. 741–755.

Cocero, M. J. (2018) 'Supercritical water processes: Future prospects', *Journal of Supercritical Fluids*, 134, pp. 124–132.

Conner, A. (1984) 'Kinetic modeling of hardwood prehydrolysis. I. Xylan removal by water prehydrolysis', *Wood and Fiber Science (USA)*, 16(2), pp. 268–277.

Du, B. et al. (2010) 'Effect of varying feedstock-pretreatment chemistry combinations on the formation and accumulation of potentially inhibitory degradation products in biomass hydrolysates', *Biotechnology and Bioengineering*, 107(3), pp. 430–440.

Falls, M. and Holtzapple, M. T. (2011) 'Oxidative lime pretreatment of alamo switchgrass', *Applied Biochemistry and Biotechnology*, 165(2), pp. 506–522.

Fort, D. A. et al. (2007) 'Can ionic liquids dissolve wood? Processing and analysis of lignocellulosic materials with 1-n-butyl-3-methylimidazolium chloride', *Green Chemistry*, 9(1), pp. 63–69.

Gao, M. et al. (2010) 'Effect of SC-CO2 pretreatment in increasing rice straw biomass conversion', *Biosystems Engineering*, 106(4), pp. 470–475.

Gao, P. et al. (2016) 'Preparation and characterization of hydrochar from waste eucalyptus bark by hydrothermal carbonization', *Energy*, 97, pp. 238–245.

García-Cubero, M. A. T. et al. (2009) 'Effect of ozonolysis pretreatment on enzymatic digestibility of wheat and rye straw', *Bioresource Technology*, 100(4), pp. 1608–1613.

García, A. et al. (2010) 'Study of the antioxidant capacity of *Miscanthus sinensis* lignins', *Process Biochemistry*, 45(6), pp. 935–940.

Garrote, G., Domínguez, H. and Parajó, J. C. (2002) 'Interpretation of deacetylation and hemicellulose hydrolysis during hydrothermal treatments on the basis of the severity factor', *Process Biochemistry*, 37(10), pp. 1067–1073.

Garrote, G. and Parajó, J. C. (2002) 'Non-isothermal autohydrolysis of Eucalyptus wood', *Wood Science and Technology*, 36(2), pp. 111–123.

Gütsch, J. S., Nousiainen, T. and Sixta, H. (2012) 'Comparative evaluation of autohydrolysis and acid-catalyzed hydrolysis of Eucalyptus globulus wood', *Bioresour. Technol.* 109, pp. 77–85.

Haddadin, M. S. et al. (2002) 'Bio-degradation of lignin in olive pomace by freshly-isolated species of Basidiomycete', *Bioresource Technology*, 82(2), pp. 131–137.

Hastrup, A. C. S. et al. (2012) 'Enzymatic oxalic acid regulation correlated with wood degradation in four brown-rot fungi', *International Biodeterioration and Biodegradation*, 75, pp. 109–114.

van Heiningen, A. et al. (2017) 'Minimizing precipitated lignin formation and maximizing monosugar concentration by formic acid reinforced hydrolysis of hardwood chips', in *Hydrothermal Processing in Biorefineries: Production of Bioethanol and High Added-Value Compounds of Second and Third Generation Biomass*, Cham: Springer International Publishing, pp. 421–441.

van Heiningen, A. (2006) 'Converting a kraft pulp mill into an integrated forest products biorefinery', in *Annual Meeting of the Pulp and Paper Technical Association of Canada (PAPTAC)*, pp. 38–43.

Himmel, M. et al. (1997) 'Advanced bioethanol production technologies: A perspective', in *ACS Symp Ser*, pp. 2–45.

Holtzapple, M. T. et al. (1991) 'The ammonia freeze explosion (AFEX) process', *Applied Biochemistry and Biotechnology*, 28(1), pp. 59–74.

Hongzhang, C. and Liying, L. (2007) 'Unpolluted fractionation of wheat straw by steam explosion and ethanol extraction', *Bioresource Technology*, 98(3), pp. 666–676.

Inoue, H. et al. (2008) 'Combining hot-compressed water and ball milling pretreatments to improve the efficiency of the enzymatic hydrolysis of eucalyptus', *Biotechnology for Biofuels*, 1(1), p. 2.

Itoh, H. et al. (2003) 'Bioorganosolve pretreatments for simultaneous saccharification and fermentation of beech wood by ethanolysis and white rot fungi', *Journal of Biotechnology*, 103(3), pp. 273–280.

Jain, S. and Sharma, M. P. (2010) 'Prospects of biodiesel from Jatropha in India: A review', *Renewable and Sustainable Energy Reviews*, 14(2), pp. 763–771.

Kaar, W. E. and Holtzapple, M. T. (2000) 'Using lime pretreatment to facilitate the enzymic hydrolysis of corn stover', *Biomass and Bioenergy*, 18(3), pp. 189–199.

Kang, S., Li, X., Fan, J. and Chang, J. (2013) 'Hydrothermal conversion of lignin: a review', *Renew. Sustain. Energy Rev.*, 27, pp. 546–558.

Kannan, K., Oblisami, G. and Loganathan, B. G. (1990) 'Enzymology of ligno-cellulose degradation by Pleurotus sajor-caju during growth on paper-mill sludge', *Biological Wastes*, 33(1), pp. 1–8.

Kim, S. and Dale, B. E. (2004) 'Global potential bioethanol production from wasted crops and crop residues', *Biomass and Bioenergy*, 26(4), pp. 361–375.

Kim, T. H. and Lee, Y. Y. (2005) 'Pretreatment of corn stover by soaking in aqueous ammonia', in *Twenty-Sixth Symposium on Biotechnology for Fuels and Chemicals*, pp. 1119–1131.

Kim, T. H., Taylor, F. and Hicks, K. B. (2008) 'Bioethanol production from barley hull using SAA (soaking in aqueous ammonia) pretreatment', *Bioresource Technology*, 99(13), pp. 5694–5702.

Kiss, A. A., Dimian, A. C. and Rothenberg, G. (2008) 'Biodiesel by catalytic reactive distillation powered by metal oxides', *Energy and Fuels*, 22(1), pp. 598–604.

Klinke, H. B. et al. (2002) 'Characterization of degradation products from alkaline wet oxidation of wheat straw.', *Bioresource Technology*, 82(1), pp. 15–26.

Knauf, M., Moniruzzaman, M. et al. (2004) 'Lignocellulosic biomass processing: A perspective.', *International Sugar Journal*, 106(1263), pp. 147–150.

Kootstra, A. M. J. et al. (2009) 'Comparison of dilute mineral and organic acid pretreatment for enzymatic hydrolysis of wheat straw', *Biochemical Engineering Journal*, 46(2), pp. 126–131.

Kruse, A. and Dahmen, N. (2015) 'Water - A magic solvent for biomass conversion', *Journal of Supercritical Fluids*, 96, pp. 36–45.

Kumar, R. and Wyman, C. E. (2009a) 'Access of cellulase to cellulose and lignin for poplar solids produced by leading pretreatment technologies.', *Biotechnology Progress*, 25(3), pp. 807–819.

Kumar, R. and Wyman, C. E. (2009b) 'Effects of cellulase and xylanase enzymes on the deconstruction of solids from pretreatment of poplar by leading technologies.', *Biotechnology Progress*, 25(2), pp. 302–314.

Kuo, C.-H. and Lee, C.-K. (2009) 'Enhanced enzymatic hydrolysis of sugarcane bagasse by N-methylmorpholine-N-oxide pretreatment.', *Bioresource Technology*, 100(2), pp. 866–871.

Lau, M.J., Lau, M.W., Gunawan, C., and Dale, B.E. (2010) 'Ammonia fiber expansion (AFEX) pre-treatment, enzymatic hydrolysis, and fermentation on empty palm Fruit bunch fiber (EPFBF) for cellulosic ethanol production.' *Applied Biochemistry and Biotechnology*, 162, pp. 1847–1857.

Li, B.-Z. et al. (2010) 'Process optimization to convert forage and sweet sorghum bagasse to ethanol based on ammonia fiber expansion (AFEX) pretreatment', *Bioresource Technology*, 101(4), pp. 1285–1292.

Li, Q. et al. (2009) 'Improving enzymatic hydrolysis of wheat straw using ionic liquid 1-ethyl-3-methyl imidazolium diethyl phosphate pretreatment.', *Bioresource Technology*, 100(14), pp. 3570–3575.

Libra, J. A. et al. (2011) 'Hydrothermal carbonization of biomass residuals: A comparative review of the chemistry, processes and applications of wet and dry pyrolysis', *Biofuels*, 2(1), pp. 71–106.

Liu, H. W. et al. (2002) 'Steam pressure disruption of municipal solid waste enhances anaerobic digestion kinetics and biogas yield', *Biotechnology and Bioengineering*, 77(2), pp. 121–130.

Liu, L. and Chen, H. (2006) 'Enzymatic hydrolysis of cellulose materials treated with ionic liquid [BMIM] Cl', *Chinese Science Bulletin*, 51(20), pp. 2432–2436.

Liu, Y., Wang, J. and Wolcott, M. P. (2016) 'Assessing the specific energy consumption and physical properties of comminuted Douglas-fir chips for bioconversion', *Industrial Crops and Products*, 94, pp. 394–400.

Lu, C. et al. (2010) 'An efficient system for pre-delignification of gramineous biofuel feedstock in vitro: Application of a laccase from Pycnoporus sanguineus H275', *Process Biochemistry*, 45(7), pp. 1141–1147.

Luterbacher, J. S., Tester, J. W. and Walker, L. P. (2010) 'High-solids biphasic CO2–H2O pretreatment of lignocellulosic biomass', *Biotechnology and Bioengineering*, 107(3), pp. 451–460.

Lynd, L. R. et al. (2005) 'Consolidated bioprocessing of cellulosic biomass: An update', *Current Opinion in Biotechnology*, 16(5), pp. 577–583.

Martín, C. and Thomsen, A. B. (2007) 'Wet oxidation pretreatment of lignocellulosic residues of sugarcane, rice, cassava and peanuts for ethanol production', *Journal of Chemical Technology and Biotechnology*, 82(2), pp. 174–181.

Marzialetti, T. et al. (2008) 'Dilute acid hydrolysis of loblolly pine: A comprehensive approach', *Industrial and Engineering Chemistry Research*, 47(19), pp. 7131–7140.

McHardy, J. and Sawan, S. P. (1998) *Supercritical fluid cleaning: Fundamentals, technology and applications*. Park Ridge, NJ: Noyes Publications.

McIntosh, S. and Vancov, T. (2011) 'Optimisation of dilute alkaline pretreatment for enzymatic saccharification of wheat straw', *Biomass and Bioenergy*, 35(7), pp. 3094–3103.

Mesa, L. et al. (2011) 'The effect of organosolv pretreatment variables on enzymatic hydrolysis of sugarcane bagasse', *Chemical Engineering Journal*, 168(3), pp. 1157–1162.

Monrroy, M. et al. (2011) 'Structural change in wood by brown rot fungi and effect on enzymatic hydrolysis', *Enzyme and Microbial Technology*, 49(5), pp. 472–477.

Morjanoff, P. J. and Gray, P. P. (1987) 'Optimization of steam explosion as a method for increasing susceptibility of sugarcane bagasse to enzymatic saccharification', *Biotechnology and Bioengineering*, 29(6), pp. 733–741.

Mosier, N. et al. (2005) 'Features of promising technologies for pretreatment of lignocellulosic biomass', *Bioresource Technology*, 96(6), pp. 673–686.

Naik, S. et al. (2010) 'Characterization of Canadian biomass for alternative renewable biofuel', *Renewable Energy*, 35(8), pp. 1624–1631.

Narayanaswamy, N. et al. (2013) 'Biological pretreatment of lignocellulosic biomass for enzymatic saccharification', in Fang, Z. (ed.) *Pretreatment techniques for biofuels and biorefineries*. New York: Springer Berlin Heidelberg, pp. 3–34.

Nguyen, Q. A. et al. (2000) 'Two-stage dilute-acid pretreatment of softwoods', in *Applied Biochemistry and Biotechnology - Part A Enzyme Engineering and Biotechnology*. United States, pp. 561–576.

Nguyen, T.-A. D. et al. (2010) 'Pretreatment of rice straw with ammonia and ionic liquid for lignocellulose conversion to fermentable sugars', *Bioresource Technology*, 101(19), pp. 7432–7438.

Nilsson, T. et al. (1989) 'Chemistry and microscopy of wood decay by some higher ascomycetes', *Holzforschung*, 43(1), pp. 11–18.

Okuda, K. et al. (2004) *Disassembly of lignin and chemical recovery - Rapid depolymerization of lignin without char formation in water-phenol mixtures, Fuel Processing Technology*. Elsevier.

Palmqvist, E. and Hahn-Hägerdal, B. (2000) 'Fermentation of lignocellulosic hydrolysates. II: Inhibitors and mechanisms of inhibition', *Bioresource Technology*, 74(1), pp. 25–33.

Pan, X. et al. (2007) 'Pretreatment of lodgepole pine killed by mountain pine beetle using the ethanol organosolv process: Fractionation and process optimization', in *Industrial and Engineering Chemistry Research*, 46(8), pp. 2609–2617.

Park, C. Y., Ryu, Y. W. and Kim, C. (2001) 'Kinetics and rate of enzymatic hydrolysis of cellulose in supercritical carbon dioxide', *Korean Journal of Chemical Engineering*, 18(4), pp. 475–478.

Perepelkin, K. E. (2007) 'Lyocell fibres based on direct dissolution of cellulose in N-methylmorpholine N-oxide: Development and prospects', *Fibre Chemistry*, 39(2), pp. 163–172.

Pettersen, R. C. (1984) 'The Chemical Composition of Wood', in Rowell, R. M. (ed.) *The Chemistry of Solid Wood: Advances in chemistry series*, pp. 57–126.

Poornejad, N., Karimi, K. and Behzad, T. (2013) 'Improvement of saccharification and ethanol production from rice straw by NMMO and [BMIM][OAc] pretreatments', *Industrial Crops and Products*, 41(1), pp. 408–413.

Puri, V. P. and Mamers, H. (1983) 'Explosive pretreatment of lignocellulosic residues with high-pressure carbon dioxide for the production of fermentation substrates', *Biotechnology and Bioengineering*, 25(12), pp. 3149–3161.

Qiu, Z., Aita, G. M. and Walker, M. S. (2012) 'Effect of ionic liquid pretreatment on the chemical composition, structure and enzymatic hydrolysis of energy cane bagasse', *Bioresource Technology*, 117, pp. 251–256.

Ruiz, E. et al. (2006) 'Ethanol production from pretreated olive tree wood and sunflower stalks by an SSF process', *Applied Biochemistry and Biotechnology*, 130(1), pp. 631–643.

Saha, B. C. et al. (2005) 'Dilute acid pretreatment, enzymatic saccharification and fermentation of wheat straw to ethanol', *Process Biochemistry*, 40(12), pp. 3693–3700.

Saha, B. C. and Cotta, M. A. (2006) 'Ethanol production from alkaline peroxide pretreated enzymatically saccharified wheat straw', *Biotechnology progress*, 22(2), pp. 449–453.

Saka, S. and Ueno, T. (1999) 'Chemical conversion of various celluloses to glucose and its derivatives in supercritical water', *Cellulose*, 6(3), pp. 177–191.

Samayam, I. P. and Schall, C. A. (2010) 'Saccharification of ionic liquid pretreated biomass with commercial enzyme mixtures', *Bioresource Technology*, 101(10), pp. 3561–3566.

Sasaki, M., Adschiri, T. and Arai, K. (2004) 'Kinetics of Cellulose Conversion at 25 MPa in Sub- and Supercritical Water', *AIChE Journal*, 50(1), pp. 192–202.

Schmidt, A. S. et al. (2002) 'Comparison of the chemical properties of wheat straw and beech fibers following alkaline wet oxidation and laccase treatments', *Journal of Wood Chemistry and Technology*, 22(1), pp. 39–53.

Shafiei, M. et al. (2014) 'Enhanced ethanol and biogas production from pinewood by NMMO pretreatment and detailed biomass analysis', *BioMed Research International*. Edited by M. Tabatabaei, 2014, p. 469378.

Sharma, R. K. and Arora, D. S. (2010) 'Changes in biochemical constituents of paddy straw during degradation by white rot fungi and its impact on in vitro digestibility', *Journal of Applied Microbiology*, 109(2), pp. 679–686.

Shuai, L. et al. (2010) 'Comparative study of SPORL and dilute-acid pretreatments of spruce for cellulosic ethanol production', *Bioresource Technology*, 101(9), pp. 3106–3114.

Silverstein, R. A. et al. (2007) 'A comparison of chemical pretreatment methods for improving saccharification of cotton stalks', *Bioresource Technology*, 98(16), pp. 3000–3011.

Singh, P. et al. (2008) 'Biological pretreatment of sugarcane trash for its conversion to fermentable sugars', *World Journal of Microbiology and Biotechnology*, 24(5), pp. 667–673.

Singh, S., Simmons, B. A. and Vogel, K. P. (2009) 'Visualization of biomass solubilization and cellulose regeneration during ionic liquid pretreatment of switchgrass', *Biotechnology and Bioengineering*, 104(1), pp. 68–75.

Song, T., Pranovich, A. and Holmbom, B. (2012) 'Hot-water extraction of ground spruce wood of different particle size', *BioResources*, 7(3), pp. 4214–4225.

Sun, F. and Chen, H. (2007) 'Evaluation of enzymatic hydrolysis of wheat straw pretreated by atmospheric glycerol autocatalysis', *Journal of Chemical Technology and Biotechnology*, 82(11), pp. 1039–1044.

Sun, X. F. et al. (2005) 'Characteristics of degraded cellulose obtained from steam-exploded wheat straw', *Carbohydrate Research*, 340(1), pp. 97–106.

Sun, Y. and Cheng, J. (2002) 'Hydrolysis of lignocellulosic materials for ethanol production: A review', *Bioresource Technology*, 83(1), pp. 1–11.

Sun, Y. and Cheng, J. J. (2005) 'Dilute acid pretreatment of rye straw and bermudagrass for ethanol production', *Bioresource technology*, 96(14), pp. 1599–1606.

Taherzadeh, M. J. and Karimi, K. (2008) 'Pretreatment of lignocellulosic wastes to improve ethanol and biogas production: A review', *International Journal of Molecular Sciences*, pp. 1621–1651.

Torry, L. A. et al. (1992) 'The effect of salts on hydrolysis in supercritical and near-critical water: Reactivity and availability', *The Journal of Supercritical Fluids*, 5(3), pp. 163–168.

Tripathi, M. K. et al. (2008) 'Selection of white-rot basidiomycetes for bioconversion of mustard (Brassica compestris) straw under solid-state fermentation into energy substrate for rumen micro-organism', *Letters in Applied Microbiology*, 46(3), pp. 364–370.

Tucker, M. P. et al. (2003) 'Effects of temperature and moisture on dilute-acid steam explosion pretreatment of corn stover and cellulase enzyme digestibility', *Applied Biochemistry and Biotechnology*, 105(1), pp. 165–177.

Wan, C. and Li, Y. (2010) 'Microbial delignification of corn stover by Ceriporiopsis subvermispora for improving cellulose digestibility', *Enzyme and Microbial Technology*, 47(1), pp. 31–36. doi: https://doi.org/10.1016/j.enzmictec.2010.04.001.

Wang, G. et al. (2015) 'Production of bioethanol from rice straw by simultaneous saccharification and fermentation of whole pretreated slurry using Saccharomyces cerevisiae KF-7', *Environmental Progress and Sustainable Energy*, 34(2), pp. 582–588.

Wang, K. et al. (2012) 'Structural transformation of hemicelluloses and lignin from triploid poplar during acid-pretreatment based biorefinery process', *Bioresource Technology*, 116, pp. 99–106.

Wang, T. et al. (2018) 'A review of the hydrothermal carbonization of biomass waste for hydrochar formation: Process conditions, fundamentals, and physicochemical properties', *Renewable and Sustainable Energy Reviews*, pp. 223–247.

Wang, Z.-M. et al. (2009) 'Homogeneous sulfation of bagasse cellulose in an ionic liquid and anticoagulation activity', *Bioresource Technology*, 100(4), pp. 1687–1690.

Wayman, M. et al. (1986) 'SO2-catalysed prehydrolysis of coniferous wood for ethanol production', *Biotechnology Letters*, 8(10), pp. 749–752.

Wiedner, K. et al. (2013) 'Chemical modification of biomass residues during hydrothermal carbonization - What makes the difference, temperature or feedstock?', *Organic Geochemistry*, 54, pp. 91–100.

Xu, D. et al. (2018) 'Catalytic hydrothermal liquefaction of algae and upgrading of biocrude: A critical review', *Renewable and Sustainable Energy Reviews*, pp. 103–118.

Xu, J. et al. (2010) 'Lime pretreatment of switchgrass at mild temperatures for ethanol production', *Bioresource Technology*, 101(8), pp. 2900–2903.

Yamashita, Y. et al. (2010) 'Alkaline peroxide pretreatment for efficient enzymatic saccharification of bamboo', *Carbohydrate Polymers*, 79(4), pp. 914–920.

Yu, H. et al. (2009) 'The effect of biological pretreatment with the selective white-rot fungus Echinodontium taxodii on enzymatic hydrolysis of softwoods and hardwoods', *Bioresource Technology*, 100(21), pp. 5170–5175.

Yu, H. et al. (2010) 'Fungal treatment of cornstalks enhances the delignification and xylan loss during mild alkaline pretreatment and enzymatic digestibility of glucan', *Bioresource Technology*, 101(17), pp. 6728–6734.

Zaldivar, J., Martinez, A. and Ingram, L. O. (1999) 'Effect of selected aldehydes on the growth and fermentation of ethanologenic Escherichia coli', *Biotechnology and Bioengineering*, 65(1), pp. 24–33.

Zhang, M. et al. (1995) 'Metabolic engineering of a pentose metabolism pathway in ethanologenic Zymomonas mobilis', *Science*, 267(5195), pp. 240–243.

Zhao, H. et al. (2009) 'Regenerating cellulose from ionic liquids for an accelerated enzymatic hydrolysis', *Journal of Biotechnology*, 139(1), pp. 47–54.

Zhao, Y. et al. (2008) 'Enhanced enzymatic hydrolysis of spruce by alkaline pretreatment at low temperature', *Biotechnology and Bioengineering*, 99(6), pp. 1320–1328.

Zhu, J., Wan, C. and Li, Y. (2010) 'Enhanced solid-state anaerobic digestion of corn stover by alkaline pretreatment', *Bioresource Technology*, 101(19), pp. 7523–7528.

Zhu, Y. et al. (2014) 'Techno-economic analysis of liquid fuel production from woody biomass via hydrothermal liquefaction (HTL) and upgrading', *Applied Energy*, 129, pp. 384–394.

15 Corporate Social Accountability in Waste Production and Management

K. N. Ajoykumar[1], Gurudatta Singh[2] and A. M. Shackira [3]

[1] Department of Plant Science, Kannur University, Mananthavady Campus, Kerala, India-670645.

[2] Institute of Environment and Sustainable Development, Banaras Hindu University, Varanasi, India – 221005

[3]Department of Botany, Sir Syed College, Kannur, Kerala, India – 670142.

CONTENTS

DOI: 10.1201/9781003196358-15

15.1 INTRODUCTION

Waste is one of the major environmental concerns in the world nowadays. We are living in an Anthropocene era in which the world is being irreversibly moulded by human activities. The unresolvable happenings in the world environment lead to a situation that, our planet is not any longer safe for humanity. The new pandemics such as COVID-19 are emerging from the intrinsic problems associated with the development and agribusiness of corporates which creates the primary route of disease (Foster et al., 2020). The rising amount of waste, and its management and consequent effects, is just one feature of this Anthropocene era (Davis, 2016). Human actions and deviations in lifestyles and consumption patterns have caused an increase in waste generation rates. The wastes produced by industries are different and they may be solid, liquid, gaseous or radioactive materials. Solid waste, especially plastic waste is an inevitable component of present-day life throughout the world. We cannot imagine a world without plastic (WEF, 2016). According to the World Bank report, there will be a 70% increase of solid waste in the coming thirty years from the 2.01 billion tonnes produce today. This is because of the population growth and the pattern of development we follow. We expect an annual generation of 3.4 billion tonnes of waste in the coming years (World Bank Report, 2018).

The high accumulation of plastic waste is of great concern in recent years because of its harmful influence on the environment and human health. It takes thousands of years to decompose from organic waste. Plastic waste is triggering floods by creating an obstruction in drains, producing respiratory problems when burned and polluting water bodies when discarded into rivers and oceans (World Bank Report, 2018). In oceans, plastic is accumulating in spinning whorls that are much wider. It converts plastic into 'microplastics' under UV light, that is practically not possible to repair and that are disturbing food chains and troubling natural habitats. Recent reports states that there will be more plastic in the oceans than fish by 2050 if nothing is done (The Ellen MacArthur Foundation, 2016, 2017; European Commission, 2017).

Plastics are widely used for the production of a variety of components and have evacuated other materials that were formerly used for many applications. Plastics now dominate wood, metal and glass. It can be designed into polyesters in the textile industry, polycarbonates for eyeglasses and compact discs and polyvinylidene chloride for food packing along with many other usages. In 2019, the global production of plastics reached 368 million metric tons, around 100 million metric tonnes of it is produced in China alone and 58million metric tons produced in Europe (Garside, 2019). We use plastics throughout the economic sectors such as construction, the automobile industry, the electric and electronic packaging industry, and the medical field. Moreover, now plastic makes up roughly 15% of an automobile by weight, about 50% of jet airliners, and contributes much in the economic sector which is low-cost, durable, and has a strength-to-weight ratio. They are inexpensive and its lightweight reduces fuel consumption in transportation.

Microplastics create a serious risk to the environment as they get into the bodies of organisms such as fish, turtles, shrimps, sea birds, and thereby enter the food chain, affecting humans as well. Several studies describe the presence of microplastics in water (Wan et al., 2019; Wang et al., 2020). Some 50% of the plastics produced are of single-use and only 9% is recyclable. The rest of the plastic accrue on the shoreline. The usage of plastic tremendously increased after the 1950s and the production of plastic increases in a larger amount compared to other materials. The major source of plastic products is oil, natural gas and coal. These are non-renewable. If the present trend continues in the future, it is expected that by the year 2050 we may utilise 20% of the world's oil for plastic production. There is an increase in plastic pollution because of the manufacture of more and more plastic products. The usage of plastic bags has increased to 500 billion, and the number of plastic bottles discarded each year is around 35 billion.

Industries excrete solid waste, chemicals, and other liquid and gas waste, which are degradable or persistent poisonous contaminants. Landfills, backyards, rivers, and the ocean are filled with solid waste. The waste as molecules blows out through gases move to the atmosphere and water bodies from there it moves to plants, animals, microorganisms, and other living bodies. Many of these are imperceptible or only partially observable. The world has already accepted these facts and taken them for granted. A small fraction of these materials are re-utilised for household items required to maintain our standard of living. It requires a bigger quantity for constructions, transportation, and infrastructure (Venezuela and Steffen Bohm, 2017). Appropriate control of waste production is possible by uniting different knowledges in a system. Governments and other international agencies should see that industries and corporates should take steps towards this goal line. A wide-ranging practice is necessary to tackle the environmental issues created by waste production.

15.2 WASTE GENERATION AND MANAGEMENT

15.2.1 The Extractive Industry and Waste

Analysing the textile industry, Salas-Molina (2020) proposed that we are facing an inverse Malthusianism in the case of resources, as the usage of textiles reaches about 13 kg per individual. The environment impacts of reverse Malthusianism due to the exponential growth of consumption of resources are having serious repercussions. Technological advancement does not minimise the use of resources.

A typical example is the more advanced vehicles that are not fuel efficient due to profitability reasons (Alcott, 2005).

Greater quantities of natural resources are extracted from nature as part of this exponential production process. J.B. Foster called this method of exploitation of natural resources as 'extractive industry' (Foster and Clark, 2012). Many of these resources are non-renewable, like fossil fuels and minerals. These resources are utilised over a long period and cannot replace. The waste which is produced because of this exponential production pollutes the non-renewable resources too making them unusable. Natural resources once used for manufacturing commodities are not reusable once it is formed as waste. It cannot convert the product into its parts. Yates (2015) called it a natural waste.

Usually, waste is as something that is rejected after usage. But, in the present world situation, there is an increase in material waste discarded after production but before consumption. Textile materials which are not in par with fashion or season are rejected regularly from the market. The Food and Agricultural Organisation (FAO) stated that roughly one-third of all food produced (1.3 billion tons) for human consumption is lost and wasted every year across the entire stream (Ishangulyyev et al., 2019). USA rejects 2.5 tonnes of unsaleable food each year and many of the products are removed from the market after expiry. We also reject even cars and electronics as pre-consumer waste (Yates, 2015). One billion tonnes of such discarded waste is produced in USA and Europe each year (Trenkle and Lohoff, 2012). Waste is produced at the time of the extraction of resources, during transportation and at the production process (Laurenti et al., 2017).

Pre-consumer waste is a unique feature in the era of late capitalism (Yates, 2015). The economic development is extracting natural resources in an uncontrolled manner, and the monetary wealth depends upon this kind of exponential production oriented towards the accumulation of capital and profit. They discard the commodities because of non-fulfilling of its primary purpose of profit-making. Yates (2015) argues that "waste is a historically specific social category for delineating the ecological limits of capitalism's drive for valorisation and seemingly limitless capital accumulation". Savini (2019) coined a new term 'regime of accumulation' to show the changing role of waste in contemporary economic development and role of the existing system on waste accumulation. He put forward the concept that the changing approaches are because of the reorganisation of capitalism in a socio-economic crisis.

15.2.2 THE THROWAWAY SOCIETY AND THE ACCUMULATION OF WASTE

In 1958 John Kenneth Galbraith had introduced the concept of 'The Affluent Society'. He predicted that affluence would lead to oversaturation, which would reduce consumer demand, thus diminishing economic growth. The classical book *The Waste Makers*, written by Vance Packard (1960), formed an extensive discussion over the development and extraction of nature of a throwaway society. It was the first widely discussed book to expose the throwaway society proposition.

The fundamental issue with a throwaway culture is societal. The alternative lifestyle not only establishes constant exhaustion of consumer goods but also considerable changes in the society. Thus, people of a throwaway society worth merely what pays to their self-worth, neglecting everything outside it. Everything progresses around individual happiness, nothing else merits more importance. It associates everything with consumerism and consumption. It transforms the complete life into a means of self-possession and social status demonstration. So many authors have used the concepts such as 'throwaway culture' or 'throwaway ethic' (Slade, 2009; Cooper, 2010; O'Brien et al., 2013).

This is the consumer culture, which is not inherent in individuals but is deployed from outside. It is through the advertisement which shares a major part of the consumer goods in the world. What is there in a long-lasting mobile phone or laptop? The business promotion depends on the reduction in the durability of a product. What is the benefit, if a laptop lasts for ten years instead of two? Naturally, business profit becomes declined. This is a crucial issue with business and corporates. Waste management should also focus on the prevention of unnecessary waste products. People in the United States abandon 7 billion tons of PVC (polyvinyl chloride) plastic-the most dangerous plastic product-annually. Many of the unnecessary waste carry huge prices to the environment and human health. Many of these chemicals are cancer-causing and inducing mutation (Brandt-Rauf et al., 2012).

But nowadays consumers are substantially more aware of intentional desuetude, and different governments have already started plans to change this extensive throwaway behaviour. However, these programmes are still insufficient. If we started rejecting plastic goods at the consumer level, it is good. But the real improvement requires a significant sterner industry level control of production and marketing directives. So, it continues to be a significant and central challenge for policymakers to tackle the throwaway society. The more successful an economy becomes, the more it turns into a throwaway society (Hellmann and Luedicke, 2018).

Schor and White (2010) called it as 'materiality paradox', which proposes that "people in our society are not too materialistic, but are not materialistic enough". We don't keep, reuse, and repair products. They learn us to discard products as soon as possible as our product becomes useless, He says that modern marketing has created a kind of psychological obsolescence in people, inspiring them to throw away what they have only just purchased because it is not new. Dawson (1991) called it as a consumer trap.

15.2.3 WASTE MANAGEMENT

Waste managing is one of the fundamental steps of sustainable development (Wan et al., 2019). It seeks to complement economic, social, and environmental improvements by reusing, treating, and managing in other ways by-product flows of economic activities by businesses and society (Escalante et al., 2007). If there is a suitable method of waste management system, different waste can be used for tapping resources. A comprehensive waste management system includes the entire process of waste of origin, transference and treatment and clearance.

Waste management is one of the crucial issues throughout the world (Ferronato and Torretta, 2019). There are no specific management systems in municipalities and industries in most of the countries (Romero-Hernández and Romero, 2018). There are different programmes to help institutions in the waste management sector. Peoples' involvement is one of the key features of waste management. The role of industries and corporate firms are important here. We see most of the waste deposits are in poor countries. The electronic wastes of developed countries are usually dumped in the southern part of the world. The waste generation of e-waste in India is four times that of the United States. India produces about 1.85 million tonnes of e-waste annually: a study by the World Bank says that the waste deposit considerably increased throughout the world during the social distancing days of COVID-19. The increases of single-use products and continuous buying have increased production and consumption, hence uncomfortable exertions towards dropping plastic pollution. So many countries have introduced strategies to ensure sustainable management of waste while caring for the security of waste handlers (World Bank, 2019).

15.2.4 PLASTIC

It was calculated that out of the total plastic produced forever 8.3 billion tonnes, the plastic waste calculated out of this is 6.9 billion tonnes and about 91% of this is not recycled. Some 12% of the total is burned and 79% goes to the environment as waste. Predictably, for the coming 20 years the plastic production is likely to double. The amount of plastic refuse that flows into the oceans every year is expected to nearly triple by 2040 to 29 million tonnes (Parker, 2018). The major obstacle of the recycling of plastic is its high production cost. COVID-19 has enhanced the plastic pollution problem due to the use of 65 billion gloves and 129 billion face masks every month. The situations become worsen due to the reduction of recycling programmes because of budget strains. By 2050 there will be more plastic waste than the weight of fish in ocean (Ford, 2020). Plastics should remain at the top of the political agenda across the world, to minimise plastic pollution and also to promote a circular economy, and thereby ensuring sustainable growth (Silva et al., 2020).

15.2.5 E-WASTE

The formation of e-waste and management is one of the foremost ecological challenges in the world (UNEP, 2016). The e-waste flow has reached 53.6 million tonnes annually on a global basis (Kirkpatrick, 2020) and the global e-waste will increase by 30% by 2030 (Mishra et al., 2020). Electronic and electrical waste contains both dangerous and valuable components and is a quickly increasing waste flow (Debnatha et al., 2016). The recycling and reuse of e-waste reduce the harm to the environment and enables the retrieval of valuable constituents (Salhofer et al., 2016), including rare earth elements and other precious raw materials that are of vital importance for modern industry (Cucchiella et al., 2015). The recycling of e-waste is a multifaceted task causing very good technical knowledge and complex managerial skills (Rasnan et al., 2016). The recycling leads to economic profit and recovery of very rare important metal components (Alsheyab and Kusch, 2013). It is calculable that the recycling retrieval from the printed circuit board of the household material alone will give a profit of 6 billion euros by 2030 (profitability as net present value) (D'Adamo et al., 2016). The important part of the management of e-waste is to predict the amounts of e-waste produced in present and future (Zeng et al., 2016)

15.2.6 RECYCLING

Sustainable development is a slogan put forward throughout the world. It became the central agenda to all national policies. Even though the entire world is talking about sustainability, waste is not a crucial issue for nations. Still, it is considered as a management issue. It is still a matter of economic governance (Gregson et al., 2015). A 'war-on-waste' was put forward by the European Green movements (Cooper, 2009). Anti-consumerist protestors pointed out waste accumulation's negative, environmentally damaging consequences.

The waste can be recycled back into further production, theoretically. Mechanical, chemical, and incineration to recover inherent energy are the methods used in recycling. It may be primary, secondary, tertiary, and quaternary recycling. The most important is the composition of the waste. Only a small amount is recycled internationally. Most of the waste is composted or disposed of. Recycling is more expensive than the extraction of natural resources (Economist, 2009; Schor, 2011). The continuous production of commodities from the recycled waste led to a situation where the products cannot be marketed or the products are discarded shortly after consumption. Goods produced from recycled waste still end up as waste. The amount of waste generated as a by-product in the manufacturing process is production waste. The conflict between the environment and the economy can be a major barrier in recycling (Goodship, 2007; Ghosh et al., 2016). The selection of recycled material as a suitable alternative in the manufacturing industry is important (Salas Molina et al., 2020).

15.3 CORPORATE RESPONSIBILITY AND SUSTAINABLE DEVELOPMENT

15.3.1 RECYCLING AND THE ROLE OF CORPORATIONS

Several companies have become involved in environment-friendly projects because they came to know that there are competitive advantages in an environmentally oriented design. There are benefits for the company when they display responsibility in the areas of waste management and other environmental issues. The companies themselves recognise that they may get a reputation as a business firm concerned about the environment, and many people prefer companies with environmental concern.

There are many case studies for different companies. For example, Philips is a multinational company in the Netherlands. As a part of its circular economic model, the company was focusing on its operational designs and upgrading their products. It promotes the repairing of materials (Ellen MacArthur Foundation, 2013) and creates awareness about the circular economy in the corporate community. Philips claims that the company's circular economic model helped to recover its resource efficiency and financial appeal. The company ensures the longevity of its products and it maintains the product after installation. Moreover, Philips takes back its products to recycle, upgrading its infrastructure for reuse, in a different place (McKinsey, 2014). It has devoted itself to comprehensive ecological management as it disposed of the street lighting infrastructure and its constituent parts at their end of life. The ecological footprint of Philips shows that it is committed to sustainable development. It has also applied the circular economy principle to medical equipment. Most Philips products are recyclable, upgradable, and maintainable (McKinsey, 2014). Likewise, Vodafone gives incentives to customers to return their old phones as part of the circular economy global platform formed by different companies. The returned mobile phones are renovated and sold again. It has been calculated that they could reduce the cost of manufacturing of mobile phones from used phones by 50% (WEF, 2014).

Canon is a Japanese-based company with a commitment to corporate responsibility. The company is active in recycling. For instance, it started a project joining with other companies in a cartridge recycling scheme. A new design of Canon equipment which consumes less energy in their manufacture, transportation and use was launched. The method adopted by Canon has decreased CO_2 emissions by roughly 11 million tonnes and saved 350 billion Yen in electricity costs for consumers between 2003 and 2010 (Torres et al., 2012).

Another example from the textile industry is the recycling of the used clothes to make fabrics that save new fibres. Different companies such as Marks and Spencer, Hennes, and Mauritz (Hand M) Denim started old clothes collection programmes to encourage customers to bring in end-of-use clothes in exchange. It could downstream be the processing of over 18,000 tonnes of unwanted clothing from its global customers (Figure 15.1) (Gold, 2015; H&M, Statista, 2019). They have been working on increasing upcycling and functional recycling for re-wear, reuse, recycling (WEF, 2014). A major portion (about 30%) of the used clothes cannot be reused as textile fibres. They could convert these materials into damping and insulating materials for the auto industry. The collected old jeans can produce new jeans after crushing and transformation into fibres. This reserve has substituted up to 25% of new materials (Gold, 2015). This will minimise the textile waste and also preventing the tapping of new natural resources. Another example is the Ricoh Company, producing hardware and software and famous for recirculating resources. The company is targeted in renovating pre-owned machines and declared that they will reduce the resource input by 87.5% by 2050 (WEF, 2014).

Levi Strauss has also taken part in an environmentally friendly programme of minimising the use of natural resources. It had a trademark for their 'Water Less' campaign by using less water when manufacturing their products (Mazurkiewicz,

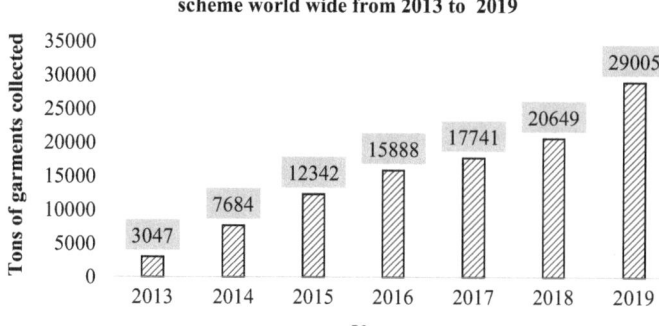

FIGURE 15.1 H & M-recycling scheme (Data Source: Hand M Group Sustainability Report 2019).

2004). They declared the goal of reducing its cumulative water use for manufacturing by 50% in water-stressed areas by 2025 (Holbrook, 2019). Still, taken all these measures, waste disposal and management are unsettled issues.

In the last released environmental responsibility report (2020), Apple placed several proposals for minimising waste by introducing several environmentally friendly programmes such as recycling. When the products are no longer functioning, the parts are again utilised for new manufacturing. The company claims that even if the parts are not reusable; they design the new technology in a way to unlock the useful materials inside them. Apple constantly ranks top among companies trying to lessen the environmental harm that comes with electronic equipment (Apple Environmental Progress Report, 2020). Apple is aiming to decrease the amount of plastic in their packaging by 2025. The company claimed that it has reduced the quantity of plastic in their packaging by 58% for the last four years (Table 15.1). Apple is aiming to set a new standard in the industry.

TABLE 15.1
Types of Waste Generation at Apple

Waste generation at Apple- Kilograms (Million)

Year	Land filled	Recycled	Composted	Hazardous waste	Waste to Energy
2015	7.0	9.0	2.0	1.0	0
2016	9.0	13.5	7.0	1.5	0
2017	9.0	32.0	7.20	2.0	1.0
2018	14.0	30.0	5.0	3.0	1.0
2019	17.5	32.5	5.0	3.0	1.0

Data Source: Apple Environmental Progress Report 2020.

The activities of companies in the field of waste management are to be monitored closely.

15.3.2 CORPORATE RESPONSIBILITY AND IRRESPONSIBILITY

Corporate social responsibility (CSR) has been a widely discussed and researched topic for the last several years. The concept of corporate responsibility now becomes a managerial tool to build up the reputation of corporations. This becomes an indicator of the company's standard of functioning. This was a topic of large corporations in the past, but nowadays it is accepted by all types of small-scale and medium-sized companies too. There are many examples of corporate responsibility and irresponsibility. Corporate social irresponsibility (CSI) refers to the opposite concept, as it examines the illegal and destructive corporate choices that the companies might take (Popa and Salanta, 2014).

Development and environment usually limit each other because of their position as two separate entities. But the concept of sustainable development helps to understand the limitations of ecological efficiency which is intended to satisfy the needs of future generations too (She et al., 2006; Karlaite, 2013). The factors considered for sustainable development are many and can be broadly summarised as social, economic, and environmental. Integral to the concept are social equity and responsible consumption, responsible marketing and business, conservation of natural resources and biodiversity, mitigation of climate change, waste management, and preventing pollution. The scopes of sustainable development and CSR are the same. Discussions on sustainability always refer to environmental policies, while the focus on corporate social responsibility is on social issues (Uddin et al., 2008).

Rapid industrialisation and development in the age of globalisation. There are a lot of social and environmental problems so that waste management and conservation of resources become essential for sustainable development. The increase in production leads to widespread exploitation of natural resources. As corporations are focusing on their profit, they are less sensitive to environmental problems (Figure 15.2). But a law prevailing in most of the countries concerning corporate responsibility in conserving environment binds them to involve in the environmental management strategies. The responsibility lies not solely with governments but with private companies too. It has become compulsory for large companies to spend a small percentage of their profit percentage to corporate social responsibility activities.

Water conservation, reducing the waste, waste management, tree plantations to compensate carbon dioxide emission, etc. are the different activities undertaken by companies for environment management under corporate social responsibility. There should be proper litigation to see that companies take care of the waste produced by industries and other firms. Now it is agreed that there should be a waste management strategy in all types of industries and every manufacturing firm should implement it as its social responsibility. The consensus is that as far

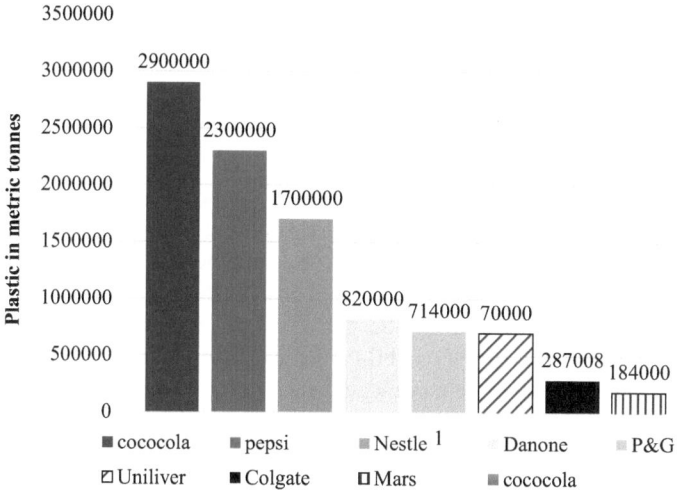

FIGURE 15.2   The world most corporate offenders for plastic pollution in metric tonnes based on companies that have disclosed their packaging figures (Data Source: Changing Market Foundation-STATISTA-The World's Worst Offenders for Plastic Pollution-Statista Sep 18, 2020).

as possible they should recycle the waste. The company or industry gets its good-will in the field if it maintains its environmental interest and has an excellent mechanism of governing the environmental apprehensions. Social responsibility is an important feature of maintaining corporate business status. Companies face extreme pressure from stakeholders and try to respond with by varied approaches, including corporate social responsibility announcements and corporate polit-ical activity (Greiner and Kim, 2021). A proper waste-dealing system helps the industry to remove or reuse waste materials. A manufacturing company must implement the best appropriate technique of waste treatment as part of corporate social responsibility so its waste products cannot damage the society (Gangwar and Yamini, 2016).

A manufacturing company realises that corporate social responsibility is a support to other organisations or individuals by transferring their financial help in varied fields such as medical, social, cultural, or ecological causes and other charitable activities (Falck and Heblich, 2007; Camilleri, 2013). There are signifi-cant gains to the business community for corporate socially responsible activities (Orlitzky et al., 2003; McWilliams et al., 2006; Ameer and Othman, 2012; Wang and Choi, 2013). Companies could influence themselves through social perform-ance and ecologically wide-ranging practices (Camilleri, 2017). At the emergence of COVID-19, Greenpeace reported that there is a propaganda campaign urging by many companies and plastic industries to stop using re-usable bags because there is the chance of dispersal of corona virus. Many companies, including

McDonald's, stopped their reusable systems and asked to use single-usage plastic bags (Schlegel, 2020).

15.3.3 COMPANIES AND PROFIT-SEEKING EVENTS

Most Fortune 75 companies publish corporate social responsibility reports. The publications generally highlight their efforts to encounter sustainability challenges. While the vast majority of these reports cite improved waste management initiatives, many of them actually gained financially from good solid waste management. Many companies implement waste management purely to boost a corporate image (Omar and Sergio, 2018). Some observers note that corporate responsibility and irresponsibility are usually discussed in terms of Western-centred views of corporate damages and usually ignore the realities of the developing and underdeveloped southern part of the globe (Alcadipani and Rodrigues, 2019; 2020).

Corporate firms harm environment leading to the death of people at a different magnitude, to both internal and external stakeholders of company. Many analyses of corporate responsibilities do not challenge the heart of present-day corporation profit-seeking actions that eventually create damage to the environment and people (Alcadipani and Rodrigues, 2020).

The Bhopal gas tragedy is a notorious example in India in which the corporate Union Carbide India Ltd. showed high irresponsibility towards a disaster which happened because of the absence of a proper management system in the gas plant. The methyl isocyanate leaking was the result of anomalies and lack of a proper management system of the plant. The company was reluctant to address the question of responsibility for the tragedy, led to 2,000 deaths and 200,000 injuries. The plant authorities have declined to discuss the wrong doings of the accident or the circumstances that formed it, as shown by The *New York Times* study. The International Labour Organisation (ILO) labelled the 1984 Bhopal Gas Tragedy as one among the world's 'major industrial accidents' of the 20th century in the report released in April 2019. At least 30 tonnes of methyl isocyanate gas, released from the Union Carbide pesticide plant, had affected more 600,000 employees and local populations. Highly toxic methyl isocyanate (MIC) gas can cause death within minutes of inhalation if its concentration exceeds 21 ppm. That is the reason for the number of deaths. Moreover, thousands of survivors have faced the devastating after-effects for the last four decades.

There are several examples where companies do not follow proper environmental laws so that the waste materials are channelled into rivers and landfills, creating creates pollution. Popular protests arise because of this in various parts of India. There are several cases of closing down of industries because of corporate social irresponsibility and lack of a specific waste disposal system. The pulp and paper factory of Grasim Industries Mavoor Gwalior Rayons (Kerala) is a classic example: the waste chemicals produced from the factory were dumped in the nearby

Chaliyar river. The company did not take proper dispensation measurements. The factory was eventually closed down because of the people's protest.

15.4 THE EXCESS PRODUCTION AND EXCHANGE VALUE OF THE PRODUCT LEADING TO WASTE

15.4.1 EXCESS PRODUCTION

One of the major problems with the creation of waste is the excess production of goods by big corporations. This is very much related to the profit-seeking mode of economic development. Marketing trends and competition between companies lead to sudden changes in the quality and design of products. The central problem of today's economic policy is not the scarcity of resources standing against the needs of the society, but it is the surplus production conflicting against the saturated consumption and speculation of marketplaces. The economic growth of society now depends on persuading existing customers to buy new commodities one by one. Economic survival in the present system depends on the continuous extension of market share and reach. People's requirements are artificially created so that customers no longer need what they want or want what they need, while the most important requirements of people continue to be unsatisfied (Dawson and Foster, 1991).

Most of the products sometimes become status symbols and markers of respectability nowadays. It is not the use-value which decides marketing (McMillan Cottom, 2014). Their value goes beyond the market place extending to social and cultural institutions. For example, the vehicle you use decides the success of your life as pointed by Ehrenreich (2010). The new market promotes you to pursue purchase things beyond your need. It is endorsed by credit cards and the society becomes highly consumerist. Advertising efforts and broader cultural consolidation have stimulated consumption on credit cards and in the housing market beyond individuals' immediate means (Rudel et al., 2011).The turnover rate of advertising industry is always high across the world as captivation of the monetary excess through publicity has strengthened marketing economics.

Every product we consume has a concealed history, an unrecorded list of its constituents, resources, and influences. It also has associated waste produced by its use and nature. The resources utilised for manufacturing a semiconductor chip are over one 100,000 times in weight. For a laptop, it is 4,000 times. The production of one tonne of paper requires the use of 98 tonnes of various resources (Hawkens et al., 2013).

Excess production in a society is the difference between what a society produces and the society reproduces in the next period. The economic surplus applies to all societies and production forms. The production outside the level of output of the society is surplus production. This is the hallmark of the present economic system.

(Wrenn, 2016). Marketing and selling products create a major part of the waste. The integration of workmanship and salesmanship is the quality assigned to a modern industry worker. Most of the money spent on a product goes to advertising and market research. This includes maintenance of disproportionate numbers of sales outlets and the salaries and bonuses of sales associates. It is not the quality of product that is paramount: how to sell a commodity is the central question. Public relations and lobbying go along with this. These entire processes contribute greatly to waste production.

15.4.2 Use Value and the Exchange Value of the New Economy

John Bellamy Foster explains the concept of production of the new economy in terms of the use-value and exchange-value concept of Karl Marx. He articulates the relationship between money and a commodity change into a new paradigm in the new economy of big corporates. The relationship between money and commodity goes like this. In the beginning, it was C-M-C, i.e., a commodity is traded for money and then this money used for purchasing another commodity. But later this order converts into M-C-M in which money comes first, purchasing a commodity and then this exchange for money and this continues like M-C-M', M'-C-M'', M''-C-M'''.... like that, leading to a financial economy. But now a new concept of capitalistic use-value (C^k) has been introduced into the system, especially in packaging industry where a wide usage of plastic is predominant. Instead of the former use-value of the product, it introduces a new type of use-value only for marketing. The concept of M-C-M' then becomes one of M-C^k-M', in which the material with no use-value for the consumer is formally incorporated in the system, leading to the major part of the dumping plastic waste. That is, the use-value C has gradually given way, to specifically capitalist use-value, C^k—incorporating many socially unproductive features. In the packaged goods, the package, intended to market the product and merged into its manufacture costs, is now the bigger part of the product (Foster, 2012). So it is necessary to interfere in the present model of development and marketing to resolve waste management.

The distinction between use-value and exchange value of a product is also important here. Most of the products are manufactured for the market; hence the use-value of the product is not important. To create a product in a marketable form, the product should be attractive. Attractive packing is a prerequisite for selling a product. The packaging industry is flourishing along with any production. Most of the materials available in the market feature a colourful and intact packet made of plastic. A packet is sometimes costlier than the pen. The package sells the product, which must also incorporate the package cost in the production cost. It becomes a major share of the cost value of the products. Plastic becomes the wrapping material for even a loaf of bread. Some 40% of the total plastic waste is produced by the packaging industry. But this has no use-value and creates all these things for the market value of the product. This becomes the major share of waste in the world.

15.5 THE CONCEPT OF CIRCULAR ECONOMY AND WASTE

The growing perception of sustainable development led to the advent of the concept of a circular economy (CE). Many environmental groups and policymakers put forward this idea for shifting towards an ecologically viable society. It encompasses reclamation of organic debris such as biomass and biogas; an alternative method of product design to enable reuse; integrated heat set-up in building construction; aquathermic in residential areas; digital sharing platforms are the most common examples (Reike et al., 2018). The circular economy is entirely different from the 'linear', economy, founded on a 'produce-use-dispose', consumption paradigm (Figure 15.3).

Environmental economists emphasise that a circular economy put forward a pathway to a 'post-extractive society' in which it recovers the material needs of economic growth from prevailing societal activities (Ellen MacArthur Foundation, 2015; Geissdoerfer et al., 2017; Ghisellini et al., 2016; Mathews, 2011; Murray et al., 2017). Different concepts are seen in the literature concerning the circular economy (Cooper, 1999, 2012; Yuan et al., 2006). Environmental marketing (Baker et al., 2003; Leonidou and Leonidou, 2011), ecological marketing (Lindridge et al., 2013), green marketing (Cronin et al., 2011; Prakash, 2002), and sustainable marketing (Hunt, 2011) are closely related terms appears in literature. The circular economy is a term used to give a balance between environment and development (UNEP, 2006). Many researchers have observed that circular waste management practices are more cost-saving and revenue-generating than the linear pathway for many companies. CE practices create new values from the formerly discarded materials. These methods emphasise new designs and production methods to eliminate waste, and the end products come back as raw materials for further production. Circular economy principles encompass more than traditional waste management improvement practices (Omar and Sergio, 2018).

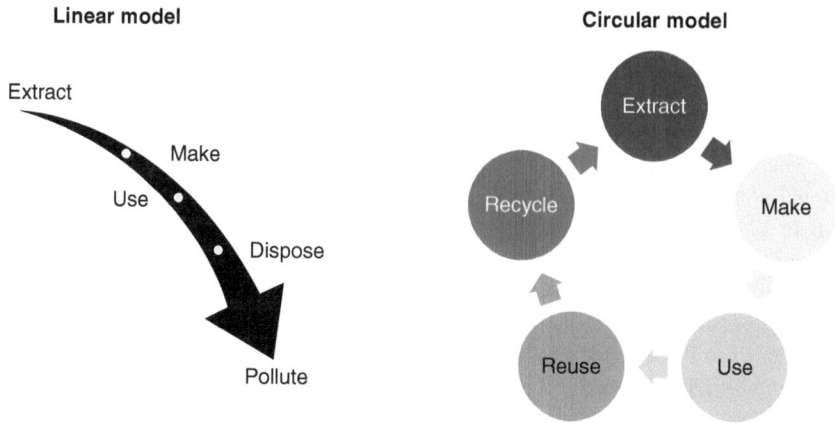

FIGURE 15.3 Linear and circular economies (Source: Catherine Weetman, CC 3.0, 2016).

As a part of a circular economic model, it is proposed to increase the life period of a product so that the user can use it for a longer period. By increasing the longevity of products through better engineering and maintenance, the rate of replacement declines, and there is a reduction in the use of resources. Individual consumers would prefer to use longer-lasting products. For a customer it is profitable. But the durability in product design could not continuously be effective in environmental terms. The problem is that durable long-lasting products require more energy when compared to short-lived ones. Plastic is durable, but paper is not durable but sustainable (UNEP, 2006). Long-lasting materials may in due course prove harder to break down into components for further recycling. The emphasis of the circular economy is on the productivity of the resources and ecological efficiency through reduce, reuse, and recycle. So, it significantly reduces the quantity of energy and the raw material requirement (Cooper, 1999). Repairing and redesigning also feature in the circular economy. The circular economy focuses on the positive restoration of the ecosystem within the industry (Cooper, 1999).

A new method of waste handling has been practised. Growth and development can go together considering the environmental and social aspects. The term green economy is widely used today. Waste treatment now focuses on converting the waste to useful substances. Decentralised waste control is important, especially of household waste. Large amounts of food waste are transferred to treatment plants, which require high cost and management. There should be adequate knowledge of the waste operation, particularly at the local level.

Several studies have suggested that the circular economy activates social interchange, environmentally accountable consumerism, product sharing, and eco-manufacturing (Girardet, 2014). Moreover, the concept of CE represents one of the most mentioned frameworks trying to integrate economic activity and environmental wellbeing sustainably (Murray et al., 2015). CE concepts have introduced a systematic approach to discussing the environmental problems related to economics (Schneider, 2014). CE puts forward many alternative models for business and the sustainability concept becomes the balanced state of the economics, environment and social anxieties (Dresner, 2002). This is a much debated concept in the academic literature, but has not yet progressed far in its practical analysis (Murray et al., 2015; Garcés-Ayerbe et al., 2019).

There are many criticisms of the CE model, specifically because of its unintentional consequences. For example, the substitute oil that is produced from palm oil or soybeans has unavoidably led to the destruction of a wide area of forest throughout the world. Equally, green energy production often exploits large stretches of arable land and puts huge pressures on the food supply chain, particularly in the poorest countries. For example, the production of ethanol is yet another example that requires more fossil fuel than it produces. In addition, environmentally friendly technologies, including wind farms and solar panels, rely on certain minerals that are also difficult to recycle (Giurco et al., 2019). These green structures will invariably require servicing and replacement.

15.5.1 Zero Waste Circular Economy—Contradictions

It was in the 1970s Paul Palmer coined the term 'zero waste' to reduce the number of chemicals going to waste in industrial processes. 'Zero waste' was later accepted as a lifestyle movement, which proposing that people and societies reduce consumption, reuse, and repair; aiming at throwing nothing away and reprocess as slight as possible. The Zero Waste International Alliance (ZWAI) developed strategies to create awareness in the society and industries about the profits gained when the waste is considered as a resource (ZWIA, 2015). The principal aim was to reduce the amount of waste produced. However, the world isn't set up for zero waste living. The ZWIA defined "Zero Waste as a goal that is ethical, economical, efficient and visionary, to guide people in changing their lifestyles and practices to emulate sustainable natural cycles, where all discarded materials become resources for others to use", featuring similar goals to that of Circular Economy. "Zero waste means designing and managing products and processes to systematically avoid and eliminate the volume and toxicity of waste and materials, conserve and recover all resources and not burn or bury them. Ideally, implementing zero waste will eliminate all discharges to land, water or air that are a threat to planetary, human, animal or plant health" (ZWIA, 2015).

But there are criticisms against the very concept of zero waste and CE. It is reported that transitioning waste to circularity faces important barriers across several categories and it is not currently economically viable (Dieckmann et al., 2020). It has been argued that this concept itself is something impossible, but a justification of the prevailing economic model of development. Some observe that CE proposals are a structural adaptation of the present economic system, safeguarding the capitalist system against problems of waste accumulation and the scarcity of natural resources. The sustainable development programme itself was criticised on this ground. There are several citations in literature that question the risks of constructing a paradigm for green economic growth based on waste recovery (Savini, 2018).

15.5.2 Green Economy and Greenwashing

'The limits to growth' report (Meadows et al., 1972) proposed the concept of sustainability and was set against the then present economic development and the over-consumption of natural resources (Kidd, 1992). Initially, the concept of sustainability was formed against the negative social and environmental effects of industrialisation and the lack of regulation of economic policies (Haque, 1999; Crouch, 2012). It was a negative concept because it was pointing towards the developmental strategies of economic growth. But the term sustainability has since been used by economists claiming that economic growth can be achieved in a way that succeeds to remain accountable to the environment and society and is positive (Magretta, 1997). The report of the United Nations Environment Programme (UNEP 2011; 2018) has substantiated the concept of sustainability as positive construction of a 'green economy', delivering a 'green development' solution (Fay,

2012). Many corporations acquired this concept of the green economy as part of their greenwashing business strategies and reputation-saving initiatives (Dauvergne and Lister, 2013). For example, companies promise to change the packaging system with biodegradable plastics. Usage of recycling alone cannot keep up with the vast production of plastic. All these promises are a kind of greenwashing. What we need far more is a total shift in the way products are packaged Publicover et al., 2019).

The argument is that the idea of sustainability has been subsumed under the sense of 'eco-business', a mode of producing sustainable goods and services to help companies secure 'competitive advantage and increase sales and profit' and 'enhance their growth and control within the global economy' (Dauvergne and Lister, 2013). Now, the term sustainability has been used not only for marketing to pacify the socio-environmental concerns of shareholders but also as a vital business driver (Seuring et al., 2008; Dauvergne and Lister, 2013). Banerjee (2008) contended that CSR initiatives are really a kind of window dressing. CSR reports are usually noticed with a phenomenon entitled CSR decoupling, which occurs when corporate social responsibility practices do not reflect what is publicised in corporate CSR reports. The linguistic analysis study conducted with special reference to Volkswagen CSR reports, as well as other corporation, shows that decouplers use more words and articles with negative emotional tone and very few words connected with risk factors involved. These linguistic characteristics would help identifying the corporate deceptions (Holtbrugge et al., 2020).

15.5.3   APPLE IPHONE AND WASTE

Valenzuela and Bohm (2017) elucidated the classical example for the argument against the zero waste concept and the role of corporations with Apple iPhone. The company claims that it is reusing the resources to keep electronic devices out of the landfill. It argues that these devices are recycled properly and that it has developed recycling assemblage events, take-back inventiveness, and new plans like Apple Renew, a global programme that can give used Apple phones to any Apple Store for reuse or accountable recycling. Apple claims that it has more than 160 recyclers around the world (Apple Inc., 2016). The Apple advertisement goes like this: "Through our efforts, we've kept over 270 million kilos of equipment out of landfill since 1994" (Apple Inc., 2016). The company sends its 'zero waste' message. It claims that it has created Liam, a series of robots that can quickly disassemble an iPhone 6, categorising its components and eliminating the need to excavate more resources from the Earth. With two Liam series running, Apple claims that it can take apart up to 2.4 million phones a year (Apple Inc. 2016). The company is also "committed to making sure all the waste created by our supply chain and by us is reused, recycled, composted or, when necessary, converted into energy. It's an ambitious goal that requires collaboration among multiple Apple teams, local governments and speciality recyclers, but we've already seen great success" (Apple Inc., 2016). Apple claims that Apple iPhone 11 Pro uses 40% less energy than the energy conservation standard. "First-ever smartphone made with 100% recycled rare earth elements in the Taptic Engine" is the slogan (Apple Environmental Progress Report, 2020).

It has created a new logo and a brand for the new 'circular' Apple Inc.: 'Renew', a greened apple made of the dart of continuous circuitousness. Apple's iPhone device sales produce millions of tons of waste, which is a major concern for the consumer electronics industry (Grant and Oteng-Ababio, 2012). The packing material itself has many covering sheets of plastic waste, which is non-degradable, but the company claims that the packing of iPhone-11Prois made with recyclable, fibre-based materials (Apple Environmental Progress Report, 2019). It is noticeable that on the Apple iPhone the 'green-and-lean' ZCWE certification stamped as a logo on the plastic that encases the Apple commodity. A person who is ecologically committed to waste management can see that his ethic-political commitment to sustainability is maintained by the company and thrilled to purchase the product. Here the commitment to the environment is reduced to the consumption of the Apple brand.

On the one hand, Apple is arguing about its wastelessness while going on with 'planned obsolescence': a policy of designing products with an artificially limited useful life period so they become rapidly outdated, forcing the customer to replace them with newer types that are already in early stages of manufacture. The company says this is 'innovation', but it is more about the company prompting consumers to reduce the time between repeated purchases and replace devices that have become obsolete for a variety of reasons. Apple may or may not deliberately create their products to run more slowly after a certain period.

How can we explain Apple's 'planned obsolescence' and the circularity concept of waste management? The consumer says: "I need to buy the latest iPhone or MacBook because it is more efficient and optimal; it has been designed better; it is lean and wastes less". Some critics contend that the company is sidestepping the dangerous issues of overproduction and worker rights by emphasising industrial innovations on resource management.

The zero-waste practices hide the actual situation of deliberate uselessness that has been built into the manufacture and selling of products created by companies like Apple (Herod et al., 2013). The Apple company displays its attainments in conforming with design for-recycling standards of production (Underwriters Laboratory, 2016). The public remains unaware of the environmental consequences of Apple's competitive business strategy, aiming for the customers to dispose the old products and to purchase the new arrival. The consumer feels that the old product is slow, but the newer one is fast and 'green by design'. Therefore, Apple products become more persistent than ever.

The present economic system of development reminds us of certain facts:

i. Waste will keep growing faster than our ability to handle it (Hoornweg et al., 2015).
ii. The present way of waste management cannot destroy waste but only transport to marginalised regions (Gidwani and Reddy, 2011; Gregson et al., 2015).

iii. Since the product is of 'green' sustainable and with a 'zero-waste' quality, the desire for the consumption increases.

This results in the consumption of such waste-fewer commodities (Jones 2010). It is proposed that waste should not be regarded as an error (Dhingra et al., 2014). Instead, the waste should be conceptualised critically, as the inherent by-product of a rule that thrives on the disproportionate exploitation of the environment (Yates, 2011). The planned obsolescence and improved disposable products relieve the economic surplus fund, all the while threatening the ecological balance of the planet (Dawson and Foster, 1992). The planned obsolescence of mobile phones encourages a culture of consumerism. This leads to a remarkable increase of waste generation (ICMR-IBS, 2019).

15.6 THE WASTE GENERATION AND MANAGEMENT— REQUIREMENT OF A COMPREHENSIVE METHOD

Plastic is cheap to manufacture. Why go for recycled plastic when the new plastic itself is very inexpensive? This is a question from the business community. So, to hang on reclaimed plastic is not advantageous for a good business. When this is the argument on plastic usage, how can we look forward to the waste management of plastic? Plastic management poses a range of problems. There are many technical problems to solve in maintaining a standard method for plastic reduction and re-usage. Biodegradable plastics may not degrade in oceans but convert instead to component microplastics. The labelling itself is confusing, and the consumers cannot identify the actual type of plastic. Most of them are contaminated. Chemical recycling is the method adopted to convert plastic to its molecules. Many companies follow this path of the recycling process. Since it is expensive, profit-motivated companies still look for new plastic production. Methods of production and management in the plastic industries should focus on multi-use plastics to reduce the remnants. To reduce the increasing waste production, the circular model of the economy is a plausible for society. It also leads to reduce the usage of raw materials. There is a need for a systematic approach which moves beyond the existing step-by-step modifications. Many industries are not at all aware of a circular model of the economy (Whiteley, 1987). We cannot measure the use and values of green technologies in terms of the actual prices of resources and raw materials. Long-term savings in sustainable development could cause substantial improvements in the economy.

The concept of the CE addresses economic and environmental issues together. It has the minimum potential to conserve the global ecosystem. But this model itself is not fully realised, considering the social and ecological characteristics of the current system. Future attention of research and practice should emphasis on this aspect. Even though the circular economy and zero waste model put forward the idea of sustainable development, in practice it is not well adapted by the corporate business with very few exceptions. In the present situation of depletion of natural resources, the linear model of economics should divert to a sustainable approach to operation. the society, as a part of CSR and keep silence over the environmental problems. With natural resources currently being depleted, the linear model of economics should divert to a sustainable approach to operation.

Environmental, political, economic, and social problems could also influence responsible corporate behaviour. Policymakers may not necessarily take the transition towards the circular economy seriously. Business people would probably feel bitter about any compulsory changes in their traditional methods. They may stick to the linear model and be against the paradigm shift of realigning economic and management practice because of their profit motives (WEF, 2016). But this paradigm shift is not plausible alone through the effort of a single entity. Participation and commitments at the company, government, and individual levels are mandatory (Upadhyay and Alqassimi, 2018). As there is a growing anti-plastic mindset throughout the world, many corporations and plastic manufacturers have responded by making impressive promises of enhanced sustainability. They promise to recycle and reduce waste by advanced chemical recycling or using compostable plastics. Companies claim that significant advancement has occurred in plastic waste management such as integration of recycling method in plastic packaging and the removal of hazardous plastics like PVC from packing and also the banning of single-use plastics (Ellen Macarthur Foundation, 2020). But they do not see the genuine issue of plastic production for no use value as in package industry. The observations of Winiwarter in understanding waste are important here. Waste created in society depends upon the mode of production of society. The type of use of natural resources and input determines the final product and the composition and nature of waste (Winiwarter, 2002). The overproduction of products has created an excess of goods on the market, most of them are not saleable and some of them less valuable than the material value embodied in it (Yates, 2015). Odum called it as Emergy—*Embodied energy* (Odum and Scienceman, 2005).

Very strict guidelines and laws are necessary for controlling plastic waste. At present, there is no mandatory international convention regulating environmental problems caused by plastic pollution (Loges and Jakobi, 2020). The existence of global, national, international, and regional laws will be beneficial for reducing and supervising plastic waste (Abril Ortiz, 2020). Legal regulation is essential to restore the marine and land environment (Pramudianto, 2019). Governments have endorsed laws to reduce the damaging effects of plastic in various countries such as the UK, Ireland, the Netherlands, China, the Philippines, and Australia. These actions include levies, bans, and other legislation. Strict laws restrict single-use plastic bags (Nyathi and Togo, 2020).

Companies should take this problem seriously since it is a life-and-death problem for the planet. Of course, most of the discourse on waste concentrates on waste management and disposal, with very little attention on the production-level management of waste. Plastic packaging is an important part of the global economy today. The concept of exchange value in the present economic mode of production is disputed. Excess production is another component which is aimed at a profit-oriented market and the main supplier of plastic waste. It is not the

A change in production and marketing is a prerequisite for managing waste. It is not only the scale of production, but a change in the product's design that is required. These characteristics of production are unnoticed, while the world focuses on the campaign against the use of plastic containers and plastic bottles. We should focus on developing strategies to ensure that the corporate world acts in accordance with technical and safety ideals while assigning funds for risk evaluation and monitoring. The most important part aim is to obey environmental rules. Individual consumers can also implement many changes that can help them reduce the amount of waste they produce. There should be changes in the consumption practices of seeking alternative material usage instead of plastic packaging. Waste management is one of the keystones of development and sustainability. The proper management of waste is possible by combining different technologies in a system. Governments and other international agencies should see that industries and corporates should take steps towards this goal.

15.7 CONCLUDING REMARKS

The method of production decides the nature of waste. The type of use of natural resources and input determines the final product and the composition and nature of waste. The present reductionist approach towards economics and the environment should be dismantled to manage waste. In order to face the future challenges, sustainability factors should be incorporated in the production process. The most important element is the selection of recycled material as a suitable alternative in the manufacturing industry. A new culture of practice of the utilisation of natural resources is a prerequisite for this.

A multifaceted comprehensive procedure is necessary to accomplish waste management. To sum up, we should focus on the following aspects of waste management in future:

i. There are barriers inherent in the system because of corporate business targets and operations. Driven by the profit motive, private business people never consider environmental issues seriously. Excess production and incorporating exchange value in the product are examples. As the concept of the exponential growth of corporations remains, the accumulation of waste will also rise exponentially in the future.

ii. There are no coordinated efforts to link different players such as local governments, people and corporates together.

iii. The non-implementation of the existing laws of nations are concerned with the environment and waste management.

iv. The economic influence of big corporations over national governments is a serious issue for framing laws.

v. The lack of commitments of governments and corporations reduces the effectiveness of international agreements on the environment.

vi. The consumerist attitude of a throwaway society continues, so that it becomes a barrier for waste management. A change in consumers' attitude is important.

vii. CE should be promoted, despite all its limitations, in a coordinated way and there should be proper monitoring.

viii. The planned obsolescence of electronic equipment boosts a culture of consumerism, leading to an increase of waste generation. We should see that corporate responsibility should focus on waste generation and environmental considerations instead of focusing other social problems.

REFERENCES

Abril Ortiz, A., Sucozhañay, D., Vanegas, P. and Martínez-Moscoso, A., 2020. A regional response to a global problem: single use plastics regulation in the countries of the Pacific alliance. *Sustainability*, 12(19), p. 8093.

Alcadipani, R. and de Oliveira Medeiros, C.R., 2019. When corporations cause harm: A critical view of corporate social irresponsibility and corporate crimes. *Journal of Business Ethics*, 167(2), pp. 1–13.

Alcott, B., 2005. Jevons' paradox. *Ecological Economics*, 54(1), pp. 9–21.

Alsheyab, M. and Kusch, S., 2013. Decoupling resources use from economic growth – Chances and challenges of recycling electronic communication devices. *Journal of Economy, Business and Financing*, 1(1), pp. 1615–1619.

Ameer, R. and Othman, R., 2012. Sustainability practices and corporate financial performance: A study based on the top global corporations. *Journal of Business Ethics*, 108(1), pp. 61–79.

Apple Environmental Progress Report (2019) www.apple.com/environment/pdf/Apple_Environmental_Progress_Report_2019.pdf

Apple Environmental Progress Report (2020) www.apple.com/environment/pdf/AppleEnvironmentalProgress_Report_2020.pdf

Apple Inc. 2016. Environment (Resources). www.apple.com/environment.

Baker, W.E. and Sinkula, J.M., 2005. Environmental marketing strategy and firm performance: Effects on new product performance and market share. *Journal of the Academy of Marketing Science*, 33(4), pp. 461–475.

Banerjee, S.B., 2008. Corporate social responsibility: The good, the bad and the ugly. *Critical Sociology*, 34(1), pp. 51–79.

Brandt-Rauf, P.W., Li, Y., Long, C., Monaco, R., Kovvali, G. and Marion, M.J., 2012. Plastics and carcinogenesis: The example of vinyl chloride. *Journal of Carcinogenesis*, 11.

Camilleri, M.A., 2017. The corporate social responsibility notion. In *corporate sustainability, social responsibility and environmental management*. Cham, Switzerland: Springer Nature, pp.3–26.

Cooper, T., 1999. Creating an economic infrastructure for sustainable product design. *Journal of Sustainable Product Design*, 8(8), pp. 7–17.

Cooper, T., 2009. War on waste? The politics of waste and recycling in post-war Britain, 1950–1975. *Capitalism Nature Socialism,* 20(4), pp. 53–72.

Cooper, T., 2010. The significance of product. *Longer lasting products: Alternatives to the throwaway society* (p. 3).

Cooper, T. ed., 2016. *Longer lasting products: Alternatives to the throwaway society* (pp. 1–393). CRC Press..

Cronin, J.J., Smith, J.S., Gleim, M.R., Ramirez, E. and Martinez, J.D., 2011. Green marketing strategies: An examination of stakeholders and the opportunities they present. *Journal of the Academy of Marketing Science*, 39(1), pp. 158–174.

Crouch, C., 2012. Sustainability, neoliberalism, and the moral quality of capitalism. *Business and Professional Ethics Journal*, 31(2), pp. 363–374.

Cucchiella, F., D'Adamo, I., Koh, S.L. and Rosa, P., 2015. Recycling of WEEEs: An economic assessment of present and future e-waste streams. *Renewable and Sustainable Energy Reviews*, 51, 263–272.

D'Adamo, I., Rosa, P. and Terzi, S., 2016. Challenges in waste electrical and electronic equipment management: A profitability assessment in three European countries. *Sustainability*, 8(7), 633.

Dauvergne, P. and Lister, J., 2013. *Eco-business: A big-brand takeover of sustainability.* MIT Press.

Davies, J., 2016. *The birth of the anthropocene.* University of California Press.

Dawson, M. and Foster, J.B., 1991. The tendency of the surplus to rise, 1963–1988. *Monthly Review*, 43(4), pp. 37–51.

Debnath, B., Roychowdhury, P. and Kundu, R., 2016. Electronic components (EC) reuse and recycling – A new approach towards WEEE management. *Procedia Environmental Sciences*, 35, pp. 656–668.

Dhingra, R., Kress, R. and Upreti, G., 2014. Does lean mean green? *Journal of Cleaner Production*, 85, pp. 1–7.

Dieckmann, E., Sheldrick, L., Tennant, M., Myers, R. and Cheeseman, C., 2020. Analysis of barriers to transitioning from a linear to a circular economy for end of life materials: A case study for waste feathers. *Sustainability*, 12(5), p. 1725.

Dresner, S., 2008. The principles of sustainability. *Earthscan.* pp. 1–197.

Economist. 2009. A special report on waste. February 28th 2009.

Ehrenreich, B., 2010. *Smile or die: How positive thinking fooled America and the world.* Granta books.

Ellen MacArthur Foundation. 2013. Towards the circular economy. Ellen MacArthur foundation rethinking the future. www.ellenmacarthurfoundation.org/assets/downloads/publications/TCE_Report-2013.pdf

Ellen MacArthur Foundation. 2015. Why the circular economy matters. Delivering the circular economy: A toolkit for policymakers.

Ellen MacArthur Foundation. 2016, 2017. A circular economy of plastic in which it never becomes a waste – 2016 and 2017 report

Ellen Macarthur Foundation. 2020. UNEP, Global commitment 2020 progress report.

Escalante, N. et al. 2007. Environmental evaluation of household waste management system in Southern Germany. *Waste management and landfill*, 11th Symposium: 53–62.

European Commission (EC). 2017. The role of waste-to-energy in the circular economy, communication from the commission to the European parliament, the Council, the European economic and social committee and the committee of the regions; European Commission: Brussels, 26.1.2017 COM (2017) 34.

Falck, O. and Heblich, S., 2007. Corporate social responsibility: Doing well by doing good. *Business Horizons*, 50(3), pp. 247–254.

Fay, M., 2012. *Inclusive green growth: The pathway to sustainable development* (pp. 1–169). World Bank Publications.

Ferronato, N. and Torretta, V., 2019. Waste mismanagement in developing countries: A review of global issues. *International Journal of Environmental Research and Public Health*, 16(6), p. 1060.

Ford. D., 2020. COVID-19 has worsened the ocean plastic pollution problem. *Scientific American*, 17.

Foster, J.B. and Clark, B., 2012. The planetary emergency. *Monthly Review*, 64(7), pp. 1–25.

Foster, J.B. and Suwandi, I., 2020. COVID-19 and catastrophe capitalism commodity chains and ecological-epidemiological-economic crises. *Monthly Review – An Independent Socialist Magazine*, 72(2), 1–20.

Gangwar, N. and Saraswat, Y., 2016. Waste management: A corporate social responsibility or a legal obligation. *Imperial Journal of Interdisciplinary Research*, 2(11), 2454–1362.

Garcés-Ayerbe, C., Rivera-Torres, P., Suárez-Perales, I. and Leyva-De La Hiz, D.I., 2019. Is it possible to change from a linear to a circular economy? An overview of opportunities and barriers for European small and medium-sized enterprise companies. *International Journal of Environmental Research and Public Health*, 16(5), p. 851.

Garside, M., 2019. Global plastic production 1950–2018. *Statista*. www.statista.com/statistics/282732/global-production-ofplastics-since-1950.

Geissdoerfer, M., Savaget, P., Bocken, N.M. and Hultink, E.J., 2017. The circular economy – A new sustainability paradigm?. *Journal of Cleaner Production*, 143, pp. 757–768.

Ghisellini, P., Cialani, C. and Ulgiati, S., 2016. A review on circular economy: The expected transition to a balanced interplay of environmental and economic systems. *Journal of Cleaner Production*, 114, pp. 11–32.

Ghosh, S.K., Debnath, B., Baidya, R., De, D., Li, J., Ghosh, S.K., Zheng, L., Awasthi, A.K., Liubarskaia, M.A., Ogola, J.S. and Tavares, A.N., 2016. Waste electrical and electronic equipment management and Basel Convention compliance in Brazil, Russia, India, China and South Africa (BRICS) nations. *Waste Management & Research*, 34(8), pp. 693–707.

Gidwani, V. and Reddy, R.N., 2011. The afterlives of "waste": Notes from India for a minor history of capitalist surplus. *Antipode*, 43(5), pp. 1625–1658.

Girardet, H., 2014. *Creating regenerative cities* (pp. 1–202). Routledge.

Giurco, D., Dominish, E., Florin, N., Watari, T. and McLellan, B., 2019. Requirements for minerals and metals for 100% renewable scenarios. *In achieving the Paris climate agreement goals* (pp. 437–457). Springer, Cham.

Goodship, V., 2007. Plastic recycling. *Science Progress*, 90(4), pp. 245–268.

Gould, H., 2015. Waste is so last season: recycling clothes in the fashion industry. *The Guardian*.

Grant, R. and Oteng-Ababio, M., 2012. Mapping the invisible and real "African" economy: Urban e-waste circuitry. *Urban Geography*, 33(1), pp. 1–21.

Gregson, N., Crang, M., Fuller, S. and Holmes, H., 2015. Interrogating the circular economy: The moral economy of resource recovery in the EU. *Economy and Society*, 44(2), 218–243.

Greiner, M. and Kim, J., 2021. Corporate political activity and green washing: Can CPA clarify which firm communications on social & environmental events are genuine? *Corporate Social Responsibility and Environmental Management*, 28(1), 1–10.

H and M Group Sustainability Report (2019). Hennes and Mauritz – published on April 2020. www.statista.com/statistics/961998/quantity-of-apparel-collected-by-handm-s-reuse-and-recycling-scheme-worldwide/

Haque, M.S., 1999. The fate of sustainable development under neo-liberal regimes in developing countries. *International Political Science Review,* 20(2), pp. 197–218.

Hawken, P., Lovins, A.B. and Lovins, L.H., 2013. *Natural capitalism: The next industrial revolution.* Routledge.

Hellmann, K.U. and Luedicke, M.K., 2018. The throwaway society: A look in the back mirror. *Journal of Consumer Policy,* 41(1), pp. 83–87.

Herod, A., Pickren, G., Rainnie, A. and McGrath-Champ, S., 2013. Waste, commodity fetishism and the ongoingness of economic life. *Area,* 45(3), pp. 376–382.

Holbrook, E., 2019. Environment plus energy leader, August.

Holtbrügge, D. and Conrad, M., 2020. Decoupling in CSR reports: A linguistic content analysis of the Volkswagen Diesel gate scandal. *International Studies of Management & Organization,* 50(3), pp. 253–270.

Hoornweg, D., Bhada-Tata, P. and Kennedy, C., 2015. Peak waste: When is it likely to occur?. *Journal of Industrial Ecology,* 19(1), pp. 117–128.

Hunt, S.D., 2011. Sustainable marketing, equity, and economic growth: a resource-advantage, economic freedom approach. *Journal of the Academy of Marketing Science,* 39(1), pp. 7–20.

ICMR-IBS., 2019. *Planned Obsolescence: Undermining Apple`s Commitment to Sustainability.* Centre for Management Research.

Ishangulyyev, R., Kim, S. and Lee, S.H., 2019. Understanding food loss and waste – Why are we losing and wasting food? *Foods,* 8(8), p. 297.

Jones, C., 2010. The subject supposed to recycle. *Philosophy Today,* 54(1), pp. 30–39.

Karlait, D., 2013. *Waste management – The future of prosperous socially responsible business* (pp. 82–89).

Kidd, C.V., 1992. The evolution of sustainability. *Journal of Agricultural and Environmental Ethics,* 5, pp. 1–26.

Kirkpatrick, K., 2020. Reducing and eliminating e-waste. *Communications of the ACM.* 63:17–19.

Laurenti, R., Moberg, Å. and Stenmarck, Å., 2017. Calculating the pre-consumer waste footprint: A screening study of 10 selected products. *Waste Management & Research,* 35(1), pp. 65–78.

Leonidou, C.N. and Leonidou, L.C., 2011. Research into environmental marketing/management: a bibliographic analysis. *European Journal of Marketing,* 4, pp. 68–103.

Lindridge, A., MacAskill, S., Gnich, W., Eadie, D. and Holme, I., 2013. Applying an ecological model to social marketing communications. *European Journal of Marketing,* 47, pp. 1399–1420.

Loges, B., Jakobi, A.P., 2020. Not more than the sum of its parts: De-centered norm dynamics and the governance of plastics. *Environmental Politics,* 29, 1004–1023.

Magretta, J., 1997. Growth through global sustainability. *Harvard Business Review,* 75(1), pp. 78–89.

Mathews, J.A., 2011. Naturalizing capitalism: The next great transformation. *Futures,* 43(8), pp. 868–879.

Mazurkiewicz, P., 2004. Corporate environmental responsibility: Is a common CSR framework possible. *World Bank,* 2, pp. 1–18.

McKinsey. 2014. Toward a circular economy: Philips CEO Frans van Houten. Retrieved, 20, 2015.

McMillan Cottom, T., 2014. The logic of stupid poor people. tressiemc http://tressiemc.com/2013/10/29/the-logic-of-stupid-poor-people. Accessed March 25.

McWilliams, A., Siegel, D.S. and Wright, P.M., 2006. Corporate social responsibility: Strategic implications. *Journal of Management Studies*, 43(1), 1–18.

Meadows, D.H., Meadows, D.L., Randers, J. and Behrens, W.W., 1972. The limits to growth. *New York*, 102(1972), p. 27.

Mishra, A.R., Rani, P., Pandey, K., Mardani, A., Streimikis, J., Streimikiene, D. and Alrasheedi, M., 2020. Novel multi-criteria intuitionistic fuzzy SWARA–COPRAS approach for sustainability evaluation of the bioenergy production process. *Sustainability*, 12(10), p. 4155.

Murray, A., Skene, K. and Haynes, K., 2017. The circular economy: An interdisciplinary exploration of the concept and application in a global context. *Journal of Business Ethics*, 140(3), pp. 369–380.

Nyathi, B. and Togo, C.A., 2020. Overview of legal and policy framework approaches for plastic bag waste management in African Countries. *Journal of Environmental and Public Health*, 2020.

O'Brien, E., Parati, G., Stergiou, G., Asmar, R., Beilin, L., Bilo, G., Clement, D., De La Sierra, A., De Leeuw, P., Dolan, E. and Fagard, R., 2013. European Society of Hypertension position paper on ambulatory blood pressure monitoring. *Journal of Hypertension*, 31(9), pp. 1731–1768.

Odum, H.T. and Scienceman, D.M., 2005. An energy systems view of Karl Marx's concepts of production and labor value. In *Emergy Synthesis 3: Theory and Applications of the Emergy Methodology*, Proceedings from the Third Biennial Emergy Conference, Gainesville, Florida, January 2004. Gainesville, Florida: Center for Environmental Policy, pp. 17–43.

Omar, R., Sergio, R., 2018. Maximizing the value of waste: From waste management to the circular economy. *Thunderbird International Business Review*, 60, 10.1002/tie.21968.

Orlitzky, M., Schmidt, F.L. and Rynes, S.L., 2003. Corporate social and financial performance: A meta-analysis. *Organization Studies*, 24(3), 403–441.

Parker L., 2018. In a first, microplastics found in human poop. *National Geographic*, Oct. 22. https://tinyurl.com/y8x76hyk.

Popa, M. and Salanta, I., 2014. Corporate social responsibility versus corporate social irresponsibility. *Management & Marketing*, 9(2), 137.

Prakash, A., 2002. Green marketing, public policy and managerial strategies. *Business Strategy and the Environment*, 11(5), pp. 285–297.

Pramudianto, A., 2019. The role of International Law and National Law in handling marine plastic litter. *Lampung Journal of International Law*, 1(2), pp. 43–54.

Publicover, J.L., Wright, T.S., Baur, S. and Duinker, P.N., 2019. Engaging with environmental issues as a musician: Career perspectives from the musicians of the playlist for the planet. *Popular Music and Society*, 42(2), pp. 167–187.

Rasnan, M.I., Mohamed, A.F., Goh, C.T. and Watanabe, K., 2016. Sustainable e-waste management in Asia: Analysis of practices in Japan, Taiwan and Malaysia. *Journal of Environmental Assessment Policy and Management*, 18(04), 1650023.

Reike, D., Vermeulen, W.J. and Witjes, S., 2018. The circular economy: New or refurbished as CE 3.0? Exploring controversies in the conceptualization of the circular economy through a focus on history and resource value retention options. *Resources, Conservation and Recycling*, 135, pp. 246–264.

Romero-Hernández, O. and Romero, S., 2018. Maximizing the value of waste: From waste management to the circular economy. *Thunderbird International Business Review*, 60(5), pp. 757–764.

Rudel, T.K., Roberts, J.T. and Carmin, J., 2011. Political economy of the environment. *Annual Review of Sociology*, 37, pp. 221–238.

Salas-Molina, F., 2020. Risk-sensitive control of cash management systems. *Operational Research*, 20(2), pp. 1159–1176.

Salas-Molina, F., Pla-Santamaria, D., Vercher-Ferrándiz, M.L. and Reig-Mullor, J., 2020. Inverse Malthusianism and recycling economics: The case of the textile industry. *Sustainability*, 12(14), 58–61.

Salhofer, S., Steuer, B., Ramusch, R. and Beigl, P., 2016. WEEE management in Europe and China – A comparison. *Waste Management*, 57, pp. 27–35.

Savini, F., 2019. The economy that runs on waste: accumulation in the circular city. *Journal of Environmental Policy & Planning*, 21(6), 675–691.

Schlegel, I., 2020. How the plastic industry is exploiting anxiety about COVID-19.WWW Document. Greenpeace . www. greenpeace.org/usa/how-the-plastic-industry-isexploiting-anxiety-about-covid-19/. April 7, 2020.

Schneider, A., 2015. Reflexivity in sustainability accounting and management: Transcending the economic focus of corporate sustainability. *Journal of Business Ethics*, 127(3), pp. 525–536.

Schor, J. and White, K.E., 2010. Plenitude: The new economics of true wealth. New York: *Penguin Press*.

Schor, J.B., 2011. *True wealth: How and why millions of Americans are creating a time-rich, ecologically light, small-scale, high-satisfaction economy*. Penguin.

Seuring, S., Sarkis, J., Müller, M. and Rao, P., 2008. Sustainability and supply chain management – An introduction to the special issue. *Journal of Cleaner Production*, 16(15), 1545–1551.

Shi, L., Xing, L., Bi, J. and Zhang, B., 2006, July. Circular economy: A new development strategy for sustainable development in China. In *Proceedings of the 3rd World Congress of Environmental and Resource Economists, Kyoto, Japan* (pp. 3–7).

Silva, A.L.P., Prata, J.C., Walker, T.R., Duarte, A.C., Ouyang, W., Barcelò, D. and Rocha-Santos, T., 2020. Increased plastic pollution due to COVID-19 pandemic: Challenges and recommendations. *Chemical Engineering Journal*, p. 126683.

Slade, G., 2009. *Made to break: Technology and obsolescence in America* (pp. 1–316). Harvard University Press.

Torres, C.A.C., Garcia-French, M., Hordijk, R. and Nguyen, K., 2012. Four case studies on corporate social responsibility: Do conflict affect a company's corporate social responsibility policy. *Utrecht Law Review*, 8, p. 51.

Trenkle, N., Lohoff, E., 2012. Die grobe Entwertung: Vom finanzkapitalistischen Krisenaufschub zur globalen Notstandsverwaltung. [The Great De-Valorization: From the Finance Capitalist Lurch into Crisis to the Global Management of Emergency] Germany: Unrast Verlag.

Uddin, M.B., Hassan, M.D. and Tarique, K.M., 2008. Three dimensional aspects of corporate social responsibility, *Daffodil International University Journal of Business and Economics*, 3(1), pp. 200–212.

Underwriters Laboratory. 2016. Apple's pursuit of zero waste validation with UL environment. http://industries.ul.com/blog/apples-pursuit-of-zero-wastevalidation-thulenvi ronment]UnitedStatesEnvironmentalProtectionAgency; www.epa.gov/ems.

UN Environment-India. 2018. www.unenvironment.org/interactive/beat-plastic-pollution.

UNEP. 2006. *Circular economy: An alternative model for economic development.*United Nations Environment Programme, Paris. www.unep.org/ resourceefficiency/Portals/ 24147/scp/nap/circular/pdf/prodev-summary.pdf October 25, 2015.

UNEP. 2016. *United Nations Economic Commission for Europe. GEO-6 Assessment for the pan-European Region; United Nations Environment Programme and United Nations Economic Commission for Europe.* UNEP, Nairobi, Kenya. http://web.unep.org/geo/ assessments/regional-assessments/regional-assessment-pan-european-region.

UNEP, 2011. *Towards a green economy: Pathways to sustainable development and poverty eradication: A synthesis for policymakers.*UNEP, Nairobi.

Upadhayay, S. and Alqassimi, O., 2018. Transition from linear to circular economy. *Westcliff International Journal of Applied Research*, 2(2), pp. 62–74.

Valenzuela, F. and Böhm, S., 2017. Against wasted politics: A critique of the circular economy. *Ephemera*: *Theory & Politics in Organization,* 17(1), 23–60.

Vance Packard, 1960. *The waste makers by David Mckay Company, Inc,* New York. pp. 306.

Wan C., Shen G.Q., Choi S. 2019. Waste management strategies for sustainable development. In Leal Filho W. (eds) *Encyclopedia of sustainability in higher education* (pp. 1–9). Springer, Cham..

Wang, H. and Choi, J., 2013. A new look at the corporate social–financial performance relationship: The moderating roles of temporal and interdomain consistency in corporate social performance. *Journal of Management,* 39(2), pp. 416–441.

Wang, Q., Hernández-Crespo, C., Santoni, M., Van Hulle, S. and Rousseau, D.P., 2020. Horizontal subsurface flow constructed wetlands as tertiary treatment: Can they be an efficient barrier for microplastics pollution?. *Science of the Total Environment*, 721, p. 137785.

WEF. 2014. *Towards the circular economy: Accelerating the scale-up across global supply chains.* World Economic Forum. http://www3.weforum.org/docs/WEF_ENV_ TowardsCircularEconomy_Report_2014.pdf

WEF, 2016. The new plastics economy: Rethinking the future of plastics.

Whiteley, N., 1987. Toward a throw-away culture. Consumerism, 'style obsolescence' and cultural theory in the 1950s and 1960s. *Oxford Art Journal*, 10(2), pp. 3–27.

Winiwarter. V., 2002. History of waste. Waste in ecological economics. Katy Bisson and John Proops, Eds.Edward Elgar Publishing, Northampton, MA.

World Bank, 2019. IDBR-IDA - Solid Waste Management: September 23, 2019.

World Bank IBRD-IDA, 2018. What a waste: An updated look into the future of solid waste management. September 20, 2018.

Wrenn, M.V., 2016. Surplus absorption and waste in neoliberal monopoly capitalism. *Monthly Review,* 68(3), p. 63.

Yates, M., 2011. The human-as-waste, the labor theory of value and disposability in contemporary capitalism. *Antipode,* 43(5), pp. 1679–1695.

Yates, M., 2015. Labor as "nature," nature as labor:"Stay the course" of capitalism in WALL-E's Edenic recovery narrative. *Interdisciplinary Studies in Literature and Environment,* 22(3), pp. 525–543.

Yuan, Z., Bi, J., Moriguichi Y., 2006. The circular economy: a new development strategy in *China, J. Ind. Ecol*, 10, 4–8.

Zeng, X., Gong, R., Chen, W.Q. and Li, J., 2016. Uncovering the recycling potential of "New" WEEE in China. *Environmental Science & Technology*, 50(3), pp. 1347–1358.

ZWIA, 2015. Zero waste international alliance. http://zwia.org/aboutus/. Accessed 1 Sept 2017–2016 & 2017 Reports.

Index